NATURAL RESOURCES, SOCIO-ECOLOGICAL SENSITIVITY AND CLIMATE CHANGE IN THE VOLTA-OTI BASIN, WEST AFRICA

Natural Resources, Socio-Ecological Sensitivity and Climate Change in the Volta-Oti Basin, West Africa

Editors

Jürgen Runge

Department of Physical Geography & Centre for Interdisciplinary Research on Africa (ZIAF), Goethe-University, Frankfurt am Main, Germany

Assogba Guézéré

Department of Geography & PREDES Research Centre, Faculty of Arts and Humanities, University of Kara, Kara, Togo

Laldja Kankpénandja

Department of Geography & PREDES Research Centre, Faculty of Arts and Humanities, University of Kara, Kara, Togo

CRC Press
Taylor & Francis Group
Boca Raton London New York

CRC Press is an imprint of the
Taylor & Francis Group, an **informa** business

A BALKEMA BOOK

Front cover: Artisanal sand extraction for housing and infrastructure construction in the Volta-Oti catchment (Mô River) between Sokodé and Bassar (Togo). Population growth and economic development are in need of natural resources which causes environmental degradation of sensitive tropical landscapes (Photo: J. Runge, 2019).

Financially supported by Volkswagen Stiftung, Hannover, Germany, funding line 'Knowledge for Tomorrow'

VolkswagenStiftung

CRC Press/Balkema is an imprint of the Taylor & Francis Group, an informa business

Typeset by Integra Software Services Pvt. Ltd., Pondicherry, India

Library of Congress Cataloging-in-Publication Data

Applied for

Published by: CRC Press/Balkema
Schipholweg 107C, 2316XC Leiden, The Netherlands
e-mail: Pub.NL@taylorandfrancis.com
www.routledge.com – www.taylorandfrancis.com

ISBN: 978-0-367-61821-6 (Hbk)

ISBN: 978-0-367-61822-3 (pbk)

ISBN: 978-1-003-10670-8 (eBook)
DOI: 10.1201/9781003106708
https://doi.org/10.1201/9781003106708

Contents

Foreword

Ecosystem fragmentation and deforestation in West Africa are mainly triggered by humans. Peasants' agriculture and widespread forest disturbances for charcoal and firewood production cause severe landscape degradation. Increasing population pressure, reduced carrying capacity, ecological perturbations by demand for agricultural land lead to the decline of land conservation capacities, even in protected areas like in the Pendjari Biosphere Reserve in Benin. The complexity of interactions between environmental and sensitive socio-ecological systems, often with unexpected and unsustainable side effects, has raised international awareness in light of ongoing climate change. By the example of natural resources, land use and stakeholders' perceptions within the Volta-Oti Basin this collection of papers and proceedings presents, discusses and distributes new findings that will sustainably stimulate the international debate. This volume also intends to overcome national borders and language barriers between Anglophone and Francophone research communities, and supports better West African cooperation and networking. Young as well as established researchers formed new collaborations, and the workshop at the University of Kara (Togo) was a truly unique opportunity for all involved, not only to discuss science, but also to assess applied and best future management practices for the Volta-Oti Basin in West Africa.

The book presents 18 papers of an interdisciplinary and international workshop supported by the German Volkswagen Stiftung (funding line 'Knowledge for Tomorrow') on the topic of 'Natural Resources, Socio-Ecological Sensitivity and Climate Change in the Volta-Oti Basin' in West Africa. All the articles are in principle of regional interest, but they are embedded in the wider global environmental context.

The workshop with almost 100 participants was jointly organised by Goethe-University Frankfurt (Germany) and the University of Kara (Togo), held from March 6 to 8, 2019 in northern Togo. It was a visible sign of the 2016 signed partnership agreement between Frankfurt and Kara. Naturally, the overall quality of such proceedings was quite diverse. However, the event aimed to strengthen capacities of junior scientists from the West African sub-region, exchange and mobilise theoretical and methodological background from various scientific fields (Botany, Construction, Geology, Geography, Infrastructure, Politics, Remote Sensing, Sociology and Urban Planning). One goal was to deliver reliable elements for ongoing and profound environmental analyses that lie outside the common questions of the academic and civil society stakeholders.

We would like to express our gratitude to Volkswagen Stiftung for generously funding the event. It helped to overcome traditional boundaries between disciplines. Particularly we like to thank Linda Carstens for reliable support of the conference office, and for her tireless assistance in editing conference contributions. Sincere thanks also to the administration of the University of Kara, more particularly to the President Professor Komla Sanda for premises and logistic support. Representatively for all contributors to the project, we like to mention and thank the assistance of Nanoini

Damitonou, Tignoati Kolani, Zakariyao Koumoï, Yentougle Moutoré, and Kadouza Padabo, and the entire PREDES team. Linguistic editing was kindly done by Elise Piazza (Sonoma State University, CA, USA) to whom we are most grateful.

Jürgen Runge, Assogba Guézéré, Laldja Kankpénandja
Frankfurt am Main (Germany)
Kara (Togo)
May 2020

Contributors

Afo, B.
University of Lomé, P.O. Box 1515, Lomé, Togo. Email: abidjo81@gmail.com, Phone: +228 90183733.

Alla, K.
University Peleforo Gon Coulibaly, P.O. Box 1328, Korhogo, Côte d'Ivoire.
Email: adjoucrou@yahoo.fr, Phone: +225 58055696.

Bawa, D.
University of Lomé, P.O. Box 1515, Lomé, Togo. Email: dangnissobawa@hotmail.fr, Phone: +228 90150990.

Ouro Bitasse, E.
University of Kara, P.O. Box 43, Kara, Togo. Email: ourobitasse2001@yahoo.fr, Phone: +228 90721274.

Blivi, A.B.
Professor, University of Lomé, P.O. Box 1515, Lomé, Togo.
Email: blimblivi1955@gmail.com, Phone: +228 90053914.

Carstens, L.
Goethe-University Frankfurt, Altenhöferallee 1, 60438 Frankfurt am Main, Germany.
Email: linda.carstens@stud.uni-frankfurt.de, Phone: +49 69 26945880.

Da, C.E.D.
Professor, Joseph Ki-Zerbo University, P.O. Box 7021, Ouagadougou, Burkina Faso.
Email: dadapola432@gmail.com, Phone: +226 70283293.

Dangui, N.P.
University Félix Houphouët Boigny, P.O. Box 715, Bingerville, Côte d'Ivoire.
Email: danguinadi@gmail.com, Phone: +225 05391207.

Dansoip, G.
University of Kara, P.O. Box 43, Kara, Togo. Email: gdansoip@gmail.com, Phone: +228 90081193.

Djangbedja, M.
Associate Professor, University of Lomé, P.O. Box 1515, Lomé, Togo.
Email: paulinminkilabe@yahoo.fr, Phone: +228 90219654.

Djeri, I.
University of Kara, P.O. Box 43, Kara, Togo. Email: idrissoudjeri@gmail.com, Phone: +228 91308480.

Flevi, K.M.
University of Lomé, P.O. Box 2801, Kara, Togo.
Email: flevikossimensah@yahoo.fr, Phone: +228 90265953.

Gbénahou, H.B.M.
University of Abomey-Calavi, P.O. Box 237, Cotonou, Benin.
Email: gmb_rvpp@yahoo.fr, Phone: +229 95860574.

Gnongbo, T.Y.
Professor, University of Lomé, P.O. Box 1515, Lomé, Togo.
Email: gnongboty@gmail.com, Phone: +228 90053914.

Guézéré, A.
Associate Professor, Dean, University of Kara, P.O. Box 43, Kara, Togo.
Email: guezereassogba@gmail.com, Phone: +228 90280603.

Hombre, S.
University of Lomé, P.O. Box 1515, Lomé, Togo. Email: hombre1er@gmail.com, Phone: +228 90831660

Issa, M.S.
University of Abomey-Calavi, P.O. Box 922, Abomey-Calavi, Benin.
Email: issa.42@live.com, Phone: +229 97878676

Kambire, B.
University Félix Houphouët Boigny, P.O. Box 7152, Abidjan, Côte d'Ivoire.
Email: bekambire@yahoo.fr, Phone: +225 05467097.

Kambire, S.
University Peleforo Gon Coulibaly, P.O. Box 1328, Korhogo, Côte d'Ivoire.
Email: kambire_sambi@yahoo.fr, Phone: +225 58127054.

Kankpénandja, L.
University of Kara, P.O. Box 43, Kara, Togo. Email: kankpenang1@gmail.com, Phone: +228 90973307.

Koumoï, Z.
University of Kara, P.O. Box 43, Kara, Togo. Email: zakarietg@yahoo.fr, Phone: +228 90846016.

Lamboni, T.
University of Lomé, P.O. Box 1515, Lomé, Togo. Email: itinguedame@gmail.com, Phone: +228 91908059.

Moutoré, Y.
University of Kara, P.O. Box 43, Kara, Togo. Email: moutorey@yahoo.fr, Phone: +228 90893394.

Nantob, M.
Associate Professor, University of Lomé, P.O. Box 1515, Lomé, Togo.
Email: nantob_bobo@yahoo.fr, Phone: +228 90 36 41 62.

Ogouwale, E.
Professor, University of Abomey-Calavi, P.O. Box 922, Abomey-Calavi, Benin.
Email: ogkelson@yahoo.fr, Phone: +229 97129800.

Ouedraogo, J.
Joseph Ki-Zerbo University, P.O. Box 7021, Ouagadougou, Burkina Faso.
Email: ouedraogojoa@gmail.com, Phone: +226 78183316.

Ouorou Barre, I.F.
University of Parakou, P.O. Box 123, Parakou, Benin. Email: oubarre@yahoo.fr, Phone: +229 97727732.

Rouamba, S.
Norbert Zongo University, P.O. Box 376, Koudougou, Burkina Faso.
Email: songanaba@gmail.com, Phone: +226 71107787.

Runge, J.
Professor, Goethe-University Frankfurt, Institute of Physical Geography, Altenhöferallee 1, 60438 Frankfurt am Main, Germany.
Email: j.runge@em.uni-frankfurt.de, Phone: +49 69 79840160.

Sarpong-Nsia, G.
Open University of Malaysia & Accra Institute of Technology, P.O. Box AN-19782, Accra, Ghana. Email: ingnsiah@gmail.com, Phone: +233 0205885502.

Sokemawu, Y.K.
Professor, University of Lomé, P.O. Box 81206, Lomé, Togo.
Email: yves.soke@yahoo.fr, Phone: +228 90273021.

Souberou, K.T.
University of Abomey-Calavi, P.O. Box 4178, Cotonou, Benin.
Email: soubekate@gmail.com, Phone: +229 67903590.

Soussou, T.
Associate Professor, University of Kara, P.O. Box 404, Kara, Togo.
Email: denis.soussou@gmail.com, Phone: +228 90274191.

Vodounou, J.B.K.
Associate Professor, University of Parakou, P.O. Box 123, Parakou, Benin.
Email: vjeanbosco@gmail.com, Phone: +229 95561627.

Yabi, I.F.
Associate Professor, University of Abomey-Calavi, P.O. Box 526, Abomey-Calavi, Benin. Email: yafid2@yahoo.fr, Phone: +229 97476828.

CHAPTER 1

Workshop on natural resources, socio-ecological sensitivity and climate change in the Volta-Oti Basin – a brief outline

J. Runge

Centre for Interdisciplinary Research on Africa (ZIAF), Goethe-University, Frankfurt am Main, Germany

A. Guézéré & L. Kankpénandja

Faculty of Arts and Humanities & PREDES Research Centre, University of Kara, Kara, Togo

ABSTRACT: The international workshop 'Natural Resources, Socio-Ecological Sensitivity and Climate Change in the Volta-Oti Basin', funded by the Volkswagen Stiftung (Germany), was organised by Goethe-University Frankfurt and University of Kara from March 6-8, 2019. It aimed to strengthen capacities of junior scientists from the sub-region, exchange and mobilise theoretical and methodological background from various scientific fields, and to deliver reliable elements for a profound environmental analysis that lies outside the common questions of the academic and civil society stakeholders. In West Africa, increasing population pressure and demand for agricultural land caused the loss of arable land and the decrease of land conservation capacities in protected areas. The complexity of interactions between human and environmental systems and subsequent effects (climate change) has raised international awareness. By the example of natural resources within the Volta-Oti Basin, the participants presented, discussed and distributed new findings that sustainably stimulate the international debate. Young as well as established researchers formed innovative collaborations, and the event at the University of Kara was a truly unique opportunity for all involved not only to discuss science but also to assess best practices for the Volta-Oti Basin.

1.1 INTRODUCTION

The African continent will be hardest hit by climate change. Local farmers are very closely coupled with the climate system. The majority of stakeholders depend on sufficient rainfall to grow their food. West Africa especially is still vastly understudied, and the degree of expected climate and environmental change is large. Capacity for adaptation to climate change is low. Poverty equates to reduced choice at the individual level while governance often fails to prioritise and act rapidly enough on impairments due to climate change.

Long and short-term changes of tropical landscapes and ecosystems are continuous and ongoing processes. They are fully maintained by both natural and human-related drivers. Human imprints on terrestrial tropical ecosystems are of large extent and of major environmental concerns related to habitat fragmentation, decline of land and ecosystem services, reduced carrying capacity, biodiversity loss, and livelihood decrease. Climate variability and climate change exacerbate this situation. Ecosystem fragmentation and deforestation are mainly triggered by humans, like smallholder agriculture and forest disturbances for charcoal and firewood production.

In the meantime, the complexity of interactions between human and environmental systems and its subsequent effects has raised broad international awareness (IPCC 2013). A growing concern is how much land has been changed and how this dynamic will affect the future of land resources. Spatial patterns of land degradation surveyed by remote sensing, its causal factors and potential impacts of climate change are investigated through various approaches, scales and dimensions (Kankam-Yeboah et al. 2013, Mul et al. 2015).

Although scientific research on climate and climate change, hydrology, land cover, agriculture and landscape dynamics has been intensified in West Africa (UNEP-GEF Volta Project 2003, GLOWA Impetus 2011, www.impetus.uni-koeln.de, WASCAL www.wascal.org), consideration of local, indigenous communities, actors and stakeholders within the sub-region is still limited (Anoumou & Runge 2016). The international workshop in March 2019 in northern Togo in collaboration with Goethe University and researchers from the sub-region aimed to strengthen a sustainable and transdisciplinary dialogue between the availability of natural resources and feedback effects on traditional rural as well as on modern urban West African societies ('socio-ecological sensitivity') against the background of climate change.

Using the example of the Volta-Oti Basin and its riparian states (Benin, Burkina Faso, Côte d'Ivoire, Mail, Ghana, Togo, Figure 1) the joint workshop tended to deepen the environmental and socio-ecological debate. The event aimed to involve junior and senior scholars from the region, also strongly supporting the participation of women junior scientists. Finally, it brought together about 100 researchers, PhD students and local stakeholders with an interest in carrying out cross-border cooperation on environmental, natural and human driven climatic changes, including socio-economic aspects.

1.2 BACKGROUND

For many years (since 1983) the first editor, Professor Runge, has been working on West African regional planning, landscape development, ecosystem degradation, former and recent ecosystem dynamics as well as on biodiversity in West Africa (www.biota-africa.org). In July 2016 an official Memorandum of Understanding and cooperation agreement was signed between the Universities of Frankfurt and Kara. They decided to intensify joint field research and academic education in the years that follow. In February/March 2018, Professor Runge visited the Togolese partner university in Kara as a DAAD short term lecturer (German Academic Exchange Service, www.daad.de). During his stay, the idea of the Volkswagen Stiftung workshop (funding line 'Knowledge for Tomorrow') in Kara 2019 targeting on environmental and socio-ecological aspects of the Volta-Oti catchment was developed. In December 2018 the Volkswagen Foundation (www.volkswagenstiftung.de) granted funds (file

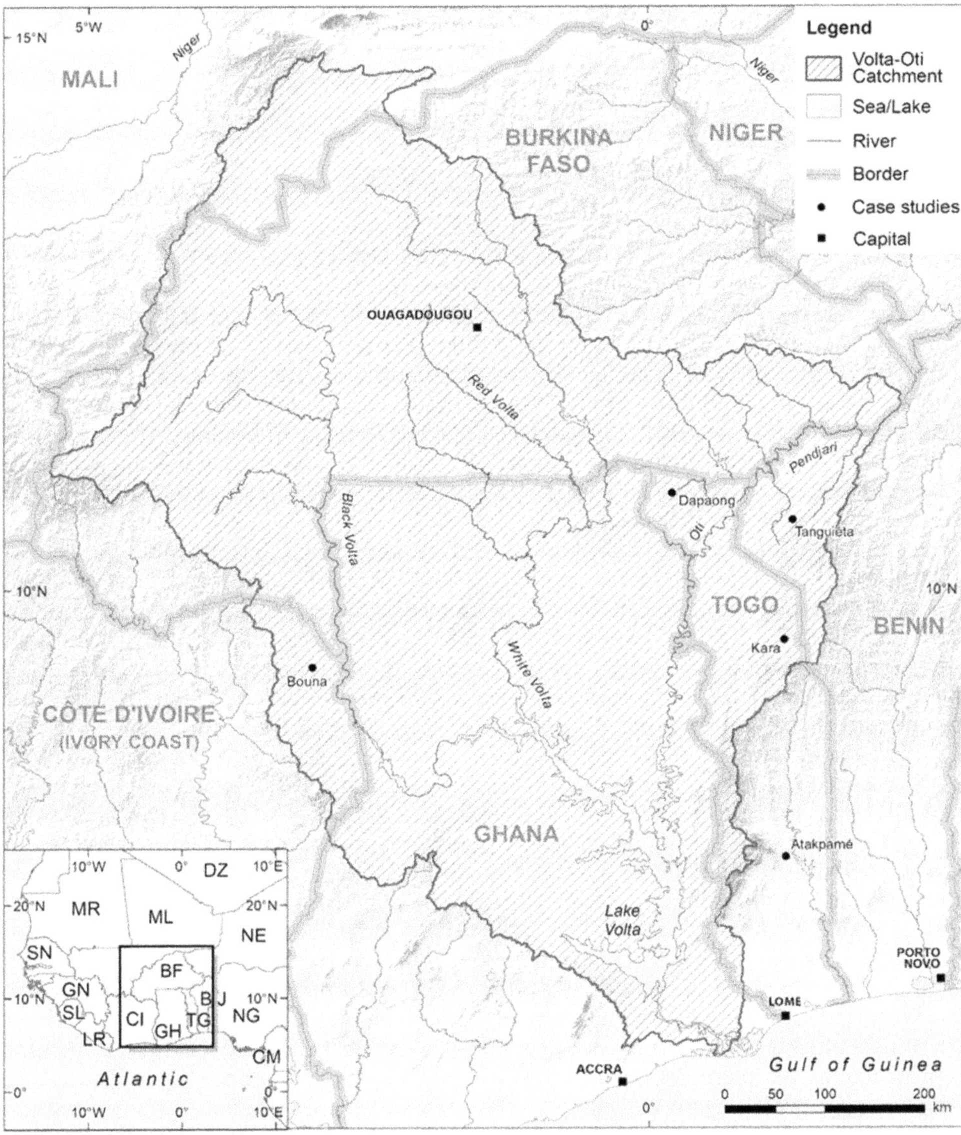

Figure 1. Location map of the Volta-Oti Basin in West Africa with some research localities mentioned in the book (Draft: J. Runge, Cartography: J. Eisenberg).

reference 95057) for planning and implementation of such a conference with the objective of having a scientific book on the outcomes in 2020.

1.3 STUDY AREA, RESEARCH CONTEXT AND OBJECTIVES

About 20 million people in the countries of Benin, Burkina Faso, Côte d'Ivoire, Ghana, Mali and Togo make their homes within the Volta-Oti River Basin. The Volta-Oti catchment is highly depended upon by these countries for the purposes of agriculture (irrigation), hydro-energy, and industrial and domestic water usage. The most significant hydrological structure within the basin is the Akosombo Dam, built between 1961-1966, which holds back the water of the White Volta, Black Volta and the Oti-Pendjari River for the generation of hydro-electric power. The dam formed the largest man-made lake in the world (Lake Volta surface area 8500 km^2) (Figure 1). The total area of the Volta basin is about 407,093 km^2 (i.e. significantly greater than Germany, surface of 357,386 km^2). Climate, landscape and soils, vegetation and the ethnical composition are quite diverse over the catchment. Most streams within the basin are ephemeral; it is thus extremely difficult to find sufficient potable water in most parts of the basin during the dry season (Ferreira et al. 2012).

The Oti sub-catchment is located between 7°48′-12°13′N and 0°15′-2°16′E, and covers an area of 73,900 km^2 (Barry et al. 2005). It stretches over four West African countries (Figure 1), and is influenced by a Guinean to Sudanese unimodal climate (alternating of rainy and dry season). These climatological conditions effect the development of multifold forest and savannah ecosystems (UNEP-GEF Volta Project 2013). It is home of several protected areas, like the Flora and Fauna Reserve of Oti-Kéran in Togo, the Pendjari Biosphere Reserve (PBR) in Benin, and the Integrated Nature Reserve of Pama in Burkina Faso.

The Oti-Pendjari River catchment owns an enormous natural potential. However, it is densely populated counting more than 7,000,000 inhabitants of which 80% rely as subsistence farmers directly on these natural resources (poverty rate is up to 75%) (Barry et al. 2005). The traditional economy is a slash and burn agriculture. Land use and related issues such as ecosystem degradation went up strongly during the past decades. Other economic activities within the catchment are industry, trade, mining and energy production (Awotwi et al. 2015, Kasei 2010).

The sub-catchment represents less than 20% of the main Volta basin surface but it provides up to 30% of runoff water to the Volta system. The water balance of the Volta River is strongly influenced and dependent on the Oti River (Kankpénandja 2016). The aim in producing more food to nourish the continuously growing riparian rural populations, results in a rapid increase of agricultural areas. This anthropogenic pressure leads to a declining carrying capacity of ecosystems, and in context with climatic change, it causes an accelerated degradation of the environment (Opoku-Ankomah 2000, Fricke 2004). Erosion of arable land, urbanised surfaces, roads and mining sites, silting up of rivers and water reserves, and eutrophication of lakes and ponds are manifesting the degradation of the catchment. In consequence, one notes great difficulties and challenges of providing drinking water and energy, associated with problems of unemployment, transport and mobility, undersupply of rural and urban centres, which again triggers poverty ('vicious circle').

It is essential that a multi-causal problem needs an inter- and transdisciplinary approach to exchange data, receive new perceptions, learn from best practices, and develop proposals for sustainable solutions (Andah et al. 2003, Runge & Eisenberg 2016). A reasonable and rational use of natural resources and landscape potentials, facing socio-ecological sensitivity and climate change in the wider Oti-Pendjari River catchment was the long-term objective of the workshop. Informatics services ('tools') like Geographic Information systems (GIS) and remote sensing approaches have evolved increasingly to support monitoring of landscapes' dynamics at different scales for better planning and management.

Researchers in various fields of science (Geography, Geosciences, Environmental and Economic Sciences, Political Sciences, Sociology, Anthropology, Psychology, History) have been invited to attend the workshop in Kara/Togo in March 2019. Oral and poster presentations were reviewed and selected for presentation by an international scientific committee. Proposals for contributions and articles had to respect four (4) thematic focuses.

The first research axis (1) was targeting on data collection and diagnostic studies as well as proposing sustainable management practices of tropical ecosystems (integrative, interdisciplinary approach). The second axis (2) considered former and recent ecosystem dynamics, present climate change, and questions of resilience and sensitivity of the Volta-Oti basin against natural and human induced change. Palaeoenvironmental history evidenced by thick alluvia allows conclusions on former ecosystem dynamics and sensitivity against natural change unaffected by humans (Runge & Tchamiè 2000, Runge 2019). Beside the established and indirect methods (GIS, remote sensing), it was expected that empirical field work also plays a substantial role within this research objectives (Badjanaa et al. 2017).

Socio-economic and socio-ecological sensitivity is mirrored by statistical data and empirical case studies on the role and individual perception of stakeholder groups inside the Volta-Oti catchment (research axis 3). Such factors are enabling the West African research community to better understand problems and challenges of rural and urban societies, and to develop stakeholder based sustainable development scenarios. Finally, research axis (4) primarily focused on economic aspects, transport and mobility of people, sustainable road and rail infrastructure for circulation of basic agricultural goods as well as items of industrial export between the Guinea coastal area into the Sudan-Sahelian hinterland.

The workshop intended to overcome national borders and language barriers between the Anglophone and the Francophone research communities, and stimulated sustainable West African cooperation and networking. The young as well as the established partners set up new scientific collaborations. The best two junior scientists were awarded a prize. In conclusion the event was a truly unique opportunity for all involved not only to discuss science but also to assess best practices for the Volta-Oti Basin (Figure 2).

Figure 2. Group picture of delegates on the campus of the University of Kara in March 2019 (Photo: A.E. Tchazim, 2019).

1.4 CONTENT OF THIS BOOK

The 18 papers gathered together in this volume are mainly contributions with a regional focus on the Volta-Oti catchment. However, they are all embedded in a global environmental context. Naturally, the quality of such papers in a proceedings volume is quite diverse. All incoming articles have been reviewed by an anonymous reviewing procedure to improve quality. For the French-speaking colleagues it sometimes remained difficult to produce good results ('language barrier'). Therefore, much linguistic editing had to be done on some articles. The same applied to illustrations (graphics, figures, photos) where a lot of reworking was required.

The contributions of Runge and Carstens (2+3) deal with infrastructure, geology, mining and landscape scenery on the Oti-Mono watershed in Togo. The challenge of international road infrastructure ('Gateway to the Sahel') and sustainable economic development (transparency and mining) faces many obstacles due to the lack of maintenance and economic value for residents.

One new concept as well as multiple aspects of selecting climate friendly (fewer CO_2 emissions) building material in Ghana are introduced in the paper by Sarpong-Nsiah (4). It is followed by two contributions by S. Kambire, B. Kambire, Dangui and Alla (5+6) on the geomorphological setting interlinked to urban erosion in the margins of the Black Volta River by example of the City of Bouna in Côte d'Ivoire (Ivory Coast, Figure 1). Perceptions of different stakeholders and climate change resilience are subsequently studied (7) by Ouro Bitasse, Moutoré and Dansoip in the northern Togolese towns of Kara and Dapaong (Figure 1). They highlight the importance of better understanding traditional and cultural attitudes on the complexity of climate variability and change.

Vodounou (8) examines by dendrometric measurements and remote sensing techniques the distribution and importance of traditional agroforestry and the role of the Shea butter tree (*Vitellaria paradoxa*) in northern Benin and Togo for climate resilience. Soil degradation by land use connected to strong population growth of the Moba ethnic group in the environs of the sandstone escarpments of Bombouaka and Dapaong in northern Togo (9) are studied by Kankpénadja, Bawa, Afo, Gnongbo and Blivi. Aside from direct field observations, they carry out multitemporal NDVI measurements from earth exploring satellites to conclude on vegetation and biomass dynamics.

A further botanical, dendrometric and stakeholders' approach is used by Djangbedja in the Kpendjal Prefecture of northern Togo (10). He looks at processes of plant adaptation, floristic composition and biodiversity (Shannon and Pielou index) under the impact of humans. He gives evidence that Combretacea and Mimosaceae are most abundant in the study area of the Oti catchment. The next contribution (11) of Flevi and Nantob examines by a comparative analysis of two agricultural cooperatives in the Plateau Region of Togo (Atakpamé, Figure 1) the success or failure ('socio-cultural burdens') of such initiatives when dealing with climate change topics from a differentiated stakeholder perspective.

Unsupervised spatiotemporal, satellite-based analysis of vegetation cover and land use in the Assoli Prefecture (south of Kara, Figure 1) between 1987 and 2015 is presented in the contribution (12) of Djeri, Koumoï and Soussou. They show that general pressure on land resources, ecosystems and environmental impoverishment increased over decades, and caused a significant loss of dry forests whereas savannah woodland ecosystems and agricultural fields strongly increased over almost 30 years.

Extensive research on climatological and agronomical data is presented by Lamboni and Hombre (13). They study rainfall variability between 1965 and 2014 in the Oti basin, and highlight agro-ecological feedbacks on the agricultural system by studying selected crops. The re-greening tendency (Ouedraogo et al. 2014) of the Sudano-Sahelian zone due to more annual rainfall becomes especially evident in the time frame 2005-2014.

Remote sensing and GIS data of land and vegetation cover is equally illustrated for the Oti-Pendjari catchment in Benin by Souberou, Yabi and Ogouwale (14). They also discuss two landscape scenarios for 2050 on ecologically and socio-economically driven development paths with different environmental consequences on biodiversity, hydrological cycle and soil quality.

Aspects of governance in nature conservation and resident's perception on the term 'climate variability' and/or 'climate change' in the Pendjari Biosphere Reserve (PBR) are intensively discussed by Gbénahou and Sokemawu (15). Focus on climatic factors for soil degradation and stakeholders' adaptation measures in the Oti plain in Benin are subsequently discussed by Ourou Barre, Issa, Yabi and Oguwale (16). Finally, two case studies from Burkina Faso illustrate land degradation and food insecurity in the urban and peri-urban areas of Ouagadougou (Figure 1). Ouedraogo and Da (17) apply LANDSAT data to survey vegetation cover dynamics between 1992-2018 for the suburb Koubri, whereas the study of Ouedraogo (18) looks by empirical social research on the economically difficult residents' situation (water and food supply) in the suburb of Polesgo.

1.5 CONCLUSION

Natural resources exploitation, changes and degradation in vegetation, endangerment of protected areas, smallholder agriculture, subsistence economy and stakeholders' perceptions of environmental and climate change have been introduced and discussed by numerous regional case studies from the Volta-Oti Basin (West Africa) with the aim to stimulate the international climate debate on future sustainable management practises. Junior and established scientists from Benin, Burkina Faso, Côte d'Ivoire (Ivory Coast), Ghana and Togo strengthened their cooperation and networking in this area, home of about 20 million of people.

REFERENCES

Andah, W.E.I., van de Giesen, N. & Biney, C.A. 2003. *Water, climate, food, and environment in the Volta Basin. Contribution to the project ADAPT. Adaptation strategies to changing environments.* URL: http://www.weap21.org/downloads/ADAPTVolta.pdf. (accessed 15/06/2018).

Anoumou, A.C.N. & Runge, J. 2016. Von Menschen und Kakao: sozial-ökologische Aspekte des Landnutzungswandels in Badou-Tomégbé, Togo (Westafrika). *Zbl. Geol. Paläont.* I, 1/2: 5–21.

Awotwi, A., Kumi, M., Jansson, P.E., Yeboah, F., & Nti, I.K. 2015. Predicting hydrological response to climate change in the White Volta catchment, West Africa. *J. Earth Science & Climatic Change* 6(1): 1–7.

Badjanaa, H.M., Benjamin Renard, B., Helmschrot, J., Edjamée, K.S., Afouda, A., & Wala, K. 2017. Bayesian trend analysis in annual rainfall total, duration and maximum in the Kara River basin (West Africa). *J. Hydrol., Regional Studies* 13: 255–273.

Barry, B., Obuobie, E., Andreini, M., Andah, W. & Pluquet, M. 2005. The Volta River Basin, Comprehensive Assessment of Water Management in Agriculture, Comparative study of river basin development and management. Accra: International Water Management Institute (IWMI), Comprehensive Assessment of water management in agriculture.

Ferreira, V., Gong, Z. & Ato Andam-Akorful, S. 2012. Monitoring mass changes in the Volta River basin using GRACE satellite gravity and TRMM precipitation. *Bol. Ciênc. Geod., sec. Artigos, Curitiba* 18(4): 549–563.

Fricke, W. 2004. Population shifts and migration in the Sahel and Sub-Sudan Zone of West Africa reviewed from the aspect of human carrying capacity. In D. Werner (ed.), *Biological Resources and Migration*: 297–316. Berlin, Heidelberg: Springer.

IPCC, International Panel on Climate Change 2013. Summary for Policymakers. In T. F. Stocker, D. Qin, G.-K. Plattner, M. Tignor, S.K. Allen, J. Boschung, A. Nauels, Y. Xia, V. Bex & P.M. Midgley (eds.), *Climate Change 2013: The Physical Science Basis. Contribution of Working Group I to the Fifth Assessment Report of the Intergovernmental Panel on Climate Change*: 1–27. Cambridge, New York: Cambridge University Press.

Kankam-Yeboah, K., Obuobie, E., Amisigo, B. & Opoku-Ankomah, Y. 2013. Impact of climate change on streamflow in selected river basins in Ghana. *Hydrolog. Sc. J.* 58(4): 773–788.

Kankpénandja, L. 2016. *Morphogenèse et gestion des terres dans les sous-bassins versants de l'Oti au Togo*. PhD thesis. Lomé: Université de Lomé.

Kasei, R.A. 2010. Modelling impacts of climate change on water resources in the Volta Basin, West Africa. *Ecology and Development Series* 69: 1–183.

Mul, M., Obuobie, E., Appoh, R., Kankam-Yeboah, K., Bekoe-Obeng, E., Amisigo, B., Logah, F. Y., Ghansah, B. & McCartney, M. 2015. Water resources assessment of the Volta River Basin. Colombo, Sri Lanka: International Water Management Institute (IWMI). *IWMI Working Paper* 166: 1–83.

Opoku-Ankomah, Y. 2000. *Impacts of Potential Climate Change on River Discharge in Climate Change Vulnerability and Adaptation Assessment on Water Resources of Ghana*. Accra: Water Research Institute (CSIR).

Ouedraogo, I., Runge, J., Eisenberg, J., Barron, J. & Sawadogo-Kaboré, S. 2014. The Re-Greening of the Sahel: Natural Cyclicity or Human-Induced Change? *Land* 3(3): 1075–1090.

Runge, J. & Eisenberg, J. 2016. Interdisziplinäre Afrikaforschung - Umwelt und Mensch im Kontext der Millenniumsziele. *Zbl. Geol. Paläont.* I, 1/2: 1–151.

Runge, J. & Tchamiè, T. 2000. Inselberge, Rumpfflächen und Sedimente kleiner Einzugsgebiete in Nord-Togo: Altersstellung und morphodynamische Landschaftsgeschichte. *Zbl. Geol. Paläont.* I, 5/6: 497–508.

Runge, J. 2019. Quaternary landscape sensitivity evidenced by the colluvial and alluvial sedimentary record in river catchments of tropical Africa. *Frankfurt Archaeological Studies* 37: 307–319.

UNEP-GEF Volta Project 2003. *Benin, Ghana, Togo and Burkina Faso Countries Report for the Integrated Management of the Volta River Basin Project*. Accra: UNEP.

CHAPTER 2

Geological and geomorphological aspects of infrastructure projects in Togo

J. Runge

Centre for Interdisciplinary African Studies (ZIAF), Goethe-University, Frankfurt am Main, Germany

ABSTRACT: Togo is dominated by S-N and SSW-NNE (north of 9°N) striking tectonic structures of Late Proterozoic age (Togo Mountains, in Benin: Atacora chain). This zone is interpreted as a collision event by continental plate convergence and overthrusting with a western and north-western direction affiliated with the West African Craton (Panafrican Orogeny, ~610-550 Ma). For many years the crossing of the predominately quarzitic escarpments (600-800 m asl) between the south and the north of Togo caused severe infrastructure problems. Due to long and narrow, steep and winding mountain roads many heavy-duty trucks accidents occured at the Alédjo fault and the Défalé Pass. As Togo became in recent years a transit country to the landlocked Hinterland ('Gateway to the Sahel' for Burkina Faso, Niger, Mali), connecting the deep water port in the capital Lomé with the international border into Burkina Faso, the National Road 1 is an important international link that has sustainably been developed. Between 2012-2015 two bypass roads were built by the China Road and Bridge Cooperation (CRBC) to ease the 'bottleneck' at the Aledjo fault and the Défalé Pass. The bypass roads were surveyed in detail and routing was critically discussed. Due to large openings and road cuts, new insights into the geology and geomorphology of the Atacora structural unit in Togo became possible. Aspects of marine morphodynamics (coastal erosion) due to the port extension in Lomé, and also perspectives of the reconstruction of railway infrastructure instead of road traffic are discussed.

2.1 INTRODUCTION

An efficient and functioning infrastructure is essential for economic development. After years of recession, stagnation and international isolation, the Republic of Togo transformed since the turn of the millennium by its' deep sea harbour ('Port Autonome de Lomé', PAL) to an important service axis for the Togolese nation and for the Hinterland of landlocked Sahelian states Burkina Faso, Niger and Mali ('Gateway to the Sahel').

Due to the lack of a railway network, the National Road 1 with transport of goods/containers by heavy-duty trucks from Togo's capital Lomé to the Burkina Faso border at Cinkansé (661 km) is of major importance. However, there are some challenges to the routing arising from the geological and geomorphological setting especially in northern Togo in the watershed area between the Mono and the Oti River (section Sokodé-Kara-Kandé). Since colonial times the S-N leading road presents a strategic link which created for years infrastructure problems for truck traffic due to long and narrow, steep and winding mountain roads at the Faille d'Aledjo and the Défalé Pass (Figure 1). To facilitate the road's task and to guarantee the

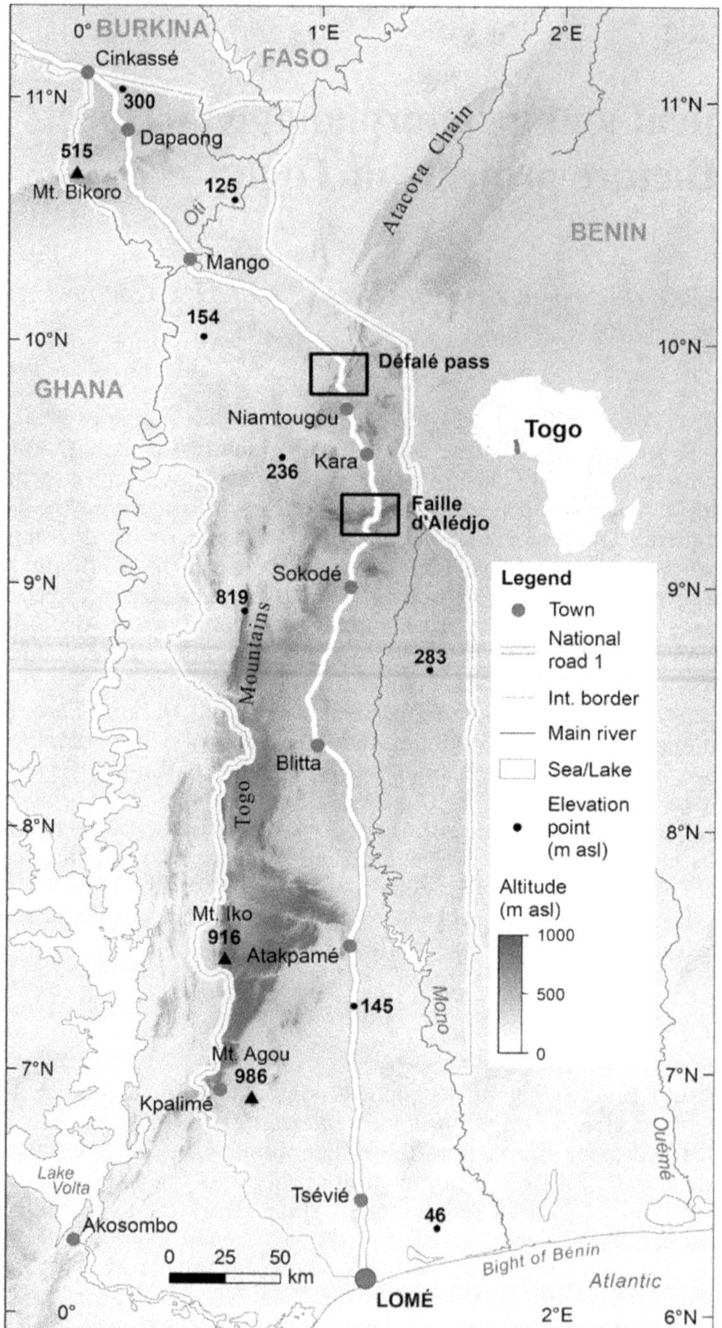

Figure 1. Topographic overview map of Togo with major towns and National Road 1 from Lomé to Cinkassé. Boxes show location of the bypass roads between Sokodé and Mango (Source: Togo, Carte Générale au 1:500.000, IGN-Paris and Direction de la Cartographie Nationale et du Cadastre, Lomé; Relief data: Shuttle Radar Topographic Mission, Global 1 arc second, v3 (SRTM), NASA, JPL 2013).

competitiveness of the PAL, Togo has solicited funds from international doners to build two bypass lanes at the Aledjo fault (18.1 km long) and at the Défalé Pass (12.1 km long).

This contribution summarises new findings from geologic fieldwork about landscape and infrastructure development in Togo over the last 35 years. Firstly, an outline of the historical framework and the unique geography and geological history is given. Secondly, the setting up of Lomé harbour at the Bight of Bénin (Atlantic) is sketched and the geoecological consequences on coastal currents and erosion are explained. Thirdly, the construction of the colonial railway and the road network is described. A regional focus is given on the new bypass roads: the routings (and alternative tracks) are described in detail and critically discussed. Due to large openings and road cuts by the road construction of China Road and Bridge Cooperation (CRBC), new insights into the geological composition and geomorphology of the Atacora structural unit in Togo became possible. Finally, an outlook on the future national development plan (2018-2022) and the socioeconomic perspectives of Togo are given.

2.2 STUDY AREA – OUTLINE OF TOGO'S HISTORY, GEOGRAPHY AND GEOLOGY

Togo was once a German colony ('Togoland' 1884-1914) divided after the First World War between the British and French governments (Metzger 1941). Since 1916 French Togoland became part of French West Africa as a trust territory of the League of Nations while the smaller British share was incorporated into Gold Coast, the later Ghana. Togo gained independence in 1960 as the République Togolaise (Cornevin 1959, 1988). The arbitrary drawn colonial-era borders of the Berlin Congo Conference (1884/85) produced abnormal shapes of new national territories dividing natural habitats and ethnic homelands such as the 'Eweland' in Togo and Ghana (Cornevin 1968).

As a consequence, Togo is one of the smallest countries in Africa. Its surface measures only 56,785 km^2 which is less than the federal state of Bavaria (Germany). From west to east the country is quite 'slim', just 50-100 km wide; this compares with a south-north spread of roughly 660 km forming a regional corridor between Ghana and Benin. In the north Togo, holds a short boundary with Burkina Faso (Figure 1).

Togo's population is rapidly expanding with almost eight million people (2018 estimate, Statista 2019) that are culturally and ethnically diverse. Even if urbanisation in the coastal Atlantic region strongly increases (~ 2,5 million inhabitants including the town of Lomé [Spekker et al. 2010]), most of the economy – aside of commodities like phosphate and iron ore (see chapter 3 in this volume) – depends on agriculture favoured by a semi-humid tropical climate. Togo belongs to the Sudano-Guinean climate zone, Aw-climate type according to Köppen with two contrasting seasons, 1000-1500 mm annual precipitation and an average yearly temperature between 25-28 °C.

Geologically, Togo is located on the south-eastern margin of the West African Craton (Neoarchean to Paleoproterozoic [Birrimien], Schlüter 2008) with a transition to slightly younger basement rocks of the Dahomeyide belt (2064 Ma [République du Togo 1984]), and a Neoproterozoic Panafrican Orogeny in between (Stein 2007). The onset of the Panafrican event took place around 610 Ma (République du Togo 1984) and lasted up to about 550 Ma (Tairou et al. 2012). It is considered as a collision event by continental plate convergence and overthrusting with a western and north-western direction affiliated with the West African Craton. Low to high grade and dislocation metamorphosis (mylonitisation) are common (Crenn

1957, Bessoles & Trompette 1980, République du Togo 1984, Runge 1990, 1993, Tairou et al. 2012). The complex structure first strikes with a N-S trend in the southern and central part of the country ('Togo Mountains'), further north it converts into a NE-SW/ENE-WSW direction in the Défalé area, and subsequently into northern Benin (Figure 1, Atacora chain). At 'Faille d'Aledjo' it is partly oriented in a W-E trend; forming a steep topography with quartzitic scarps, and at this locality it is an obstacle for road infrastructure. The Dahomeyide belt is subdivided from east to west into the structural units Plaine de Bénin ('Plaine Bénino-Togolaise', République du Togo 1984), Atacora and Buem (Affaton 1987).

The Atacora structural unit is made up of sandstones, quarzitic sandstones and green schists strongly folded and epi-metamorphically deformed to weathering resistant mica slates, quartzites and sericite slates (Figures 5-7). These highest parts of northern Togo relief are locally topped by iron rich laterites (ferricretes). Its external western part (Buem) consists of a sedimentary sequence ('para-autochthone' Bessoles & Trompette 1980), dominated by sandstones, shales and mixtites of low-metamorphic grade, which forms the transition between the Oti-Volta Basin (Dapaong and Bombouaka sandstones, Mango shales). The eastern part of the Dahomeyide belt has an ultrabasic trend along the Kara-Sotouboua-Atakpamé-Agou line (SSW-NNE). It is characterised by an association of crustal metamorphic and supracrustal rocks, an abundance of basic to ultrabasic, amphibolite and eclogitic rocks. The structural unit of the 'Plaine Bénino Togolaise' is composed of granitoids, migmatites, gneisses, metasediments, orthogneisses and ultrabasic rocks (Runge 1990, 1993, Schlüter 2008).

Due to the geological setting the physiognomy of Togo is diverse: from south to north seven geomorphic landscape types are recognisable (Petit 1981, Runge 1990): The youngest section is the Tertiary coastal sedimentary basin ('Terre de Barre', marine transgression, 0-60 m asl) with a backfilled former Last Glacial Maximum (LGM) Rias coast, followed by a Quaternary lagoon and highly dynamic coast system with longshore W-E drift (1). It is composed of marine and terrestrial clays, sands and gravel interlocked with alluvia. Eocene (56-38 Ma) phosphate deposits with open pit extraction take place between Kpogamé and Hahotoé (~15-20 km from Lomé) located in a SW-NE striking graben that contains manifold marine mammals (Gingerich & Capetta 2014).

Subsequently, a slow increase of topography follows up to about 140-150 m asl with a hardly noticeable transition into the levelled main erosion surface (2) of the 'Plaine Bénino Togolaise' (Dahomeyide belt). Drainage of this etchplain is done by the Mono River (Figure 1). Near the Togo Mountains some isolated gabbro (Kpalimé, Mont Agou, 986 m asl) and ultrabasic inselbergs occur parallel east to the quarzitic Atacora chain. Between Atakpamé and Sokodé sparsely populated areas ('Middle belt', Manson 1969) are typical, probably due to former slave raiding. The term 'Middle Belt' was used for such recurring population poor hinterland regions that have an overall lower population density than adjoining areas to the north and south (Cornevin 1959, 1988, Manson 1969). To the west and north the Togo Mountains (Atacora) separate Togo from the Volta basin by uplifted S-N striking escarpments with remnants of multiple erosion surfaces in 600-800 m asl (3) (e.g. 'Plateau de Danyi', 'Monts de Fazao'). Neotectonic uplift of the panafrican structures was caused by the fragmentation of Gondwanaland and the opening of the Atlantic with a new base level of erosion during the Cretaceous. North of Sokodé (9°N) the Togo Mountains run almost parallel of latitude into Bénin forming a continental divide that separates the catchments of the Mono and the Volta-Oti River (Figure 1). The structural unit Buem forms a western landscape zone parallel to the Togo Mountains characterised by numerous Inselbergs (4)

(Runge 2002). Northwest of the Défalé Pass the landscape merges into the flat, sedimentary Oti depression (5) (110-250 m asl) with extensive alluvia of the Oti floodplain around the town of Mango (Figure 1). Towards Bombouaka and Dapaong a cuesta relief (400-350 m asl) (6) emerges from the Oti plain with two major sandstone scarps, and north of Dapaong the Gondwana planation surface over Birrimien basement rocks (7), riddled with laterite topped tabular mountains, connects to the West African Craton into Burkina Faso (Runge 1990, 1993, 2013).

2.2.1 Lomé free port and coastal erosion

Togo has become due to its geographic position along a highly mobile, sediment transporting 56 km long compensatory coast (W-E longshore drift at Bight of Bénin), the principle gateway for trade between Europe, Asia, America and Africa. This strategic position led the Togolese government, financially supported by Germany, to construct a new seaport with ~17-18 m deep harbour basins, 8 km east of Lomé. It was officially opened on the 26th of April 1968 as the free port or 'Port Autonome de Lomé (PAL)'. 50 years later PAL became the motor drive for development in West Africa. The facility is the only deepwater port in the subregion and offers space for industrial settlement. Originally designed for 400,000 tons of traffic per year, PAL has been successively enlarged, and today handles over 15,000,000 tons/year of general cargo. The port has recently, since 2006 to date, completed the expansion and modernisation of its infrastructure. In addition to its two existing jetties and the fishing port, it now offers two modern container terminals up to a length of 950 metres at a storage area of 36 hectares. The 'Togo-Terminal' is operated as a Public Private Partnership (PPP) between the Togolese government and the French Bolloré Group.

Bypassing the city of Lomé from the harbour and linking to a new W-E motorway connecting Ghana and Bénin, as well as the installation of the 'Terminal du Sahel' in the outskirts of Lomé reduced the circulation of heavy traffic of general cargo and containers which became over the years an increased burden for the urban population. The four-lane expansion of the national road 1 as a modern motorway has meanwhile reached the town of Tsévié some 40 km from Lomé in the hinterland towards Atakpamé (Figure 1). The challenging road situation in northern Togo between Sokodé, Kara and Mango was partly relieved and improved by the construction of two bypasses (Figures 3-5).

The Togolese coastline is a sensitive geomorphological system (Runge & Anoumou 2017). Along the Bight of Bénin an around 300 km long, sandy 'hybrid' beach barrier-lagoon system was formed during the Holocene (Blivi & Anthony 1999). While the Last Glacial Maximum (LGM, ~26-18 ka BP) an eustatic regression with lowering of the Atlantic sea level of more than 100 metres (Peltier & Fairbanks 2006) gave a steeper fluvial gradient to the Volta-Oti, Ouemé and Mono River by the longer distance to the base level and estuary that subsequently caused strong fluvial incision (Rias coast). At the onset of the Holocene and especially between 6-5.5 ka BP (due to global warming and melting of northern hemisphere ice sheets), the elongated and bay-shaped Rias coast was flooded by rebound of the Atlantic's sea level ('drowned rivers' mouths) leading to fluvial-marine sedimentation and formation of a lagoon coast with longshore drift of material (Blivi et al. 2002).

From 2.0-1.5 ka BP the coastline between Ghana, Togo and Bénin developed a stable equilibrium between accumulated and eroded coastal sediments. In Togo this balance has already been reached by 3.8 ka BP due to the shorter distance to the main source of sediment in the Volta delta (Blivi et al. 2002). This natural balance has been disturbed by anthropogenic interventions: the construction of Akosombo dam in Ghana (1960) that led to a drastic fluvial sediment reduction and input into the Atlantic, and the setting up of harbours at Cotonou (Bénin, 1962) and Lomé (Togo, 1968) disturbing coastal geometry, followed by new dynamics of coastal marine erosion (Blivi et al. 2002, Runge 2013, Runge & Anoumou 2017).

The Bight of Bénin is not shaped by tropical cyclones, and slow to moderate waves with heights of 1.5 to 4 metres occur from July to October (rainy season) and from 0.5 to 1.5 metres in the remaining months. A striking feature is the permanence of the wave system which lands on the coast every 10 to 15 seconds with a strong, eastward ocean current and a refraction angle of 4 to 9 degrees. The tidal range is around one meter to a maximum of 1.95 metres (Blivi et al. 2002). The strong eastward and even flow produces a continuous transport of terrestrial sediment load (1.2 to 1.5 x 10^6 m^3/a) which is also strongly displaced eastwards and deposited. As a result of the high sedimentation rates, a compensatory coastline is developing (Blivi et al. 2002: 188).

The jetties and quay facilities of the 'Port Autonome de Lomé' (PAL) hamper and interrupt the coastal W-E oriented longshore drift. Increased sediment accumulation in the west (windward) and increased coastal erosion in the east (lee) have been the consequences (Blivi 1993, Laibi et al. 2014). Within 30 years only almost one kilometre of sandy deposits have been accumulated west of the main jetty. The Hotel Sara Kawa, built in the 1970s close to the port, used to be on the seashore. Today it lies 600-800 metres inland. East of PAL, coastal erosion of the same amount took place reaching up to the shores of neighbouring Bénin. Near to coast settlements were lost to the ocean, and the international road to Cotonou and Lagos disappeared as 'victim' to the waves, and was subsequently relocated twice into the hinterland. A Holocene outcrop of a Beach rock forms a natural breakwater parallel to the shore and east of the port of Lomé, was exposed by coastal erosion in 1975. It lowers kinetic wave energy and partly protects the eastern coast (Blivi 1993). However, the huge freshwater lagoon (Lake Togo) near Anecho, part of the former Rias coast system, was repeatedly salinated by marine waters intruded by coastal erosion. In addition, the phosphate loading station (jetty at Kpémé) leads to environmental and ecological problems by eutrophication of the ocean.

2.2.2 Pre-colonial, colonial and modern infrastructure: rail and road network

The infrastructure development and colonial exploitation of Togo started at the Atlantic coast with commercial outposts and trading houses of European merchants. At that time (1884) Togo was part of two distinct trading networks: the Arabic Haussa trade in the Sudano-Sahel, and the Atlantic, colonial trade in South (Guinea). Even before colonial and modern infrastructure had been established, German penetration succeeded in extending the Atlantic trading network further north (de Haan 1983). General orientations of pre-colonial tracks, footpaths between villages, and caravan trails showed an overall direction from east to west or parallel to the coast. By setting up interior colonial stations and military posts (e.g. Atakpamé, Bismarckburg, Sokodé, Mango) the derived historical pattern transformed into a linear, more south-north to radial structure connecting the coast with the hinterland. As motor vehicles in Africa were hardly common at the beginning of the 20th century, and natural navigable waterways into the interior of Togo were lacking, the setting up of a railway network became a priority for the colonial government (Sebald 2013).

Although the German Empire was well advanced in railway technology, it took 20 years (1904) until the railway became a reality in Togo. As the West African territory was considered to be a self-sufficient 'Musterkolonie' (model colony) the colonial administration had to provide verification that a bond of the German Empire – including interest rates – can be successfully payed back by annual rates. In the end, the railway project in Togo achieved a profit of about 21% (Sebald 2013).

During the German colonial period (1884-1914) three railway lines were realised: 1) since 1905 the coastal or palm railway (44 km long) on the sandy lagoon towards Aného (Bénin border) was operational, mainly used for the exportation of palm kernals, palm oil and mais via the jetty in Lomé; 2) the cacao and coffee railway (116 km long) from Lomé to Kpalimé near the Togo Mountains, and 3) the cotton railway towards Tsévié (35 km long), up to 1911 extended to Agbonou (163 km long) south of Atakpamé (Cornevin 1959, Sebald 2013). Under the French administration this railway line was extended in 1934 further north to Blitta (281 km long, BAD 2014). However, the previously planned extension of this track up to the north (9°30′N) to exploit the iron ore in the Bassar region (Bandjeli) has never been implemented. From 2008 to 2016 the Indian company M.M. Mining Ltd. exploited haematite iron ore from Bandjeli, subsequently transported by trucks to Blitta station and then took advantage of the colonial railway infrastructure to reach the port of Lomé (see chapter 3 in this volume).

Today, Togo's rail network of 495 km is run-down. Since 1987 some sections of the colonial infrastructure were, due to lack of maintenance, no longer operational. The stretch Lomé-Kpalimé was abandoned in 1996 and the Lomé-Blitta rail ceased operation in 2012 simultaneously with the Lomé-Aného line (BAD 2014). Carriage of passengers was stopped. Currently only two coastal, private industrial railway lines, one for clinker production by the Indian based WACEM (West African Cement) and the other for raw phosphate exportation to the Kpémé jetty, operated by the Société Nationale des Phosphates du Togo (SNPT), are in use.

Figure 2. The 'bottleneck' at the Alédjo fault (Faille d'Alédjo) between Aléhéridé and Kpéwa: all the heavy-duty trucks had to go through this historic caravan trail up to 2015 (Photo: J. Runge, 2012).

As a consequence, traffic development in Togo focuses mainly on road infrastructure used by private and public transport ('taxi brousse', coaches, 'taxi moto' [in urban areas]), and heavy trucks for overland transport of cargo and container traffic. Togo's road network comprises a total length of 11,672 km (2014) from which 1793 km (15.4%) are asphalted roads ('Routes Nationales revêtues', RNR); 1294 km (11.1%) are non-paved national roads ('Routes Nationales non revêtues', RNNR). A huge share of 6802 km (58.2%) is accounted to rural pistes (lateritic tracks) which often only allows limited use for normal cars ('Reseau de pistes rurales', PR). Urban roads ('Voiries urbaines', VU, paved/non-paved) contribute with 1783 km (15.3%) to Togo's road network (BAD 2014); a large proportion of the latter belongs to the capital Lomé.

In 2012 the governments of Togo and Burkina Faso decided, as financially supported by international donors (ADB, African Development Bank, WADB, West African Development Bank, EBID, Ecowas Bank for Investment and Development, KfW, Kreditanstalt für Wiederaufbau and EU, European Union), to rehabilitate the Lomé-Cinkassé-Ouagadougou road as an international lifeline in the West African sub-region that connects the PAL with the Sahel (FAD, n.d.). Furthermore, the programme aims for strengthening local capacities of custom authorities and technical inspection of vehicles as well as the installation of monitoring systems (e.g. weighbridges) for traffic safety (BAD 2014). The critical 'bottleneck' situation along the Togo-Atacora Mountains between Sokodé and Kandé (Figures 2-5) was already investigated by a feasibility study in 2006. Two alternative bypasses at the Alédjo fault (18.1 km) and the Défalé Pass (12.1 km long) have been finally proposed. Financing of the 'contournements' came by a loan of China's EXIM Bank (~ 60-100 Mio. €, different sources), and technical execution was done by China Road and Bridge Cooperation (CRBC) between 2012-2015.

2.3 ROAD INFRASTRUCTURE CHALLENGES

2.3.1 Faille d'Alédjo

Crossing the Alédjo fault ('Faille d'Alédjo') was a critical section of the National Road 1 (Figure 1). Constituted by a section of slopes of more than 10% over a length of about 20 km, it was the main cause of serious accidents of large trucks. The bypass of this critical section should allow heavy vehicles to permanently avoid the steepness of the section between the localities of Aléhéridé, Kpéwa and Bafilo (Figures 2-3).

The Faille d'Alédjo (9°16.495′N/1°13.275′E, 689 m asl) is a narrow, elongated ~60 metres long and vertical (~8 metres) quarzitic rock fissure with a NW-SE strike. It was the steep traditional traffic connection between the Kara Region and the southern Central Region (Sokodé) (Figures 1-3). As a pre-colonial caravan track it was the passage for traders crossing, on foot or on horseback, the watershed between northern and southern Togo (Volta-Oti and Mono catchment).

Until 1955 the Alédjo road was non-paved; a lateritic single-lane road passing through the tunnel-like fault. In 1960 an additional external lane was added for upstream traffic towards Bafilo and Kara (Figure 2). At that time, it would have been impossible to imagine that 40-50 years later hundreds of trucks per day will have to pass this infrastructure bottleneck. The Alédjo road was and is dangerous: a great number of accidents occurred due to poorly maintained (brake and clutch

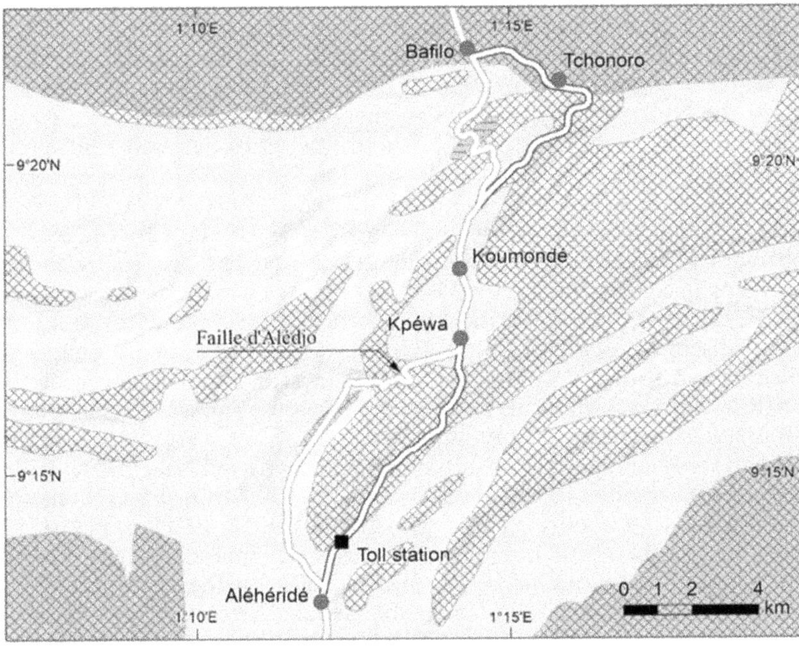

Figure 3. Geological and topographic sketch map of the Alédjo fault region with old National Road 1 (thin line) and new bypasses between Aléhéride and Kpéwa, and Koumondé and Bafilo (thicker line). For legend see Figure 5 (Source: République du Togo 1984).

problems), overloaded trucks as well as overtired drivers who were inexperienced with mountain roads.

The picturesque quarzitic fault structure of the Atacora unit is well known as a touristic attraction in Togo. Parts of the area serve as a wildlife reserve (765 ha) nominated for the UNESCO World Heritage List. Aside of biodiversity conservation its objective is the protection of local geological and physiographic formations.

The Alédjo bypass road starts within a slight topographic depression behind the village Aléhéridé to the right (Figure 3). There is a huge informal parking space at Aléhéridé where drivers prepare their trucks for the long and steady climb up to Kpéwa. The first bypass (Aléhéridé-Kpéwa) measures ~8 km and passes in NE direction parallel to the Atacora structure along the levelled ferricrete surface at Alédjo Kadara (~800 m asl). The old National Road 1 passing on to the Faille d'Alédjo was blocked by a barrier to avoid illegal use. At the first segment (~ 1 km) of the new road a toll station is located. Private cars have to pay a 'péage routier' of FCFA 500 (~1 Euro) per passage. Trucks pay a fee of 3000-4000 FCFA (~4.5-6 Euros). What is striking is that the new road has a relative narrow lane width, and a limited number of emergencies stopping points. Today, very often trucks with punctures or engine damages are blocking one lane of the newly built road!

Many geologic openings between Aléhéridé and Kpéwa provided the opportunity to newly study the petrologic and tectonic situation of the Atacora structural unit (Figure 7). The Kara amphibolitic orthogneiss crops out in a small 'geological window' (9°16.325′N, 1°12.379′ E, 515 m asl) close to the old National Road 1 before the climb to the Faille d'Alédjo (République du Togo 1984).

Geochemical analyses (Table 1) of two basement rock samples (RU-8), one taken by Runge (1990) close to Niamtougou, indicated considerable fluctuations of oxides of major elements (SiO_2 between 55.09% to 71.10%). As typical for amphibolitic gneisses the content of Al_2O_3 is high, between 15.7% to 19.85%, whereas the content of iron oxide (Fe_2O_3) is quite low and fluctuates between 1.30% to 10.54% (Table 1, Figure 6a).

On the left-hand side, before the toll gate (9°13.380′N, 1°12.075′E, 518 m asl), banked strata of quarzitic muscovite and sandstone (RU9, 90.92% of SiO_2, Table 1, Figure 6b) with inclusions of ovally garnet crystals (~ one centimetre in diameter) crop out; it dips in SE direction with 25°. Near surface weathering of strata is generally insignificant, however, quarzitic blocs, schists and 1-3 metres thick colluvial deposits of pedisediments are common. Reddish to whitish Atacora quarzites, quarzitic sandstones and mica-schistes crop out as tors or chaotic bloc accumulations with superficial Tafoni formation. Rock surfaces exposed to the air show dark-black to reddish weathering rinds 5-10 mm thick.

The analysis of sample RU-10, a banded muscovite quarzite at PK 3.275 (PK = road section point) resulted in 83.95% SiO_2 and 10.60% Al_2O_3. Fe_2O_3 content is relatively small with 1.30% only (Table 1). At PK 4.550 (9°15.203′N, 1°13.218′E, 616 m asl) the stratified and homogenous muscovite quarzite slightly modifies its strike direction with a steeper dip of 39°towards ESE. Further upwards (9°15.352′N, 1°13.336′E, 633 m asl) quarzites and mica-slates under heterogenous weathering conditions are common; general strike is NE-SW with rectangular faults on which significant waterflow (springs) is recognisable; water also runs in the dry season.

At PK 5.875 from Aléhéride (9°15.646′N, 1°13.789′E, 657 m asl) a huge outcrop of steeply dipped (>60° to SE) thin layered sandy, panafrican coastal sediments with little or no signs of metamorphic stress, but with some embedded pebbles and solid bands of greenish chlorite and sericite schist appears (sample RU11, Table 1). This whitish to slight greenish serecitic sandstone is obviously deeply weathered (white kaolinitic clays), however, chemical analysis gives only 1.06% Fe_2O_3 and 8.05% A12O3 and an expectable amount of SiO_2 (87.10%). Small sandy quartz detritus (grus) is continuously sliding from the opening (>10 m), blocking too small drainage trenches along the new National Road 1. In 2019 informal cleaning of trenches was done by simple wheelbarrows. Geotechnical slope protection measures have not been considered by the Chinese road construction.

At PK 6.658 on the right side of the road (9°16.016′N, 1°13.957′E, 704 m asl) direction Aléhéridé, a steep emergency lane covered by gravel was added to the main road in case of brake's failure of trucks. The very steep (>60°), elongated quarzitic slope shows rock falls along normal faults with slipping of huge rock blocs. Same to other sites along the new National Road 1: any (expensive) slope stabilising measures (concrete, iron nets and rock anchors) have been built in. Long term sustainability of the bypass has therefore to be questioned.

Close to the Alédjo plateau, almost a kilometre before Kpéwa, the bumpy relief of tors and quarzitic rock towers becomes dominant (9°16.362′N, 1°14.187′E, 755 m asl). A surface sample (RU12) of weathered quarzitic sandstone was analysed by a microprobe survey (Figure 6e). Aluminium (Al_2O_3), iron (FeO), TiO_2 and CaO are obviously enriched in the superficial blackish weathering coating, whereas SiO_2 values average at about 78.62% on the rock surface (Table 1).

Sporadically lateritic crusts or ferricretes can be recognized on the 'rolling' Gondwana surface in ~800 m asl. At Koumondé (Figure 3) a complete lateritisation of a glacis slope gives evidence of a former etching and down-weathering with iron enrichment of an old land surface by 'wash depressions' according to the concept of Büdel (1957). The Neogene uplifted relief of quarzitic tors is interpreted as remnants

Table 1. Geochemical X-ray analyses (percent by weight) of major oxides of elements in Aledjo fault rock samples carried out by Technical University Darmstadt, Germany. For selected handpieces see Figure 6; *) earlier taken sample, Runge 1990). Sample RU-12 (Figure 6e) was researched by microprobe analysis (H. Höfer, Goethe-University Frankfurt, Germany).

Sample rock type	RU-08 2015 Orthogneiss near Alédjo fault	Ortho- gneiss)	RU-09 2015 Quarzite-Musko- vite, with Garnet	RU-10 2015 Muscovite- quarzite	RU-11 2015 Sericite- shist, sandstone	RU-12 2015 Microprobe (Wt %) Weathered quarzitic sandstone	RU-13a 2015 Ferricrete Kpéwa	RU-13b 2015 Ferricrete Kpéwa
Oxides	**Figure 6a**		**Figure 6b**	**Figure 6c**	**Figure 6d**	**Figure 6e**	**Figure 6f**	
% by weight						70.24 80.07		
SiO_2	55.09	71.10	90.92	83.95	87.10	85.55	30.03	20.49
TiO_2	1.12	0.20	0.09	0.29	0.14	2.19 0.05 0.48	0.58	0.59
Al_2O_3	19.85	15.70	5.46	10.60	8.05	11.82 7.18 0.48	13.79	15.76
Fe_2O_3	10.54	1.30	1.87	1.30	1.06	7.20 (FeO) 3.55 (FeO) 2.24 (FeO)	53.91	61.69
MnO	0.14	0.00	0.01	< 0.003	0.03	-	0.02	0.02
MgO	3.72	0.40	< 0.3	0.34	1.00	-	0.44	0.40
CaO	1.27	0.50	< 0.05	< 0.05	< 0.05	3.83 2.16	< 0.05	< 0.05
Na_2O	1.99	3.50	0.30	0.43	< 0.1	- -	< 0.1	< 0.1
K_2O	5.56	5.10	0.99	2.92	2.48	2.42 1.29	0.78	0.68
P_2O_5	0.20	0.30	0.10	0.03	< 0.03		0.23	0.15

of the lower deep weathering front, formerly covered by a deep tropical saprolite and latosol. The terrestrial black weathering coating on rocks also contains traces of SO_3 which can be contributed to regular bush fires.

Geochemical analyses of two ferricrete samples (RU13a, RU13b, Figure 6f, Table 1) gives a high Fe_2O_3 content of 53.91% and 61.69%; the Al_2O_3 values plunge to 13.79% and 15.76% respectively. Even though the primary iron content of Atacora rocks along the Alédjo bypass road is low, it underlines the relative enrichment of iron by a deep and long-lasting tropical weathering mantle. During the Cretaceous, a new base level of erosion was formed along the Bight of Bénin due to the opening of the Atlantic in West Africa, followed by the lowering of the coastal regions. In course, the levelled hinterland ('peneplain') with low energy of relief, was tectonically uplifted, and as a consequence the older weathering mantles have been stripped and removed by denudation and fluvial erosion.

After around two kilometres on the old road behind Kpéwa, following the undulated and partly iron crust coated planation surface (ferricretes, ~800 m asl) of the Koumondé plateau (road culmination point at 9°16.719′N, 1°14.175′E, 775 m asl), the second part of the Alédjo bypass leaves the old road to the right in direction to Bafilo (Figure 3). This bypass follows the main Atacora strike (elongated quarzitic escarpment) in SW-NE. The slope gradient rises enormously on the first three kilometres of relatively straight track, and for the case of brake failure, another emergency lane was set up before the large left turn following into Tchonoro. Bypass construction deeply (> 15 m) cut through the Atacora quarzite ridge following a NW-SE fault (same tectonic direction as the 'Faille d'Alédjo'). Near to Bafilo highly metamorphised quarzites crop out, mainly thin bedded with bright mica on bedding planes. As a good fissionable rock, it was traditionally used as grindstone; today, Atacora quarzite is famous for ornamentation of buildings, and is frequently sold along the road.

At Bafilo the new National Road 1 meets the old National Road at a right angle (9°21.694′N, 1°14.427′E, 485 m asl). It marks, since the village of Tchonoro, the transition from the mountainous Atacora chain to the levelled, wavy orthogneiss basement of Kara (Figure 3). Atacora structures here dip to N and NE against the basement that gives evidence of a huge anticline structure. Main parts of Bafilo lie elevated on a flat quarzitic terrace, clearly separated from the orthogneiss through a slight depression with tabular mountains topped by ferricretes (not visible in Figure 3 due to generalisation from the 1:200,000 map, République du Togo 1984).

2.3.2 Défalé Pass

The other critical section of the National Road 1 was the routing to cross the quarizitic ridge of the Atacora structural unit between Défalé and Kandé (Figure 4). The first straight stretch of about 7 kilometres behind Défalé climbs up the escarpment in a northern direction with a general slope of more than 13%. Behind the culmination point the equally steep descent to Kandé ends with a dangerous left bend, leading into the highly metamorphised schist area of the 'Plaine de Lamba'. Overthrusting of the panafrican Atacora structural unit to the NW is well evidenced by an isolated quarzitic/sandstone 'Inselberg' Mont Béhao south of Kandé (Figure 4-5).

The bypass close to Défalé takes a routing between two Atacora escarpments with a gentle descent on the first 4 km almost parallel to the former national road. The second part of the bypass shows an increasing slope and a winding structure.

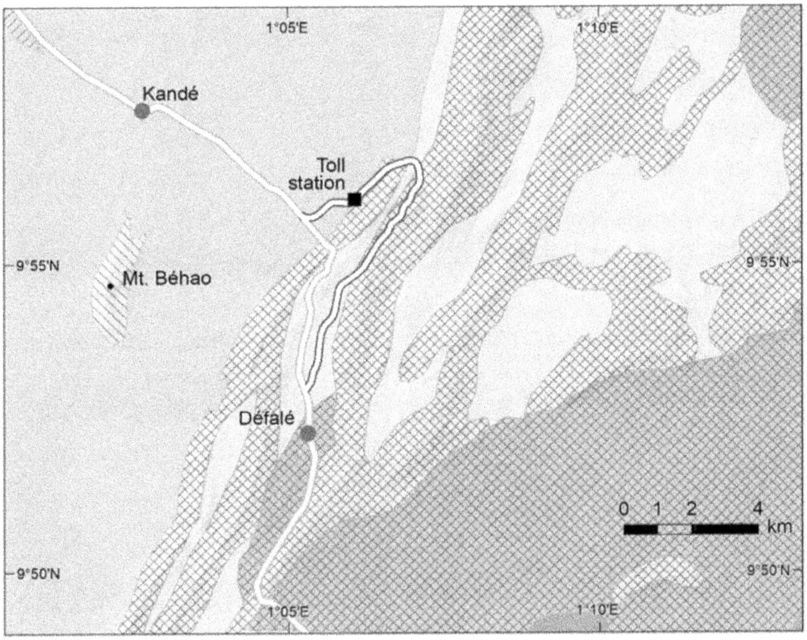

Figure 4. Geological and topographic sketch map of the Défalé Pass region with old National Road 1 (thin line) and new bypass (thicker line green). For legend see Figure 5 (Source: République du Togo 1984).

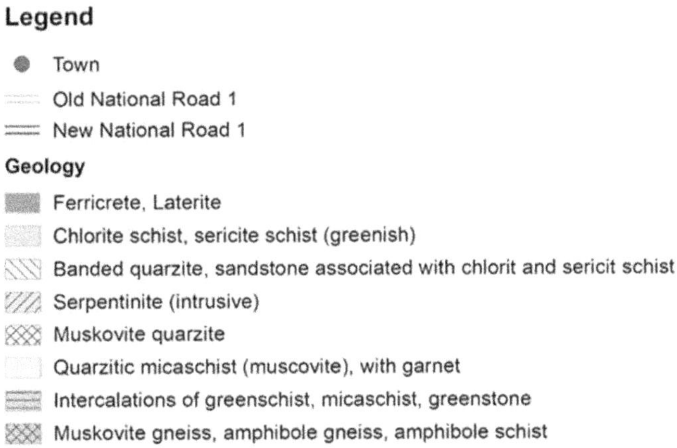

Figure 5. Legend of the geological and topographic sketch maps in Figure 3 and 4 (Source: République du Togo 1984, modified).

Finally, the new National Road 1 takes advantage of a natural geologic ‚gap‘, where obsequent drainage and incision follow a transversal fault. The road passes through a series of dominant SW-NE striking quarzitic escarpments. However, the long 180° bend (over almost two kilometres, changing from NE into SW direction, with an increasing descending slope) creates a severe risk for heavy traffic accidents in case

Figure 6. Images of rock samples and stratigraphy of the Atacora structural unit collected alongside the Aledjo bypass road and geochemical analyses (percent by weight) in Table 1; from the basement gneiss (6a) to the ferricrete crust (6f) on the Kpéwa etchplanation surface (Photos: J. Runge, 2019).

of brake failure. The second toll station ('Poste de Défalé') is located almost at the end of the bypass (Figure 4).

It remains unclear why in view of the geological and geomorphological circumstances the unfavourable Défalé bypass was finally chosen and realised. In fact, the feasibility study of 2006 proposed an alternative routing turning from Kara in the south, to the west via Sarakawa and then to the NW (via the villages of Léon, Kadjala, and Aloum), with eastern bypassing of Mt. Béhao and finally reaching Kandé from a S-N direction (Figures 4-5). Many bridges and hydraulic structures such as culverts have already been in place of this alternative track. Only the colonial German iron bridge crossing the Kara River must have been renewed to support heavy trucks. The big advantage of this alternative routing (parallel to the main

Figure 7. Finished bypass road between Aléhéride and Kpéwa in 2014 with exploration openings (Figures 1, 3 and 5); viewing direction is southwest. Today, six years later, the new road starts to degrade already (Photo: J. Runge, 2014).

Atacora escarpment) would have been avoidance of all old and new dangerous mountainous sections.

Journalist Ferdi Nando supposes on September 16, 2014 in the journal '27 Avril' (https://www.27avril.com) that the decision for the Défalé bypass finally followed political and economic interests as Défalé is the birthplace of the Togolese Minister of Economy and Finance. It is also said that the bypass road is of a local economic interest to certain Togolese companies. Moreover, it is often rumoured that the decision to realise the Défalé bypass was linked to the house construction of a new presidential villa in the Atacaora mountains. This private building should not be marooned from the international road network. Finally, Nando (2014) considers the Défalé bypass as a typical ‚white elephant', and a grave for millions that never will be in service. However, in 2016 the bypass road has already been opened and in use by international transport from Lomé to the Sahel.

Up to 2018/2019 numerous trucks accidents had occurred on the two bypass roads and significant degradation of the road surface already is visible. Deformation of the less adopted asphalt pavement, potholes and ruts are now visible at numerous places. The request of the Togolese government to the Chinese Road and Bridge Construction (CRBC) to renew the pavement was only partly considered. Meanwhile CRBC has left Togo, the budget for the two bypasses is used up and necessary road repair to sustain the roads are done in the 'traditional' way by patching single potholes where necessary. Long-term and sustainable preservation of the two new mountain bypass roads can be doubted.

2.4 CONCLUSION

Aside of Good Governance, countries in West Africa need in order to establish functioning market economies, a good network of infrastructure such as ports, railways and roads. The Republic of Togo developed since the turn of the millennium by its deep-sea harbour infrastructure a lifeline for the nation and for the landlocked Sahelian Hinterland (Burkina Faso, Niger and Mali). As the colonial railway lines had been systematically neglected and were no longer in operation, the PAL and the National Road 1 present a strategic international road link. However, the setting up of Lomé harbour caused severe problems of coastal erosion in Togo and Bénin. Due to the complicated geological and geomorphological setting in northern Togo (Sokodé-Kara-Kandé) it was essential to meet challenges for modern infrastructure planning. To facilitate reliably the transit tasks of the National Road 1, and to allow the competitiveness of the PAL in the absence of railways, Togo has solicited international funds to build and supported by China two lanes of mountain bypasses at the Alédjo fault and the Défalé Pass.

Longstanding observations in northern Togo proved that high quality feasibility studies are essential for the successful planning of infrastructure. From the geological point of view the construction of the two bypass roads allowed a deeper insight into the petrology, stratigraphy and geomorphology of the Atacora structural unit (e.g. evidence of a quarzitic anticline between Aléhéride and Bafilo). However, often political interests and decision making are more essential than geological facts and findings. Surveying and controlling technical issues of heavy trucks, respecting national laws, and combatting corruption is also of great importance to ensure safe traffic and transport of goods from the Atlantic into the Sahel.

The new national development plan ('Plan National de Développement', PND, 2018-2022) of Togo proposes as one lighthouse project the expansion of the National Road 1 from Lomé to Ouagadougou to a modern motorway with four lanes (Jeune Afrique 2019). A four lanes stretch is now realised from Lomé to Tshévié (30 km long). From today's view this plan could be feasible (if funds will be available) to the flat and levelled topographies of the south and the extreme north of Togo. The crossing of the Alédjo and the Défalé mountain section will remain a huge geotechnical problem. The recent Alédjo bypass is much too small to allow two additional lanes (Figure 7). A tunnel construction cannot be considered due to financial and logistical limitations. In the case of the Défalé Pass the discussed alternative routing deviating the escapements to the west is a conceivable scenario. However, the newly built Défalé bypass road will then be closed and for local use only.

In the future (2050) traffic volume and population in Togo will strongly increase and additional local bypasses of towns should be planned and built to protect citizens. However, this development will require agricultural land which is a limited resource in Togo. Finally, infrastructure is indeed not all for a reliable and sustainable development. The investment climate and customer confidence as well as capacity development of skilled workers are essential. It is incomprehensible why the planning of a new railway for freight trains, crossing the Togolese corridor, plays only a minor role in the recent discussion. The setting up of a new railway in Togo would be environmentally friendly which is the better link to the 'Gateway to the Sahel' in view of global emissions and global change.

REFERENCES

Affaton, P. 1987. *Le bassin des Volta (Afrique de l'Ouest): une marge passive, d'age Protérozoique supérieur, tectonisée au Panafricain (600+50 Ma)*. Doctoral thesis: 462. Marseille: Sci. Fac., St. Jérome.

BAD, Banque Africaine de Développement 2014. *Le secteur des transports au Togo. Bref Aperçu. Département des Transports, du Développement Urbain et des TIC*. Abidjan: BAD.

Blivi, A.B. & Anthony, E. 1999. Morphosedimentary evolution of a delta-sourced, drift aligned sandbarrier-lagoon complex, western Bight of Bénin. *Marine Geology* 158: 1–168.

Blivi, A.B. 1993. *Géomorphologie et dynamique actuelle du littoral du Golfe du Bénin (Afrique de l'Ouest)*. Doctoral thesis: 425. Bordeaux: Université Bordeaux 3.

Blivi, A.B., Anthony, E. & Oyédé, L. 2002. Sand barrier development in the Bight of Bénin, West Africa. *Ocean and Coastal Management* 45: 1–192.

Büdel, J. 1957. Die ‚Doppelten Einebnungsflächen'in den feuchten Tropen. *Z. Geomorph. N.F.* 1: 201–228.

Cornevin, R. 1959. Histoire du Togo. *Mondes d'outre-mer*. Paris: Berger-Levraut.

Cornevin, R. 1968. Note au sujet des limites de l'Eweland. *Journal of African History* XI(3): 501–502.

Cornevin, R. 1988. *Le Togo: des origines à nos jours*. Paris: Académie des sciences d'outre-mer.

de Hahn, L. 1983. Die Kolonialentwicklung des Deutschen Schutzgebietes Togo in räumlicher Perspektive. *Erdkunde* 37: 127–137.

FAD, Fonds Africain de Développement (n.d.). *Département de l'infrastructure: République Togolaise – Rehabilitation de la route Atakpamé-Blitta-Sokodé-Kara. Résume du Plan de gestion environmental et social*. Unpublished.

Gingerich, P.D. & Cappetta, H. 2014. A new Archaeocete and other marine mammals from lower middle Eocene phosphate deposits of Togo. *Journal of Paleontology* 88(1): 109–129.

Jeune Afrique 2019. Togo – Changement d'horizon. *Grand Format, Hebdomadaire International 3048 du 9–15 juin, Paris*: 87–112.

Laibi, A., Anthony, R., Almar, B., Castelle, N., Senechal, E. & Kestenare, E. 2014. Longshore drift cell development on the human-impacted Bight of Bénin sand barrier coast, West Africa. *Journal of Coastal Research* 70: 78–83.

Manson, M. 1969. Population density and 'slave raiding' – The case of the middle belt of Nigeria. *The Journal of African History* 10(4): 551–564.

Metzger, O.F. 1941. *Unsere alte Kolonie Togo*. Radebeul: Verlag Neumann-Neudamm.

Nando, F. 2014. Ouverture du contournement de Défalé annullée pour ne pas exposer la sompteuse demeure de Faure Gnassingbé sur le collines de Défalé. *Blog du 27 avril, L'Alternative No. 360*. URL: https://www.27avril.com/blog/transport/togo-ouverture-du-contournement-de-defale-annulee-pour-ne-pas-exposer-la-somptueuse-demeure-de-faure-gnassingbe-sur-les-collines-de-defale (accessed 01/12/2019).

Peltier, W. & Fairbanks, R. 2006. Global glacial ice volume and Last Glacial Maximum duration from an extended Barbados sea level record. *Quaternary Science Reviews* 25: 3322–3337.

Petit, M. 1981. Géomorphologie. Atlas du Togo. *Les éditions jeune afrique*: 8–9.

Port Autonome de Lomé (PAL) 2019. *Informations du port*. URL: https://www.togo-port.net/les-informations-du-port/actualite-port-autonome-lome/(accessed 21/11/2019).

République du Togo 1984. Notice explicative de la carte géologique à 1/200.000, Feuille Kara. *Mémoire* 1: 1–36.

Runge, J. & Anoumou, N. 2017. *Westafrika, Geographische Studienreisen in Bénin, Burkina Faso und Togo*. Frankfurt am Main: Institut für Physische Geographie und Frankfurter Geographische Gesellschaft (FGG), Frankfurt am Main, Selbstverlag.

Runge, J. 1990. Morphogenese und Morphodynamik in Nord-Togo (9°-11°N) unter dem Einfluss spätquartären Klimawandels. *Göttinger Geogr. Abh.* 90: 1–115.

Runge, J. 1993. Lateritic crusts as climate-morphological indicators for the development of planation surfaces - possibilities and limits. Z. *Geomorph. N.F.* 92: 201–216.

Runge, J. 2002. Morphographic analysis of Inselbergs in Northern Togo (West Africa) using high resolution DEMs derived from aerial photographs. In Yang, X. (ed.), *Desert and Alpine Environments. Advances in Geomorphology and Palaeoclimatology*: 22–34. Beijing: China Ocean Press.

Runge, J. 2013. *Republik Togo, Geographische Einblicke zwischen dem Golf von Guinea und der Sudanzone in Westafrika. Große Geographische Exkursion* 2012. Frankfurt am Main: Institut für Physische Geographie, Shaker-Verlag.

Schlüter, T. 2008. *Geological Atlas of Africa, with notes on stratigraphy, tectonics, economic geology, geohazards, geosites and geoscientific education of each country.* 2<sup>nd{{/ sup}} ed. Berlin, Heidelberg: Springer.

Sebald, P. 2013. *Die deutsche Kolonie Togo 1884–1914. Auswirkungen einer Fremdherrschaft.* Berlin: Ch. Links Verlag.

Spekker, H., Ammermann, K., Guillart,D. & Lankenau, D. 2010. Küstenschutzmaßnahmen in Togo – Stabilisierung erosionsgefährdeter Strandabschnitte und Schutz des Containerterminals Lomé durch den Neubau einer Buhne. *Bautechnik* 87(6): 311–317.

Statista 2019. Togo: Gesamtbevölkerung von 2008 bis 2018. URL: https://de.statista.com/statistik/daten/studie/415270/umfrage/gesamtbevoelkerung-von-togo/(accessed 21/11/2019).

Stein, K.-J. 2007. Vorkommen an Naturwerksteinen in Benin und deren Klassifikation nach dem UNFCR. *Z. dt. Ges. Geowiss.* 158(3): 429–446.

Tairou, M.S., Affaton, P., Sabi, B.E., Boudzoumou, N.F. & Da Cost, P.Y.D. 2012. The Aledjo quartzitic ruins in northern Togo: genesis and structural geomorphology. *Journal of Geological Sciences* 10(2): 167–174.

CHAPTER 3

Haematite iron ore deposits, environmental problems and economic development in Bandjeli, northern Togo

L. Carstens

Department of Physical Geography, Goethe-University, Frankfurt am Main, Germany

J. Runge

Centre for Interdisciplinary Research on Africa (ZIAF), Goethe-University, Frankfurt am Main, Germany

ABSTRACT: The primary haematite deposit of Bandjeli was extracted between 2008 and 2016 by several open-pit iron ore mining sites implemented by the Indian company MM Mining Ltd. Guided interviews with stakeholder groups and individuals were conducted in the canton of Bandjeli. The extraction did not lead to sustainable and increased economic development or positive long-term effects as infrastructure investments in the rural territory. Severe environmental problems occurred and peasants' agricultural yields decreased. Water sources were contaminated, and air polluted with haematitic dust. This led to health issues like pneumological diseases and abdominal discomfort. The working conditions on the mine and safety measures were not appropriate leading to numerous work accidents. The population of Bandjeli is enraged and not satisfied with the commodity exploitation and the consequences for development.

3.1 INTRODUCTION

The primary iron ore (haematite) deposit of Bandjeli have been well known for more than 1000 years. The local ethnical group of the Bassar developed a smelting furnace out of clay and smelted the iron. It was sold on the local markets and was further used for the manufacturing of tools (Hahn 1993). Due to this special smelting technique, the canton Bandjeli specialised and gained supra-regional reputation. The trade of Bandjeli's smelted iron rose until the late 19th century reaching a supra-regional level (De Barros 2012). During this time of German colonial rule, the explorer Hupfeld made in 1899 the first exploration of the iron ore of Bandjeli. His extraordinary findings made Germany to send the geologist Koert to Togo. In 1906 he first described the iron ore of Bandjeli as one of the purest haematites to be found in the world (Hahn 1993). Although, Germany planned on extracting the haematite and started building a national railway line for the transportation of the ore, the distance of more than 400 km to the coast axed the plans and the subsequent exploitation was never realised under German colonial rule (Hasselmann 2017). It took another hundred years for the iron ore of Bandjeli to become another investor's interest. In 2006, the Indian company MM Mining Ltd.

started negotiations with the Togolese government about the large-scale extraction of haematite ore, and industrial exploitation begun in 2008 (Republique Togolaise 2006).

It started as a win-win situation for both the company and the government, who were hoping to stimulate local economy and reduce regional disparities and poverty of the local population. For this purpose, MM Mining Ltd. was granted the use and maintenance of the colonial railway from Lomé to Blitta as well as port facilities in Lomé for commodity export (MME 2008, see chapter 2 in this volume). From the beginning until now, MM Mining Ltd. has shown little to no transparent business practises. There is basically no information available for the public about the mining project. This determined the growing concern about the company. The Extractive Industries Transparency Initiative (EITI, in French ITIE) confirmed the lack of transparency and lack of willingness to participate in the development of a transparent extractive industry in Togo (ITIE Togo 2017, 2019).

During the short extraction period, complaints about the environmental and working conditions started to rise. The population of Bandjeli suffered under increased dust exposition, surface and ground water pollution due to extraction. Furthermore, rumours of unpaid taxes increased the unacceptance of the Bahamas based Indian company in Togo.

In 2015, president Faure Gnassingbe visited Bandjeli while on his election campaign. He promised the canton to soon end the exploitation by MM Mining Ltd. One year later, MM Mining Ltd. officially resigned their actions due to strongly decreased world market prizes for iron ore. Although, the mine is not professionally closed, heavy machinery is still on the premise and the market prize has gone up again, MM Mining Ltd. has not yet (2019) resumed its activities. This study's aim was to examine the socio-ecological impacts of the haematite mining in Bandjeli and to investigate transparency aspects of the Indian company.

3.2 STUDY AREA

The iron ore mine is located at 9°25'29"N and 0°37'19"E in the canton Bandjeli in northern Togo. It is located in the west of the village Bandjeli and in the north-east of Byakpabé (Figure 1). The canton is part of the Kara Region and close to the Ghanaian border.

Geomorphologically, the landscape at Bandjeli is characterised by many levelled inselbergs (Runge 2002) that developed under former climates with higher humidity and warm temperatures, and therefore increased chemical weathering and erosion by levelling the surrounding rocks (Figure 2). The mountain Djohoulnal in Bandjeli with the haematite deposit forms a typical group of tropical inselbergs (Büdel 1981, Thomas 1994).

Geologically, the study area belongs to the margin of the West African Craton. The Craton was expanded during the Pan-African Orogeny (~650-500 Ma) by consolidating additional rock series to the basement. This former mobile zone is divided from East to West into the structural units of Dahomeyides, Atacora and Buem (Runge 1990). The latter hosting the iron deposit of Bandjeli (Figure 3). The Buem consists of the subunits of Byakpabé and of Katcha. These two subunits build an overlapping geological contact (Simpara et al. 1985). The haematite deposit was formed during compression and metamorphosis at this overlapping geological contact. A very strong tangential North-South-contact of the subunits led to the occurrence of ferruginous silicate rocks and mylonites. The haematite was formed during the drastic Proterozoic event as a major result of rock mylonitization: a process that occurs when high pressure due to tectonically active fault zones transforms and deforms rocks that develop

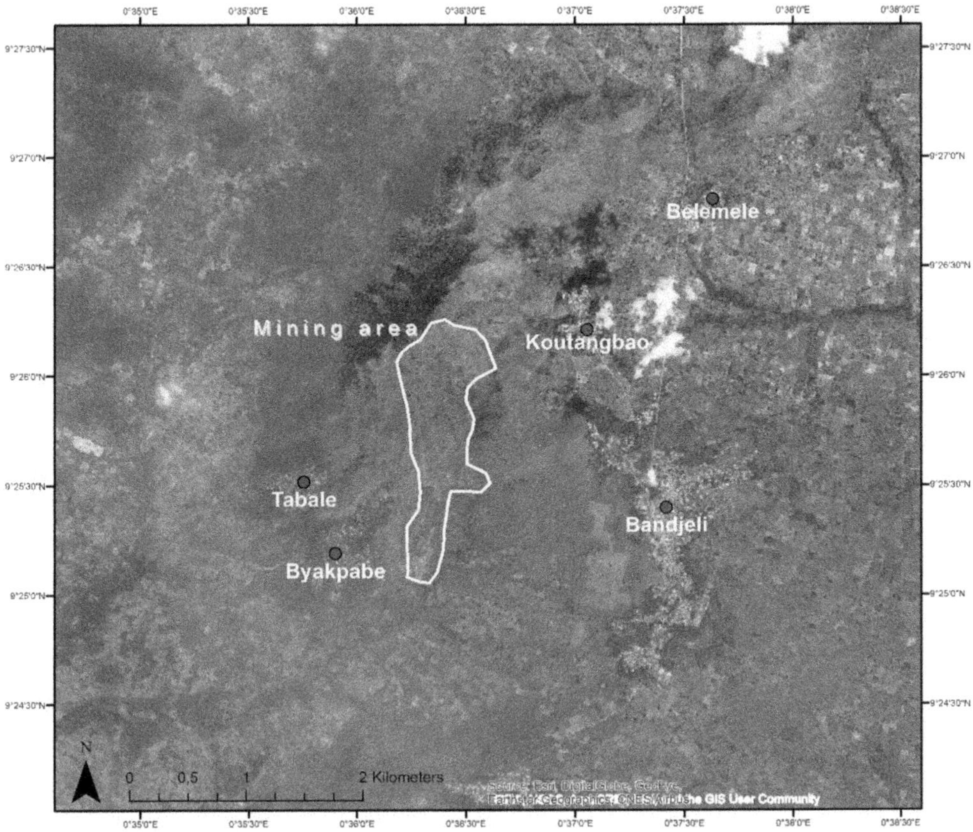

Figure 1. Study area with villages around Bandjeli that were visited and mentioned in the text. The mountain Djohoulnal is in the west of Koutangbao. The haematite iron ore mine operated by MM Mining Ltd. is located in the southern part of the mountain (status of 2012) (Source: Esri, DigitalGlobe, GeoEye, Earthstar Geographics, CNES/Airbus).

characteristic facies. It is of primary origin and mostly metamorphic (Simpara 1978). The haematite content (Fe_2O_3) of two rock samples was 79.25% and 69.42% (percent by weight) which is outstandingly high (Bessoles & Trompette 1980). The entire reserve of haematite ore is estimated at 500 Mio./tons (Lawson 1972).

The semi-natural vegetation cover is a savannah woodland that got deteriorated during the time of traditional iron smelting that was highly dependent on charcoal production and therefore led to increased deforestation. The anthropogenic influence, dominated by deforestation and agriculture, formed a savannah shrubland (De Barros 1988). The climate is dominated by a dry and rainy season. The rainy season starts in April and lasts till October; the dry season dominates the months around the turn of the year. The average annual rainfall amounts for 1400 mm and air temperature is consistently around 27 °C.

The region is characterised by a strong rurality. Most of the 18,000 inhabitants of the canton Bandjeli are small-scale (subsistence) peasants. Bandjeli is

Figure 2. Landscape with levelled, iron-rich inselbergs in the Bandjeli region. View from top of the mined haematite mountain in southern direction. Entrance of the mine is visible in the middle of image with buildings, machineries and tailings in the foreground (Photo: L. Carstens, 2019).

Figure 3. Opening at the margin of one of the iron ore open pits at Bandjeli (the scale bar on the right is about two metres long). Near surface iron rich, massive blocks crop out (with bright reddish colour) that are little weathered and can be easily mined. Blasting must be done at greater depth to loosen the solid haematite ore. Extracted blocks are subsequently crushed by nearby mining facilities (Photo: L. Carstens, 2019).

considered to be the "Yams-Capital of Togo" being the biggest yams (*Dioscorea*) producer. Apart from yams, peasants also plant maize, rice and peanuts. But other noteworthy industry or income generation is not existing. Continuous electricity is not available in Bandjeli. Most of the smaller villages have no access to power at all. The local ethnicity and the spoken language are both 'Bassar'. The population in Togo is generally very young with 60% of the population under the age of 25 and an average life expectancy of 60 years (Coulibaly 2020). No job opportunities other than agriculture, lack of educational facilities and cultural attractions result in young people leaving the villages in search for a better life.

3.3 METHODS

During fieldwork in Togo (March 2019), guided group and individual interviews were conducted (Figure 4). The guideline was designed before the field trip and adapted on-site (Mayer 2013, Kaiser 2014). It focused on ecological and social impacts of the haematite mine operated by MM Mining Ltd. but left enough space for an open conversation in which positive as well as negative aspects could be considered. In total, 20 interviews were conducted and recorded. Furthermore, additional on-site communication was taken into account by a report on memory. The main focus was on environmental and socio-ecological topics such as water, air, soil and agriculture, social conflicts, health and working conditions on the mine.

The main part of the interviewees were local authorities like the chief of the canton and local chiefs of other villages of the canton. Additionally, ordinary people on the street were recruited and interviewed, former iron mine workers

Figure 4. Example of a group interview with the representatives of the chief of Bandjeli in March 2019 (Photo: A.G. Edjabou, 2019).

included. The recorded interviews were transcribed and finally paraphrased using the qualitative content analysis by Mayring (2015).

Additional to the sociological part, the mine, that was not in operation since 2016, was visited and examined. Existing pits and tailings were inspected. This provided a better understanding of the management of the mine itself and geological outcrops and the hotspots of contamination sources concerning dust emissions could be observed.

3.4 RESULTS AND DISCUSSION

3.4.1 Economic development

Before the extraction of iron ore, the population of Bandjeli suffered from poverty and since there are no noteworthy employers or education and training opportunities. The new Indian mining project brought hope for a better economic development. MM Mining Ltd. promised to invest in the canton and to build infrastructure like schools and health centres. They wanted to rehabilitate and construct roads, wells, and most importantly, to ensure electricity supply for the whole canton.

People were hoping for new job possibilities and additional income generation through the support of local shops and facilities by foreign mine workers.

However, these expectations did not become a reality. The only positive development that was reported by the interviewees was the new school building with four classrooms that was built by MM Mining Ltd. They also reinstalled a well and repaired a small bridge in Byakpabé. The rest of the promised infrastructural measures were not implemented. Madjintéba Sertchi III, the chief of Bandjeli, summarized the change the iron ore exploitation has brought to the canton as follows:

> *"With the exploitation of iron ore nothing has changed. Everything we wanted the company to do for us... it has done almost nothing. There were a lot of things, a lot of plans. So, we asked the company and the state to provide us with electricity and until now, we don't have electricity." [Madjintéba Sertchi III, chief of Canton Bandjeli]*

While during its peak production time, there were about 300 Togolese mine workers, only 30 of them came from the canton itself. The rest was recruited in Lomé or in other parts of Togo. The company insisted on finding accommodation for the foreign workers in Kabou, where the head quarter of the Indian supervisor was located. Consequently, there was less money spent and gained in Bandjeli than expected. But people became most enraged during the interviews about the missing electricity. Only the centre of Bandjeli has access to temporarily limited electricity, that was recently installed by local authorities. The smaller villages of the canton still do not have access.

3.4.2 Air

Due to the extraction of the haematite ore that was combined with the use of explosives, red dust was constantly emitted in the environs of Bandjeli. The fine lateritic dust covered large areas of the landscape and was transported aerially over long distances. Only two people interviewed did not confirm the increased amounts of aerosols during extraction. But the majority of respondents complained about the increased amount and different colour that is associated with the reddish haematite ores (Figure 2). But there was no overall conclusion whether it was temporary during the dry season and after blastings or if it was the whole time. A teacher from Bandjeli commented on the air:

> *"Certainly, the air has changed because at the time of the operation, if you look at our rooftops, everything has turned red. You only have to travel a small distance and you will see that your own clothes have become dirty from the dust. But all I'm saying here, was in the period when they were regularly working on the mine. But today like this, no. Everything is back to normal." [Teacher in Bandjeli]*

The red dust covered houses, fields and water sources. This way the haematitic material could enter different areas. Since the red dust only occurred during the extraction, there was no possibility to examine the problem or to prove it in the scope of this study.

3.4.3 Water

The water availability and quality changed directly after the beginning of the mining activities as the chief of the canton describes:

> *"From the moment they started mining... you can see the mountain is high. The water there flows in all directions. And the water...when they started mining, they destroyed the rock and the residue got into the groundwater and everything turned red. All water sources are polluted." [Chief of canton]*

The fine haematitic material could enter the water sources via the air and through pollution of the groundwater caused by underground concussions of the explosions. From the 17 people that were asked about water conditions, 15 confirmed that the water changed because of the extraction. The water changed in different ways. Firstly, the colour became reddish and the water contained suspended material that settled to the ground after a few minutes. Secondly, some people said the taste and the smell have changed, too, describing it as "rusty". These changes involved the water from the drinking wells as well as the water from the natural stream. Since there is no other source for potable water, the population was obliged to drink the contaminated reddish water. To obtain water with less suspensions, they left the water to settle so they could extract the upper and cleaner part of the water.

The mine itself used a lot of water for the washing of the crushed iron ore. This led to a depletion of the groundwater level and consequently to less water available for the canton. It was reported that some wells close to the mine, that were used for irrigation, went dry and are still dry. The chief of Belemele stated that the stream became seasonally aquiferous and was all-season water-bearing before the onset of extraction.

Since the end of the mining activities in 2016 the water gradually returned to its usual state. According to the chief of Koutangbao the recovery of the water from the wells took about six months. The stream is still contaminated with an unnatural load of suspended iron matter. As there is no comparison to the state of the water sources before or during the mining, the conclusions are highly dependent on the statements of the population of Bandjeli.

3.4.4 Soil and agriculture

The soil and therefore agricultural activities close to the mine were affected as well. Considering the polluted water, that was partially used for simple hand irrigation, and the high amount of dust in the air, the haematitic material entered the soils. Some people said that the dust covering the buds or blossoms would lead to a decreased growth of the plant or even a loss of the fruits. In general, the majority of the people interviewed reported that their yields decreased since the exploitation started and is still not as high as it used to be. The chief of the canton blames the dust and the mud for the agricultural problems:

"The red dust that gets on the crops is not good. And even the mud that's invading the soil. The mud when they take it out of the [mining] dam, they dredge it up, they dredge the mud up there, they leave it there and it becomes hard stuff. That's not good." [Chief of the canton]

Especially the agriculture of Byakpabé suffered because of the extraction. They reported that they used to grow rainfed rice. But since their fields are directly southern of the iron ore mountain, the rain in the rainy season, that is essential for rice cultivation, brought down a lot of red mud from the open pit mine with its unsecured tailings. This led to a deeply bright reddish coloured soil. After the harvest, the grains of the rice are red. The colour is not substantial and can be washed off, but they say that the quality deteriorated and the yields became less. The rice production in Kara Region varied from 0.122 t/ha to 4.5 t/ha over the period from 1972 to 2013 (Ali 2018), implicating that yield changes may be normal and due to precipitation patterns. For some years it was not possible after all to grow rice on their usual fields. They moved most of their fields further to the west and still have problems with their agricultural production.

"They destroyed the land, it's uncultivable. The plots on which we used to grow rice are full of mud from the mine. There's so much mud that we can't cultivate anymore. We still have problems with farming." [Chief of Byakpabé]

Other fields that were directly located in the mining area had to be removed. The owners have not been compensated in any form. Only one Peul who was grazing his herd in the canton got a single financial compensation of 60,000 FCFA (~90 €).

The villages in the north-east of the mine were less affected. They reported to have had problems during the extraction as well but since the end of the mining their yields increase gradually.

3.4.5 Health

The health impacts of the haematite mine were due to the contamination of air and water. The result of the interrogations was that except of one person all claimed to have experienced negative health implications during the extraction.

The most prominent impacts were pneumological diseases caused by the high amount of haematitic dust. They suffered from pneumonia and chronic coughs. The second most frequent health impact were abdominal pains that were caused by drinking the contaminated water (Figure 5). This included stomach pains and diarrhoea, particularly for children. The water also led to skin irritations as reported by a few people. These occurred especially after having stood in the reddish water for a while. The chief of Byakpabé emphasized the skin irritations and breathing problems during the extraction:

"[...] the water has been contaminated with iron, which increases the risk of skin infection for the inhabitants. [...] Yes, there are health repercussions because the air is polluted. The only visible one at the moment is coughing. Dust also settles on the skin and could cause skin irritation." [Chief of Byakpabé]

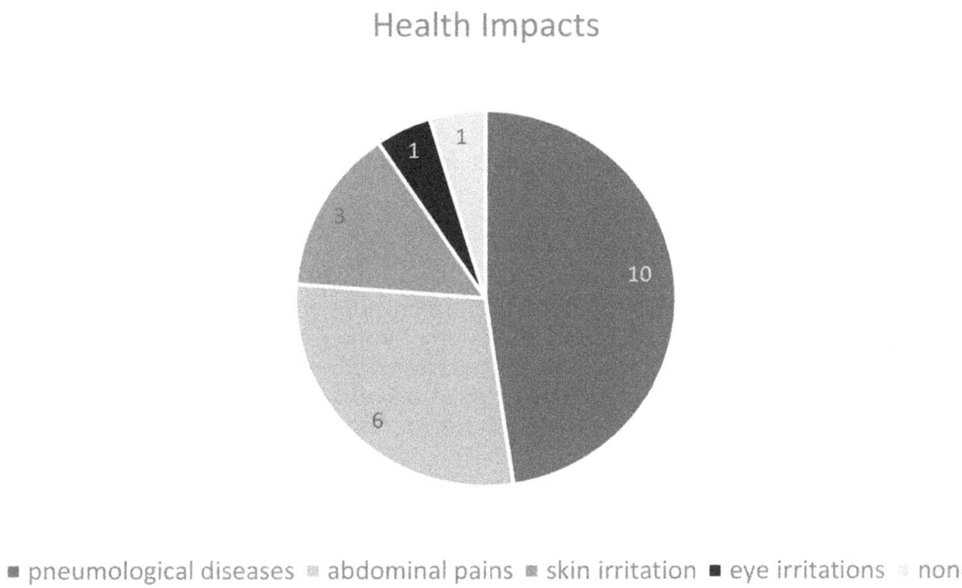

Figure 5. Health impacts on the local population in Bandjeli in relation to the haematite extraction. 16 people were asked about health conditions, multiple answers per person are presented in the diagram (Source: fieldwork 2019).

Only one person reported eye irritations.

Although there were no official medical studies implemented, the doctor of the local hospital in Bandjeli confirmed the higher abundance of the diseases mentioned above. According to these observations, he emphasized that children were more frequent to suffer from the diseases, especially the abdominal pains.

As animals are exposed to the same conditions, similar health impacts on them were observed. Especially, diarrhoea was reported as an increased animal disease during time of extraction.

Since the end of exploitation in early 2016, health impacts started to become better due to an improvement of aerial and water conditions. Some issues that became chronic like the coughs of a few people are still noticeable.

Another more indirect health impact is caused by the deterioration of the road condition. The doctor reported that there have been more road accidents since the heavy trucks of the mining company caused damage to the main road leading to Kabou and Bassar. Since the road condition has not yet been improved, the accidents continue although the mining has stopped.

3.4.6 Social conflicts

The small canton Bandjeli, besides being the biggest yams producer, played an insignificant role in Togo. This changed due to the extraction by MM Mining Ltd.. This new 'fame' made old rivalry come up. While Bandjeli is closest to the mountain and the mine and even in 1906 Koert described it as the "iron of Bandjeli" (Koert 1910: IV), MM Mining Ltd. implemented a sign on their premises saying 'iron of Kabou', a town in about 20 km distance in the east of Bandjeli. On top of that, the placement of mine workers in Kabou intensified the conflict. Until now, this conflict between Bandjeli and Kabou has not been solved.

Before the extraction MM Mining Ltd. granted the villages to conduct their religious fetishes on the mountain one last time and supported them financially so they could buy their offering. Not knowing it better, MM Mining Ltd. gave money to all of the villages, including those who do not have a fetish on the mountain and therefore could keep the money. This led to some tension in the canton as the other villages could not keep the money and felt treated unfairly.

3.4.7 Working conditions

The working conditions were reported by former mine workers and other inhabitants of Bandjeli. While some reported that the workers did not sign any official working contract, other said that the contract was imposed on them and that they had no opportunity to negotiate. They signed it without the actual salary written down. They worked in three to four different groups 24 hours from Monday to Saturday.

The daily wage for the ordinary workers were 1000 FCFA (~ 1.50 €) for a six to eight-hour work shift. The work was very physical and consisted mostly of carrying heavy rocks. Former workers complained about back pain. The chief of Belemele reported that the first shift was from 6-12 am for 1000 FCFA but after a few months MM Mining Ltd. increased the working hours from 6 am to 3 pm and keeping the 1000 FCFA wage. From their daily wage they had to buy lunch for 200 FCFA, leaving them with 800 FCFA a day. It was also reported that MM Mining Ltd. did not pay the workers the last three months, they are still waiting for their salary.

The working conditions on the mine were very poor. Although they were directly working in the dusty environment and handling heavy rocks, they did not have sufficient safety measures. MM Mining Ltd. handed them helmets and respiratory protection masks on a very unregular basis, if at all. They used their own shoes because there were no safety shoes handed by the company. After the masks wore out, MM Mining Ltd. did not replace them so the workers were obliged to use their clothes as facial protection. Only when the authorities visited did the company hand out the safety materials.

The harsh working conditions were well known in Bandjeli, even under people who did not work at the mine, like a teacher who was enraged about the situation and stated the following:

> *"The conditions under which people were working there, it's very deplorable. The financial treatment first, it's difficult. It is very, very, very, very difficult and to get their salary at the end of the month was a problem. You have to wait three months to get a month's salary. And the treatment, for example when they had taken the workers, the workers on the mines, in the dust, all that there, and he is treated with 1000 FCFA a day. It was difficult. Well, even the equipment to work. For example, they are not protected. They are in the dust, the iron deposits, but there is nothing to protect them. It's up to them to look for their own shoes, clothes, hats, even the muffler, there's none. So, they worked in very deplorable conditions. It's not right." [Teacher in Bandjeli]*

Accidents occurred regularly, reported the former workers. Some of them ended fatally, some of injured needed a hospital treatment. One former worker explained that he got hit by a rock that fell out of the truck and smashed his hip. MM Mining Ltd. paid for his first operation but not for the operation that was to follow. He has been asking for pain killers and medicine but did not receive any financial aid or compensation from the mining company. Since the roads on the mine are very rough, many workers got injured by rocks falling off the trucks. The lack of sufficient helmets increased the mortality rate of these accidents.

One reason for the frequent accidents was the condition the mine was in. Compared to international standards described in Rankin (2011) the mine was built in a seemingly provisional manner: roads and pits are allocated randomly on the mining property. Some pits are left unfinished. Tailings are spread on the whole premise, next to the pits and downhill next to the machineries. The roads are not paved or secured. Gully erosion that cut the roads leads to potholes making them more dangerous for heavy trucks.

3.5 CONCLUSION

The rich haematite iron ore deposits located in the canton of Bandjeli could help the local and national economy to grow and to insure sustainable development. However, the extraction by the Indian company MM Mining Ltd. did not meet local and national expectations. Environmental protection measures were not implemented. The increased amount of red dust in the air as well as the contamination of surface and ground water led to health issues and agricultural harvest losses. Most of the changes were temporary and are now, after the closure of the mine, improving. Although, the recent experience with a large-scale mining company was not satisfying for the local population, the majority would support a new mining project with another more transparent company that would invest in the local infrastructure and treat people well.

REFERENCES

Ali, E. 2018. Impact of Climate Variability on Staple Food Crops Production in Northern Togo. *Journal of Agriculture and Environment for International Development* 112(2): 321–341.

Bessoles, B. & Trompette, R. 1980. Géologie de l'Afrique: La chaîne panafricaine "zone mobile d'Afrique Centrale (partie Sud) et zone soudanaise". *Memoire du Bureau de Recherches Géologiques et Minières* 92.

Büdel, J. 1981. *Klima-Geomorphologie*. Berlin, Stuttgart: Gebrüder Borntraeger.

Coulibaly, N. 2020. Accélérateur de croissance. *Jeune Afrique* 3083: 122–124.

De Barros, P. 1988. Societal repercussions of the rise of large-scale traditional iron production: a West African example. *The African Archaeological Review* 6: 91–113.

De Barros, P. 2012. The rise of the Bassar Chiefdom in the Context of Africa's internal Frontier. In Monroe, J.C. & Ogundiran, A. (eds.), *Power and Landscape in Atlantic West Africa: Archaeological Perspectives*: 255–277. Cambridge: Cambridge University Press.

Hahn, H.P. 1993. Eisentechniken in Nord-Togo. *Kulturanthropologische Studien* 21. Münster: Lit.

Hasselmann, K.-H. 2017. Bassar: Les échappées de lumière a l'epoque pré-coloniale et colonial. L'Ouverture du Nord-Togo 1894–1914. *Récits Ouest-Africains* XVIII. Simmern: Pandion.

Initiative pour la Transparence dans les Industries Extractives (ITIE) Togo 2017. *Extractive Industries Transparency Initiative in Togo. EITI Report 2015*. Lomé: ITIE Togo.

Initiative pour la Transparence dans les Industries Extractives (ITIE) Togo 2019. *Initiative pour la transparence dans les industries extractives: Rapport 2017*. Lomé: ITIE Togo.

Kaiser, R. 2014. *Qualitative Experteninterviews: Konzeptionelle Grundlagen und praktische Durchführung*. Wiesbaden: Springer Fachmedien.

Koert, W. 1910. Begleitworte zur geologischen Karte von Togo. In Meyer, H. (ed.), *Das deutsche Kolonialreich: eine Länderkunde der deutschen Schutzgebiete. Teil II: Togo, Südwestafrika, Schutzgebiete in der Südsee und Kiautschougebiet*: I-V. Leipzig: Verlag des Bibliographischen Instituts.

Lawson, D.T. 1972. *Géologie et perspectives économiques de la formation ferifère de la cuvette du Buem au Togo*. Rapp. inédit. Lomé: BNRM.

Mayer, H.O. 2013. *Interview und schriftliche Befragung: Grundlagen und Methoden empirischer Sozialforschung*. München: Oldenburg Wissenschaftsverlag GmbH.

Mayring, P. 2015. *Qualitative Inhaltsanalyse: Grundlagen und Techniken*. Weinheim: Beltz.

MME, Ministère des Mines et des Energies 2008. *DECRET N°2008-021 PERMIS MM MINING*. URL: http://www.pdgm.tg/index.php?option=com_docman&view=download&alias=75-decret-n-2008-021-permis-mm-mining&category_slug=permis-recherche&Itemid=502 (accessed 2020-04–11).

Rankin, W.J. 2011. *Minerals, Metals and Sustainability: Meeting future material needs*. Boca Raton: CRC Press.

République Togolaise 2006. *Convention d'investissement entre la Republique Togolaise et MM Investment Holding Limited*. URL: http://www.pdgm.tg/index.php?option=com_docman&view=download&alias=158-convention-d-investissement-mm-mining-07-08-2006&category_slug=documents-statistiques&Itemid=502 (accessed 25/04/2020).

Runge, J. 1990. Morphogenese und Morphodynamik in Nord-Togo (9°-11° N) unter dem Einfluss spätquartären Klimawandels. *Göttinger Geogr. Abh.* 90: 1–115.

Runge, J. 2002. Morphographic analysis of Inselbergs in northern Togo (West Africa) using high resolution DEMs derived from aerial photographs. In: Yang, X. (ed.): Desert and Alpine Environments. Advances in Geomorphology and Palaeoclimatology. Beijing: China Ocean Press.

Simpara, N. 1978. *Étude géologique et structural des unites externs de la chaîne panafricaine (600 MA) des Dahomeyides dans la région de Bassar (Togo)*. Doctoral thesis: 338. Marseille: University of Marseille III.

Simpara, N., Sougy, J. & Trompette, R. 1985. Lithostratigraphie et structure du Buem unité externe de la chaine panafricaine des Dahomeyides dans la région de Bassar (Togo). *Journal of African Earth Sciences* 3(4): 479–486.

Thomas, M.F. 1994. *Geomorphology in the Tropics. A study of weathering and denudation in low latitudes*. Chichester, New York: John Wiley & Sons.

CHAPTER 4

A multi-criteria approach for CO_2 reduction through climate-friendly material selection: Housing construction in the Volta Basin, Ghana

G. Sarpong-Nsiah

Open University of Malaysia / Accra Institute of Technology, Accra, Ghana

ABSTRACT: Drawing on the concept of Sustainability in Environmental Development (SED), building stock makes a significant contribution to CO_2 emissions. The feasibility of SED lies in rethinking Locally Sourced and Recycled Building Materials (LSRBMs) selection decision making, in bringing cultural issues to the attention of civil engineers and designers in mainstream housing development, towards the wider use of Recycled Laminated Wood Flooring (RLWF) which is 55%, in CO_2 emission, as good as Fly Ash Cement Concrete Floor Slab (FACCFS). Decisions involve sociocultural factors need to be traded off. To do that, they have to be measured alongside sustainability principle indicators impacting the selection of building materials whose measurements must also be evaluated as to, how well, they serve the selection of LSRBM with the least CO_2 emissions. The analytical hierarchy process is a concept of measurement through pairwise comparisons to derive priority scales. These scales measure intangibles in relative terms. This explained a 15.9% cut back in CO_2 emissions for selecting Bamboo XL Laminated Split Paneled Flooring (BXLLSPF) instead of selecting FACCFS, the lack of informed knowledge in the wider use of BXLLSPF and RSLWF and in the Ghanaian context towards climate change mitigation in the countries situated in the Volta Basin.

4.1 INTRODUCTION

Rapid population growth and industrialization throughout the world together with increasing living standards have turned the creation of the built environment into a rising peril to the natural environment (Emmanuel 2004, Anderson et al. 2009). Building material selection is an important issue in building design and construction decision-making and environmental issues need to be incorporated into the evaluation process. Between 1960 and 2010, the population of Ghana recorded a rapid growth from 6,726,815 to 24,658,823 inhabitants with a corresponding housing deficit of 2,074,391 (Quandzie 2011). Measures are required to address the housing situation with implications for energy consumption and potential climate change impact (Ofori 2006).

As the housing construction industry faces the likelihood of increasing energy efficiency in buildings, building material selection decision-making has become more complex. Support provided to decision makers by traditional approach has evolved from simple predefined reports to complex and intelligent-based analysis and judgments (Ding 2008). Within this context, material selection tools have been developed to promote environmental awareness (Cole 1999, Ding 2008) and encourage future reduction in carbon dioxide emissions. However, while ongoing research is making positive contribution in various ways including establishing a solid base to aid *Locally*

Sourced and Recycled Building Materials (LSRBMs) selection, there have been little systematic efforts to examine the role of the sociocultural as vital resources for strengthening the material selection decision-making process (San-Jose et al. 2007, Sarpong-Nsiah & Ahadzie 2018). To this effect, bringing cultural issues to the attention of civil engineers and designers and encouraging LSRBMs selection in mainstream housing developments in the Oti-Volta basin, can potentially contribute towards future reduction in carbon dioxide emissions. Florez et al. (2010) have stated that the appropriate materials for sustainable design vary by impact priorities, regional issues, project budgets, and performance requirements. In this paper a wider use of LSRBMs, the role of sociocultural is brought to the attention of civil engineers and designers of the housing construction industry of West Africa.

4.2 TECHNOLOGY IN MATERIAL SELECTION: STATE OF ART

4.2.1 Building assessment methods

A multi-criteria decision support system for quantitative cost analysis proposed by Mahmoud et al. (1996) provides information to aid the designer with material decision-making. However, no specific information on the methodology for evaluating such materials is given. Mohamed & Celik (1998) provided data utility that enables users to intuitively choose their most preferred option from a list of materials but failed to mention the techniques used for evaluating the list of materials finally selected. Perera & Fernando (2002) acknowledged a computer-based cost modeling material management system for roofing material selection and did not encourage the integration of a broader range of factors into the material selection process. Rahman et al. (2009, 2008) described an integrated knowledge-based cost model for optimizing the selection of roofing materials and technology for residential housing designs. In the framework of their study they presented an evaluation tool to assess the cost sustainability of roofing materials with very little precise information about the technical, sociocultural and environmental issues. Ding et al. (2010) introduced a comprehensive assessment decision support system that measures the life-cycle environmental characteristics of a building product using a common and verifiable set of criteria and targets for building owners and designers, to achieve higher environmental standards. However, it appears to only direct sustainable material selection towards environmental issues.

4.2.2 Review on building materials selection tools

Comprehensive Environmental Performance Assessment Scheme for Buildings (CEPAS) is a holistic assessment tool for various building types with clear demarcation of the entire building life cycle that covers the pre-design, design, construction, demolition and operation stages. It employs an additive/weighting approach, which introduces and organizes performance criteria make a clear distinction between "human" and "physical" performance issues as well as "building" and their "surroundings (Crawley & Aho 1999). However, for the CEPAS assessment model, only single-ownership buildings are eligible for assessment. Netherland's *Environmental Preference Method* (EPM), building environmental assessment model, that is building material specific, a manual that contains a list of preferable materials and products, sorted according to their position in different components of a building.

However, it was already used as a tool for environmental evaluation in some European projects. *Building Research Establishment's Environmental Assessment Method* (BREEM) is an environmental building assessment method. BREEAM covers a range of building types including: offices, homes, industrial units, retail units, and schools. Material selection is based on awarding points for each criterion and the points are added for a total score calculated based on the credits available, number of credits achieved for each category and weighting factor. The energy performance assessment adopts the UK building regulation as a benchmark to rate the level of performance improvement, which may not necessarily apply to other regions with an entirely different assessment structure. Australia's *Green Star* evaluates the environmental merits of building products using the credit rating system based on a number of points allocated to the credits in order to determine the total scoring and hence the level of certification. The disadvantage with this tool is that it uses is limited to the evaluation of lettable areas within office buildings, hence excludes areas that are not offices or supporting the office. Moreover, the assessment structure is delineated in Australian standards and perhaps may not apply to other regions with different socio-technical background given the differing views on impact assessment. *Leadership in Energy and Environmental Design* (LEED) is another environmental building assessment method. A holistic tool for determining the environmental impact of building products and facilities from the whole building perspective. However, overemphasis on environmental benefit without an equal concern for the durability of the products employed to achieve this environmental benefit. *Sustainable Building Assessment Tool* (SBAT), environmental building assessment method, considers building as a whole, the design of a building in terms of sustainability, and explicitly introduces performance criteria that acknowledge social and economic issues, mainly assesses building performance with little recourse to material indicators. Unable to differentiate between choices of materials except for indirect consequences.

4.2.3 Carbon dioxide emissions

Carbon dioxide (CO_2) emission in the construction industry is very noticeable. This rate can be measured and quantified, both in energetic terms and in terms of carbon emissions. A great quantity of carbon dioxide is emitted into the atmosphere through the different phases of a building life cycle:

- In the exploitation,
- In the production of materials and products,
- In the setting on site,
- In the construction of the building itself,
- In the renovations,
- In the later rehabilitations
- In the final demolition.

The main sectors of energetic expense of a country are consumption for maintenance and air conditioning of buildings, transport and industry. Energy consumption for maintenance and air conditioning of buildings has a direct and immediate relation with construction. Goldenberg (1998) defines carbon dioxide emissions as a third of the energetic expense that comes directly or indirectly from the construction activity. Furthermore, Edwards & Hyett (2001) states that a participation of close to 50% of the total energetic cost in developed countries is closely linked or is a consequence of the construction industry. The energy for maintenance has already received the positive attention of state institutions, which have emphasized the implementation of the use of alternative energies, as well as the intense use of thermal insulators and responsible consumption. These decisions can drastically reduce the energetic expense of the refrigeration of buildings. With reference to the other sectors of energy consumption, those of transport and industry, there is little information as far as building construction is concerned. However, many studies (Xing et al. 2008, Monahan & Powell 2011, Shams et al. 2011) have acknowledged that the close relationship between the production of carbon dioxide and construction has the following aspects: estimation of carbon dioxide produced by all industrial activities related to the building erection and estimation of carbon dioxide emission reduction, which can be reached by an adequate selection of materials. Subsequently, drawing from the generic definition provided by Goldenberg (1998), it is contended that carbon dioxide emission measure should appropriately and explicitly be redefined as emissions from rethinking locally sourced and recycled building materials selection for mainstream housing developments. Furthermore, the emission should be linked to the conceptual phases of the project. Likewise, the deterioration of the physical environment due to housing construction activities is traceable to the choice of building products at the early design stages. For example, Shams et al. (2011), critically examined a five-floor residential building and the associated carbon dioxide emissions for different construction materials. The results of the analysis showed a 52% reduction in the total embodied energy and 45% reduction in the total carbon dioxide emissions. The achievement came from replacing cement concrete and mortar with fly-ash or blast furnace slag. Xing et al. (2008) in a comparing steel and aluminum with concrete, in residential buildings, in terms of embodied energy and carbon dioxide emissions, concluded that concrete exhibits lower energy consumption than steel or aluminum in the entire building life-cycle. The case studies revealed that concrete dominates in terms of mass, while steel and aluminum dominate in terms of carbon dioxide emissions due to the high values of carbon dioxide emissions per material mass.

4.2.4 The role of sociocultural as vital resources for strengthening the material selection decision-making process

The sociocultural perspective is a theory used to describe awareness of circumstances surrounding individuals and how their behaviors are affected specifically by their surrounding sociocultural factors (Jaramillo 1996). To this end social environment refers to the immediate physical and social setting in which people live or in which something happens or develops. It includes the culture that the individual was educated or lives in, and the people and institutions with whom they interact. San-Jose et al. (2007) argue that the sociocultural variable forms an implicit part of the design decision-making process, as it helps to define the architectural of the region, as well as promote the image of the community. The sociocultural values of mankind is known to vary from one society to another. The values of sociocultural have direct and indirect influences on humans' habitation. "Self-build houses" involves the intermittent and piece-meal acquisition and utilization of housing inputs such as plots of land, sand, various building materials in the housing construction process in tandem with building owners resource flow. Indeed, over 90% of housing stock in Ghana have been realized through this process which in effect can be simulated to gradual lock of housing capital during the housing delivery period ranging from 5-15 years. This is indeed the housing realization culture in Ghana (Ahadzie & Amoa-Mensah 2010). However, Boamah (2010) argues that majority of houses in the country are provided by the private sector, mainly by individuals through personal savings and assistance from family members. This informal way of house financing constitutes about 90% of the houses delivered in the country and construction is done incrementally.

The Ghana building code approves the use of the following building materials: Mud or swish used in a plastic state to erect an earthen wall, wattle and daub, pisé or earth rammed between wooden or other form work to make a wall in situ. Unburnt earth bricks or blocks, stabilized earth products, bricks, blocks, burnt clay products, concrete or reinforced concrete, thatch or leaves in roofing or otherwise, timber or bamboo products, asbestos-cement and metal products, glass and synthetic products, as well as stone and lime-based goods are used. In Ghana the materials for floors of dwelling units are:

- Cement/concrete (77.8%),
- Earth/mud (16.0%),
- Wood (1.0%),
- Others (5.2%)

The materials selected for roof of dwelling units:

- Metal sheet (71.4%),
- Slate/asbestos (13.0%),
- Thatch/palm leaves or raffia (8.6%),
- Other (7.0%) (UN-Habitat 2011a)

Ghana classifies dwelling units into the following categories:

- Compound house (51.5%),
- Separate house (28.7%),
- Semi-detached house (7.1%),
- Flat/Apartment (4.7%),
- Others (8%) (Owusu 2010).

To this effect, the main type of material selected depends on the part of the building, sociocultural, availability and affordability of the material.

4.2.5 Analytical hierarchy process

The *Analytical Hierarchy Process* (AHP) model offered a logical and representative way of structuring the decision problem and deriving priorities. The method is a theoretically sound and practicable approach for selecting, weighting, standardising and aggregating individual criteria into a composite index. The technique allows both quantitative and qualitative criteria to be entered into the model and offers an overall solution for the model (Singh et al. 2007). AHP compares decision factors by pairs and assigns weights to reflect their relative importance (Saaty 1986). Singh et al. (2007) describe AHP method as a multiple step analytical process of judgment, which synthesizes a complex arrangement into a systematic hierarchical structure. It allows a set of complex issues that have an impact on an overall objective to be compared with the importance of each issue relative to its impact on the solution of the problem (Raisinghani & Meade 2005).

The AHP is designed to cope with the intuitive, the rational, and the irrational when making multi-criteria materials selection decision and handle the complexities of real-world problems. AHP models a decision-making framework that assumes a unidirectional hierarchical relationship among decision levels (Presley 2006). The top element of the hierarchy is the overall goal for the decision model. The hierarchy decomposes to a more specific attribute until a level of manageable decision criteria is met. The

Table 1. Comparison scale adapted.

Degree of importance	Definition
1	Equal importance
3	Importance of one element over another
5	Strong importance of one element over another
7	Very strong importance of one element over another
9	Absolute importance of one element over another
2, 4, 6, 8	Intermediate values between two adjacent degrees of importance

method's fundamental rationality is decomposing a dataset into smaller constituent elements and then eliciting pairwise comparisons by using a fundamental (1–9) scale developed by Saaty (1980) to determine their specific priorities (Table 1). Thus, although the hierarchical structure of the AHP method does facilitate analysis, it is the method's ability to measure and synthesize the multitude of factors within the developed hierarchy that truly sets the method apart (Singh et al. 2007). In this paper, it is tested to derive weights of criteria by the prioritization of their impact to future reduction in overall carbon dioxide emission in the selection of LSRBMs in MHDs assessment. The AHP calculates the inconsistency index as a ratio of the decision maker's inconsistency and randomly generated index.

4.2.5.1 Consistency ratio

Various methods have been devised to deal with inconsistency. Saaty (2007) suggests the following consistency index (CI):

$$\mathrm{CI} = \frac{\lambda_{\max} - 1}{\mathrm{n} - 1} \tag{1}$$

where λ_{max} is the largest eigenvalue and n is the number of elements within a branch being compared. If a matrix is perfectly consistent (cardinally) than λ_{max} will be at a minimum and equal to n, producing a CI equal to zero. As inconsistency increases λ_{max} will become increasingly large, producing a larger value of CI. This consistency index can also be expressed as a consistency ratio:

$$\mathrm{CR} = \frac{\mathrm{CI}}{\mathrm{RI}} \tag{2}$$

where RI is a known random consistency index obtained from a large number of simulation runs and varies depending upon the order of matrix (Table 2).

If the value of CR is less than 10%, it implies that the evaluation within the matrix is acceptable or indicates a good level of consistency in the comparative judgements represented in that matrix. In contrast, if CR is more than the acceptable value, inconsistency of judgements within that matrix has occurred and the evaluation process should therefore be reviewed, reconsidered and improved. An acceptable consistency property helps to ensure decision-maker reliability in determining the priorities of a set of criteria (Table 2) (Saaty 2000).

Table 2. Average random index for corresponding matrix size.

Matrix size (n)	3	4	5	6
Random Index (RI)	0.58	0.9	1.12	1.24
Consistency Ratio (CR) less than or equal to	0.05	0.08	0.1	0.1

4.3 TOWARDS A CONCEPTUAL FRAMEWORK FOR MATERIALS SELECTION DECISION

The identification of the factors, strategies, drivers and barriers towards building material selection is a starting point to sketch out ideas and work out which factors were key for inclusion in the outline of a conceptual framework of multi-criteria decision-making for the mainstream housing developments of the housing construction sector of countries situated in the Oti-Volta basin. The analysis of the surveyed questionnaires identified a fifty-six (56) key influential factors as important components of the material selection process. The factors are compressed into six groupings: 1) Environmental/Health; 2) Economy/costs; 3) Sensory; 4) Socio-cultural; 5) Technical; and 6) Site conditions. For the purpose of clarity in the functions, similarities and differences in properties, the factors are compressed into four groupings: 1) Performance capabilities, 2) Sociocultural benefits, 3) Environmental impact, and 4) Economy efficiency. Socio-cultural variables are intangible and have not received much attention in the architecture literature as other groups of factors (San-Jose et al. 2007). As it can be seen from Figure 1, the visual tracking of the analysed decision factors for measuring the sustainability impacts of materials for the housing industry of Ghana in the context of climate change mitigation. The framework has been sub-divided into nine (9) areas. Each box represents a group of factors that the participants had identified as attributes that they consider when selecting building materials with sociocultural as key attribute. Building elements are: flooring, walling, doors/windows, ceiling and roofing. The sustainable principle indices are: technical factors, economy factors, general site conditions, sociocultural, sensorial and environmental factors. Table 3 describes the key influential criteria.

4.3.1 Defining problems

Building material evaluation begins with a definition of the problem, then formulating material attributes, objectives and goals (van Pelt 1994). Thereafter, the problem is structured to provide adequate specification of objectives, so that attributes can be identified. Nijkamp et al. (1990) and Ding (2008) noted that early identification of material constraints is critical to developing a more precise set of alternatives in order to optimize the best solutions.

4.3.2 Identifying alternatives

The next step of the multi-criteria assessment process is to identify alternatives, based on the identified problem. At this stage, the list of feasible alternatives is identified. Alternatives may include design alternatives, location options, material options and technology.

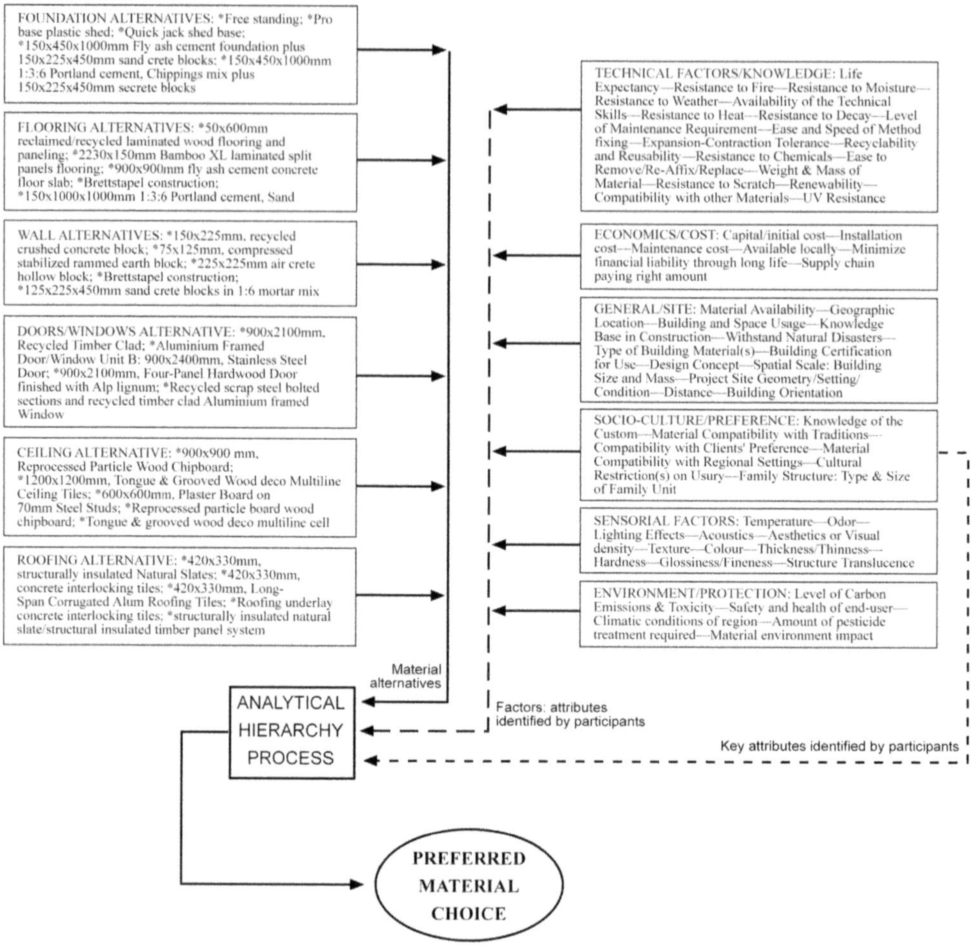

Figure 1. Conceptual framework of the analysed decision factors for measuring the sustainability impacts of materials for the housing industry of Ghana (adopted from Sarpong-Nsiah & Ahadzie 2018).

4.3.3 Identifying criteria

Next, the evaluation criteria are defined following the identification of material alternatives. The decision model eliminates those that are less important based on their weighting scores. The criteria are used as guidelines to analyze impacts from each material alternative (Nijkamp et al. 1990, Ding 2008).

4.3.4 Assessing impacts

At this stage, each criterion is measured using the most appropriate method for its nature to reflect its relative importance against each alternative (Saaty 2001). It involves expressing impacts in numeric terms and information may be presented in an evaluation matrix with alternatives set against criteria in a spreadsheet (Saaty 1994).

Table 3. Definition of key influential criteria.

Main criteria	Sub-criteria	Definition
EN: Environmental Impact	EN1: Amount of pesticide treatment required, EN2: Material environmental impact, EN3: Level of carbon emissions, EN4: Environmental statutory compliance EN5: Climatic condition of the region	General conditions promoting the completion of a project without major accidents of injuries to users
EC: Economy Efficiency	EC1: Cost of energy spent in manufacturing and transporting materials to site, EC2: Availability of material, EC3: Clients financial budget, EC4: Affordability of the materials, EC5: Life cycle cost	The long-term financial impact of housing projects when selecting building materials
SC: Sociocultural Benefit	SC1: Material compatibility with tradition, SC2: Local knowledge of the custom lifestyle, SC3: Material compatibility with client`s preference	Define the architecture of the region, as well as promote the image of the community
TN: Technical Performance	TN1: Weight/mass of material, TN2: Resistance to chemicals TN3: Resistance to heat	A rational framework for building design and construction that is flexible and amenable for accommodating innovations and change

4.3.5 Estimating weights

In material evaluation, choosing an option from a list of alternatives means that priorities must be set and weights assigned to each criterion, reflecting each criterion's priority. Nijkamp et al. (1990) and Saaty (2008) suggest that various methods of estimating criteria weighting include direct estimation which involves the expression of relative importance of the objectives or criteria in a direct way through questionnaire surveys, or the indirect approach where Weights are obtained through estimating actual previous behavior derived from ranking alternatives or through an interactive procedure of obtaining weights by questioning the decision-maker and other involved parties.

4.3.6 Reaching conclusions

Finally, decisions are made according to the score of each alternative. This stage provides further information to select the most appropriate of all ranked alternatives that meets the ultimate objective(s) (Ding 2008). Given the previous discussions on the trend towards multiple criteria in material appraisal, it is therefore necessary to identify a multi-criteria assessment technique that facilitates multiple dimensional assessments of criteria to aid informed decision-making when selecting locally sourced and recycled building materials for mainstream housing development sector of housing construction of Ghana and countries situated in the Oti watershed. In order to avoid inconsistencies across jurisdictions regarding the multi-criteria assessment techniques applied is the analytical hierarchy process. The analytical hierarchy process enables potential users to break down complex unstructured problems into component parts, and arrange these parts into a hierarchy of a much simpler and more logical judgments by simply assigning numerical values to represent each of their preferred choice of materials and variables and the formulation of the mathematical models required for computing the carbon dioxide emissions utility index (Figure 2).

4.3.7 Analysis of the material selection factors

As part of an exercise to maintain consistency in the material selection decision-making process during crucial stages of the design, and to clarify the similarities and differences in properties and functions between various factors are:

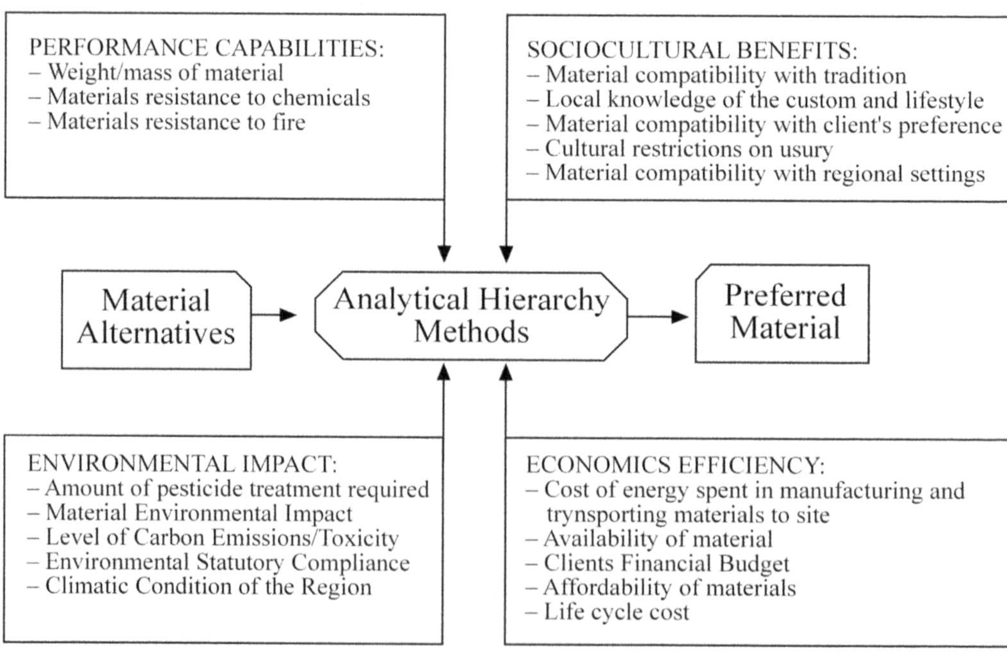

Figure 2. Conceptual framework of the analysed decision factors for measuring carbon dioxide emission of locally sourced and recycled building materials for mainstream housing development sector of the construction industry of countries situated in the Volta-Oti River catchment.

4.3.7.1 Performance capabilities

Housing project is to find trade-offs that satisfy a multitude of technical objectives that provide a rational framework for housing construction. The technical concept enables housing that are highly suitable for the functions and activities of their occupants (Wong & Li 2008). Mora (2007), maintained that undermining the issue of technical performance in housing construction has resulted in the colossal waste of material resources and underperformance of housing projects. The technical concept is an enabler to the execution of buildings that are suitable for the functions and activities of their occupants. Consequently, failure to recognise the importance of technical criteria during material selection at the early stages of the design, lead to building system incompatibility, malfunctioning and risk of obsolescence (Wong & Li 2008).

4.3.7.2 Sociocultural benefits

Sociocultural factors are much difficult to quantify and as such have not received much attention in the architecture literature as other groups of factors (San-Jose et al. 2007, Sarpong-Nsiah & Ahadzie 2018). San-Jose et al. (2007) argue that the sociocultural factors form an implicit part of the design decision-making process, as it helps to define the architecture of the region, as well as promote the image of the community. Sarpong-Nsiah & Ahadzie (2018) in a conceptual framework of material selection decision support system draws designers' attention to the major role of sociocultural in material selection decision-making at the inception phase of housing projects.

4.3.7.3 Environmental impacts

The long-term ecological footprint is important in the locally sourced and recycled building materials selection decision-making in mainstream housing development ensuring future carbon dioxide emission reduction (Behm 2005).

4.3.7.4 Economics efficiency

Financial constraint to clients in housing projects is a prime concern (Goh & Yang 2009). The cost of housing project is consequent of the choice of materials at the inception phase to ensure the objective of carbon emission reduction (Cole 1999). Often, local materials are better suited to climatic conditions, and these purchases support area economies.

4.4 AN EXAMPLE OF A MULTI-CRITERIA DECISION PROCESS

The following is a multi-criteria decision process by a direct estimation which involves the expression of relative importance of the objectives or criteria in a direct way through questionnaire surveys to determine what kind of flooring material would be best for a separate house: Bamboo XL laminated split paneled flooring, recycled laminated wood flooring and fly ash cement concrete floor slab. The goal is rethinking LSRBMs for future reduction of carbon dioxide emissions with a potential of climate change mitigation as spelled out by the criteria (Figure 3). Details for the three options of flooring materials for the proposed residential separate housing are illustrated in Table 4. The description of the three options was based on the standard practices and construction details commonly used in the housing construction industry of Ghana. These three flooring materials were analysed amongst a host of other material alternatives. This section will analyse the problem using the AHP mathematical multi-criteria decision-making technique, to identify and decide which material is the most suitable flooring material.

4.4.1 Decomposition of the decision making process

There are 14 pairwise comparison matrices in all (Tables 5 & 6). One for the criteria with respect to the goal, which is shown in Table 5, four for the sub-criteria, the first of which for the sub-criteria under environmental impact: amount of pesticide treatment required, material environmental impact, level of carbon emissions, environmental statutory compliance and climatic condition of the region (Table 5.1). For the sub-criteria under economic efficiency: cost of energy spent in manufacturing and transporting materials to site, availability of material, clients financial budget, affordability of the materials and life cycle cost (Table 5.2). Also, for the sub-criteria under sociocultural: material compatibility with tradition, local knowledge of the custom and lifestyle and material compatibility with client's preference (Table 5.3). Lastly for the sub-criteria under technical performance: weight/mass of material, resistance to chemicals and resistance to heat (Table 5.4).

Table 5 shows the main criteria listed on the left are one by one compared with each criterion listed on top as to which one is more important with respect to the goal of selecting the material with the least carbon emissions. In Table 5.1 to 5.4 the sub-criteria on the left are compared with the sub-criteria on top as to their importance with respect to the main criteria.

4.4.2 Final weights of each criterion

To find the final global weight of each sub-criterion, the results of the weighting vector for standing carbon dioxide emission criteria list were arranged (Table 6). The main criteria weighting vectors (1) are multiplied by the corresponding sub-criteria weighting vectors (2) to obtain the (global) criteria weight (3).

Table 4. Summary of the flooring option.

Description	Material A	Material B	Material C
Design Element	Paneling flooring	Laminated flooring	Concrete flooring
Building type	Residential	Residential	Residential
Material type	Bamboo XL laminated Split Paneled Flooring	Reclaimed/Recycled Laminated Wood Flooring	Fly Ash Cement Concrete Floor Slab
Size of material	230 x 150 mm	50 x 600 mm	900 x 900 mm
Rate of CO_2 Emissions	0.023 kg CO_2/m^2	0.07 kg CO_2/m^2	0.2 kg CO_2/m^2

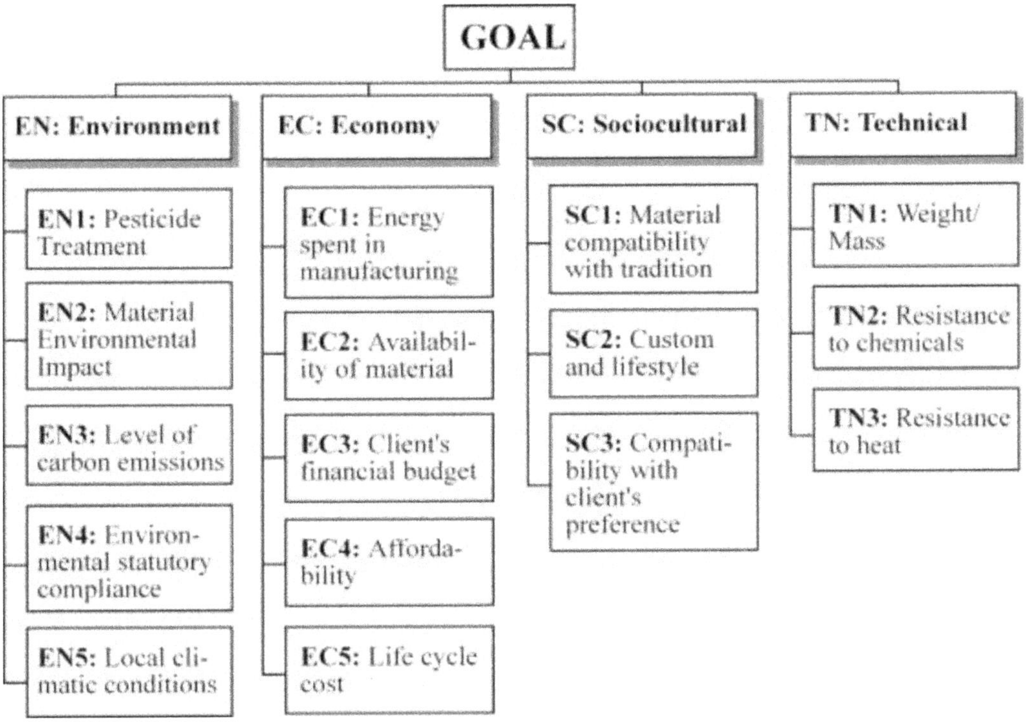

Figure 3. Appropriate material selection decision.

Table 5. Pairwise comparison matrix of the main criteria with respect to the goal (EN = Environmental impact, EC = Economic efficiency, SC = Sociocultural, TN = Technical performance).

	EN	EC	SC	TN	Priority
EN	1	3	5	7	0.5072
EC	0.3333	1	3	5	0.2959
SC	0.200	0.333	1	3	0.1437
TN	0.1429	0.200	0.333	1	0.0531
CI=0.0395		RI=0.9		CR=0.0439 <10%	

- Level of carbon emissions (Table 6.1),
- Pesticide treatment (Table 6.2),
- Material environmental impact (Table 6.3),
- Clients financial budget (6.4),
- Energy spent in manufacturing (Table 6.5),
- Material compatibility with tradition (Table 6.6),
- Availability of material (Table 6.7),
- Costume and life style (Table 6.8),
- Weight of materials (6.9).

Table 5.1 Pairwise comparison matrix for the sub-criteria with respect to Environmental impact (EN).

	EN1	EN2	EN3	EN4	EN5	Priority
EN1	1	3	1	2	5	0.2919
EN2	0.3	1	1	6	2	0.2513
EN3	1	1	1	4	6	0.3162
EN4	0.2	0.17	0.25	1	1	0.0709
EN5	0.2	0.5	0.17	1	1	0.0697
CI=0.1089			RI=1.12		CR=0.097 <10%	

Table 5.2 Pairwise comparison matrix for the sub-criteria with respect to Economy efficiency (EC).

	EC1	EC2	EC3	EC4	EC5	Priority
EC1	1	3	1	3	5	0.302845
EC2	0.33	1	1	6	2	0.240723
EC3	1	1	1	4	7	0.326141
EC4	0.33	0.17	0.25	1	1	0.064063
EC5	0.2	0.5	0.143	1	1	0.066227
CI=0.02711			RI=1.12		CR=0.06762 <10%	

Table 5.3 Pairwise comparison matrix for the sub-criteria with respect to Sociocultural (SC).

	SC1	SC2	SC3	Priority
SC1	1	3	5	0.6054
SC2	0.333	1	3	0.2915
SC3	0.200	0.333	1	0.1031
CI=0.0194		RI=0.58	CR=0.0334 <5%	

Table 5.4 Pairwise comparison matrix for the sub-criteria with respect to Technical performance (TN).

	TN1	TN2	TN3	Priority
TN1	1	3	5	0.6054
TN2	0.33	1	3	0.2915
TN3	0.2	0.33	1	0.1031
CI=0.0194		RI=0.58	CR=0.0334 <5%	

Table 6. Priority weights for carbon dioxide emission main criteria and sub-criteria.

CO_2 emissions main criterion	local weight (1)[1]	CO_2 emissions sub-criterion	local weight (2)[2]	global weight (3)[1x2]
Environmental impact	0.5072	Pesticide treatment	0.2919	0.1481
		Material environmental impact	0.2513	0.1275
		Level of carbon emissions	0.3162	0.1604
		Environmental statutory compliance	0.0709	0.0360
		Climatic condition of the region	0.0697	0.0354
Economy efficiency	0.2959	Energy spent in manufacturing	0.3028	0.0896
		Availability of material	0.2407	0.0712
		Clients financial budget	0.3261	0.0965
		Affordability of material	0.0641	0.0190
		Life cycle cost	0.0662	0.0196
Structural benefit	0.1437	Material compatibility with tradition	0.6054	0.0870
		The custom and lifestyle	0.2915	0.0419
		Compatibility with client's preference	0.1031	0.0148
Technical performance	0.0531	Weight of material	0.6054	0.0321
		Resistance to chemical	0.2915	0.0155
		Resistance to heat	0.1031	0.0055
Sum	~1		~4	1

Table 6.1 Level of carbon emissions.

	A	B	C	Priority
A	1	3	5	0.6206
B	0.333	1	3	0.3009
C	0.200	0.333	1	0.1437
CI		RI	CR	
0.0163		0.58	0.028 less than 5%	

Table 6.2 Pesticide treatment.

	A	B	C	Priority
A	1	1	9	0.5176
B	1	1	7	0.4235
C	0.11	0.143	1	0.059
CI		RI	CR	
0.0035		0.58	0.0061 less than 5%	

Table 6.3 Material environmental impact.

	A	B	C	Priority
A	1	2	9	0.5516
B	0.5	1	7	0.3907
C	0.11	0.143	1	0.0576
CI		RI	CR	
0.0109		0.58	0.0188 less than 5%	

Table 6.4 Clients financial budget.

	A	B	C	Priority
A	1	0.11	0.11	0.0514
B	9	1	2	0.5059
C	9	0.5	1	0.4427
CI		RI	CR	
0.0235		0.58	0.0406 less than 5%	

Table 6.5 Energy spent in manufacturing.

	A	B	C	Priority
A	1	1	9	0.4947
B	1	1	8	0.4497
C	0.11	0.125	1	0.0556
CI		RI	CR	
0.0008		0.58	0.0013 less than 5%	

Table 6.6 Material compatibility with tradition.

	A	B	C	Priority
A	1	1	7	0.5189
B	1	1	5	0.4036
C	0.143	0.2	1	0.0774
CI		RI	CR	
0.0063		0.58	0.0109 less than 5%	

Table 6.7 Availability of material.

	A	B	C	Priority
A	1	7	5	0.7496
B	0.143	1	1	0.1236
C	0.2	1	1	0.1269
CI		RI	CR	
0.0063		0.58	0.0109 less than 5%	

Table 6.8 Costume and lifestyle.

	A	B	C	Priority
A	1	0.2	0.25	0.0919
B	5	1	3	0.5703
C	4	0.33	1	0.3378
CI		RI	CR	
0.0421		0.58	0.0376 less than 5%	

Table 6.9 Weight of material.

	A	B	C	Priority
A	1	3	8	0.6104
B	0.333	1	5	0.3222
C	0.125	0.2	1	0.0674
CI		RI	CR	
0.0222		0.58	0.0383 less than 5%	

4.4.3 The alternatives pair-wise comparison

The final step in the pair-wise comparison involves comparing each pair of alternatives with respect to each sub-criterion. In comparing the three flooring materials, the decision-makers were asked which material is preferred with respect to each sub-criterion. They are represented by letter A, B and C. Table 6.1 to 6.9 explain the pair-wise matrix and priorities for each sub-criterion.

4.4.4 Overall carbon dioxide index score

As it can be seen from table 7, all criteria are combined, an indexing algorithm termed the 'carbon dioxide utility index' is created to rank options of competing material choices on their contribution to future reduction in carbon dioxide emissions. These priorities may also be expressed in the ideal form by dividing each priority by the largest one. The effect is to make this alternative the ideal one with the

Table 7. Overall criteria rating.

Main Criteria	Local weight (1)	Sub criteria	Local weight (2)	Local weight (3)			Global weight (2)		
				M(A)	M(B)	M(C)	M(A)	M(B)	M(C)
EN	0.5072	ENa	0.2919	0.5176	0.4235	0.059	0.0766	0.0627	0.0087
		ENb	0.2513	0.5516	0.3907	0.0576	0.0703	0.0498	0.0073
		ENc	0.3162	0.6206	0.3009	0.1437	0.0995	0.0483	0.0230
EC	0.2959	ECa	0.0896	0.4947	0.4497	0.0556	0.0131	0.0119	0.0015
		ECb	0.0965	0.0514	0.5059	0.4427	0.0015	0.0144	0.0126
		ECc	0.0712	0.0919	0.5703	0.3378	0.0019	0.0120	0.0071
SC	0.1437	SCa	0.6054	0.5189	0.4036	0.0774	0.0451	0.0351	0.0067
		SCb	0.2915	0.7496	0.1236	9.1269	0.0314	0.0052	0.3823
TN	0.0531	TN	0.6054	0.6104	0.3222	0.0674	0.0196	0.0104	0.0022
Carbon dioxide utility index (normalised priorities)							0.359	0.2498	0.4516
Carbon dioxide utility index (idealised priorities)							0.79	0.55	1

others getting their proportionate value. One may then interpret the results to mean that bamboo XL laminated split paneled flooring is about 79% as good fly ash cement concrete floor slab and recycled laminated wood flooring is about 55% as good as fly ash cement concrete floor slab.

4.5 DISCUSSION

Building stocks have been identified to make significant contribution to carbon dioxide emissions and cumulative impact of carbon dioxide emissions on climate has potentially profound economic and policy implications (Ofori 1999, Boardman 2007). Countries situated in the Oti watershed, drastic measures are required to address the acute housing situation with huge implications for cost, energy consumption and potential climate change impact (Ofori 2006). Energy consumption can be cut back through improving efficiency, which is an effective means to lessening carbon dioxide emission (Lee & Chen 2008).

Generally, the framework indicates that research towards the development of sustainability in environmental development in acknowledging the role of sociocultural as vital resources for strengthening the material selection decision-making process for future reduction in carbon dioxide emissions has not been vigorously pursued in practice (San-Jose et al. 2007; Sarpong-Nsiah & Ahadzie 2018).

The framework explores and evaluates the significance of analytical hierarchy process mode of decision-making to facilitate cut in carbon dioxide emissions in building materials selection for mainstream housing development of the construction sector of countries situated in the Oti water shed. The analytical hierarchy process, prioritize decision alternatives that would ensure the resolution of conflicts between environmental factors, economy factors, technical factors as tangible and sociocultural factors as intangible. Decisions involving the following intangibles: material compatibility with tradition, local knowledge of the custom and lifestyle, and material compatibility with client's preference need to be traded off. To do that, they have to be measured alongside tangibles whose measurements must also be evaluated as to, how well, they serve the objectives of the decision maker. The analytic hierarchy process is a theory of measurement through pairwise comparisons and relies on the judgments of experts to derive priority scales. It is these scales that measure intangibles in relative terms. The comparisons are made using a scale of absolute judgments that represents, how much more, one element dominates another with respect to a given attribute. The judgments may be inconsistent, and how to measure inconsistency and improve the judgments, when possible to obtain better consistency is a concern of the analytic hierarchy process. The derived priority scales are synthesised by multiplying them by the priority of their parent nodes and adding for all such nodes to obtain normalised priorities and idealised priorities (Table 8).

Table 8. Comparing normalised and idealised priorities.

Flooring materials	Normalised Priorities (%)	Idealised priorities (%)	Reduction of carbon dioxide
Bamboo XL laminated split paneled 0.023 kg CO_2/m^2	0.359	79	15.9
Reclaimed/recycled laminated wood 0.07 kg CO_2/m^2	0.2498	55	11.0
Fly ash cement concrete floor slab 0.2 kg CO_2/m^2	0.4516	1	0

4.6 CONCLUSION

Environmental building material assessment methods have been used by many countries to assess environmental performance of building material. These environmental building assessment methods have escalated from local programs into an international agenda and are well recognized as a way for promoting environmental awareness among building professionals. However, limitations have been identified regarding the lack of input of sociocultural factors in the context of especially developing countries. Here a multi-criteria material selection approach decision making for the selection of locally sourced and recycled building materials and components for the housing construction industry of Ghana and useful to countries situated in the Oti-Volta basin has been developed and presented.

The usefulness of the framework appears indispensable in that, there is a need to organise ways to make decisions and collect information relevant to them when a group must decide by laying out all the important factors and negotiating their understanding, beliefs and values. The framework deals with a decision from four different standpoints: the benefits that the decision brings, the opportunities it creates, the costs that it incurs and the risks that it might have to face. The effect is to make this alternative the ideal one with the others getting their proportionate value. Bamboo XL laminated split paneled flooring is about 79% as good as fly ash cement concrete floor slab (reduction of 15.9% carbon emissions) and recycled laminated wood flooring is 55% as good as fly ash cement concrete floor slab (reduction of 11.0% carbon emissions). The analytic hierarchy process model can be easily implemented using any simple and very familiar spread sheet or software application such as MS Excel, MS Access and MS Word.

REFERENCES

Ahadzie, D.K. & Amoa-Mensah, K. 2010. Management Practices in the Ghanaian House Building Industry. *Journal of Science and Technology* 30(2): 62.

Anderson, J., Shiers, D. & Steele, K. 2009. *The Green Guide to Specification: An Environmental Profiling System for Building Materials and Components*. BR 501. 4ed. UK: Building Research Establishment.

Behm, M. 2005. Linking construction fatalities to the design for construction safety concept. *Safety Science* 43: 589–611.

Boamah, N.A. 2010. Housing Affordability in Ghana: A focus on Kumasi and Tamale. *Ethiopian Journal of Environmental Studies and Management* 3(3): 1–11.

Boardman, B. 2007. Examining the carbon agenda via the 40% house scenario. *Building Research & Information* 35(4): 363–378.

Cole, R.J. 1999. Building environmental assessment methods: clarifying intentions. *Building Research and Information* 27(4/5): 230–246.

Crawley, D. & Aho, I. 1999. Building environmental assessment methods: application and Development trends. *Building Research and Information* 27(4/5): 300–308.

Ding, G.K.C. 2008. Sustainable construction: The role of environmental assessment tools. *Journal of Environmental Management* 86(3): 451–64.

Ding, L., Roberto, G.Q., Wei, L. & Ratcliffe, J. 2010. *Risky Borrowers or Risky Mortgages Disaggregating Effects Using Propensity Score Models*. Working Paper. Durham: Department of Urban Studies and Planning and the UNC Center for Community Capital.

Edwards, B. & Hyett, P. 2001. *Rough guide to sustainability*. London: RIBA Enterprises.

Emmanuel, R. 2004. Estimating the environmental suitability of wall materials preliminary Results from Sri Lanka. *Building and Environment* 39(10): 1253–1261.

Florez, L.D., Castro, D. & Irizarry, J. 2010. Impact of Sustainability Perceptions on Optimal Material Selection in Construction Projects. *Proceedings of the Second International Conference on Sustainable Construction Materials and Technologies, University Politecnica delle Marche, Ancona, Italy, Coventry University and the University of Wisconsin Milwaukee Centre for By-Products Utilization, 28-30 June 2010*: 719–727.

Goh, K.C. & Yang, J. 2009. Developing a life-cycle costing analysis model for Sustainability enhancement in road infrastructure project. In *Rethinking Sustainable Development: Planning, Infrastructure Engineering, Design and Managing Urban Infrastructure. 26 March 2009, Queensland University of Technology, Brisbane, Queensland.*

Goldenberg, J. 1998. *Energia, medio ambiente & desenvolvimento*. Sao Paulo: Edusp.

Jaramillo, J.A. 1996. Vygotsky's sociocultural theory and contributions to the development of constructivist curricula. *Education* 117(1). Alabama: Project Innovation.

Lee, W.L. & Chen, H. 2008. Benchmarking Hong Kong and China energy codes for residential buildings. *Energy and. Buildings* 40(9): 1628–1636.

Mahmoud, M.A.A., Aref, M. & Al-Hammad, A. 1996. An expert system for evaluation and Selection of floor finishing materials. *Expert Systems with Applications* 10(2): 281–303.

Mohamed, A. & Celik, T. 1998. An integrated knowledge-based system for alternative Design and materials selection and cost estimating. *Expert Systems with Applications* 14(3): 329–339.

Monahan, J. & Powell, J.C. 2011. An embodied carbon and energy analysis of modern methods of construction in housing: A case study using a lifecycle assessment framework. *Energy and Buildings* 43(1): 179–188.

Mora, E. 2007. Life cycle, sustainability and the transcendent quality of building materials. *Building and Environment* 42(3): 1329–1334.

Nijkamp, P., Rietveld, P. & Voogd, H. 1990. *Multi-criteria evaluation in physical Planning*. New York: North Holland.

Ofori, G. 1999. Satisfying the customer by changing production patterns to realize sustainable construction. In *Proceedings of the Joint Triennial Symposium of CIB Commissions W65 and 55, Cape Town, 5–10 September* (1): 41–56.

Ofori, G. 2006. Revaluing Construction in Developing Countries: A Research Agenda. *Journal of Construction in Developing Countries* 11(1): 1–16.

Owusu, G. 2010. Urban Growth, Globalization and Access to Housing in Ghana's Largest Metropolitan Area, Accra. *Institute of Statistical, Social and Economic Research (ISSER)*. Legon: University of Ghana.

Perera, R.S. & Fernando, U. 2002. Cost Modelling for Roofing Material Selection. *Built Environment: Sri Lanka* 3(1): 11–24.

Presley, A. 2006. ERP investment analysis using the strategic alignment model. *Management Research News* 29(5): 273–284.

Quandzie, E. 2011. *Ghana has 1.6 million Housing Deficit, but Policies still in Draft since 1983*. Business News: URL: http://www.ghanabusinessnews.com/2011/09/12/ghana-has-1-6-million-housing deficit-but-policies-still-in-draft-since-1983/n (accessed 21/01/2013).

Rahman, S., Perera, H. Odeyinka, H. & Bi, Y. 2009. A Knowledge-Based Decision Support System for Roofing Materials Selection and Cost Estimating: A Conceptual Framework for Cost Modelling. *25th Annual ARCOM Conference, (2009) September 7- 9,Nottingham, England.*

Rahman, S., Perera, H., Odeyinka, H. & Bi, Y. 2008. A Conceptual Knowledge-Based Cost Model for Optimizing the Selection of Material and Technology for Building Design," In A.R.J. Dainty (ed.), *24th Annual ARCOM Conference, Association of Researchers in Construction* Management, *University of Glamorgan, 1-3 September 2008*: 217–22.

Raisinghani, S. & Meade, L.L. 2005. Strategic decisions in supply-chain intelligence Using knowledge management: an analytical-network-process framework. *Supply Chain Management* 10(2): 114–121.

Saaty, T.L. 1980. *The Analytic Hierarchy Process*. New York: McGraw-Hill.

Saaty, T.L. 1986. Axiomatic foundation of the analytic hierarchy process. *Management Science* 32 (7): 841–855.

Saaty, T.L. 1994. *Fundamentals of decision-making and priority theory with the analytic Hierarchy process*. Pittsburgh: RWS Publications.

Saaty, T.L. 2000. *Fundamentals of the Analytic Hierarchy Process*. Pittsburgh: RWS Publications.

Saaty, T.L. 2001. *Decision Making for Leaders: The Analytic Hierarchy Process for Decisions in a Complex World*. Pittsburgh: RWS Publications.

Saaty, T.L. 2007. Time Dependent Decision-Making; Dynamic Priorities in the AHP/ANP: Generalizing from Points to Functions and from Real to Complex Variables. *Mathematical and Computer Modeling* 46: 860–891.

Saaty, T.L. 2008. Relative Measurement and Its Generalization in Decision Making Why Pairwise Comparisons are Central in Mathematics for the Measurement of Intangible Factors. *The Analytic Hierarchy/Network Process RACSAM* 102(2): 251–318.

San-Jose, J.T., Losada, R., Cuadrado, J. & Garrucho, I. 2007. Approach to the quantification of the Sustainable value in industrial buildings. *Building and Environment*: 916–3923.

Sarpong-Nsiah, G. & Ahadzie, D.K. 2018. Multi-Criteria Material Selection Decision Support System for the Housing Construction Industry: A Conceptual Framework. *ICCF-Psycon Conference, University of Woverhampton, United Kingdom.*

Shams, S., Mahmud, K. & Al-Amin, M. 2011. A comparative analysis of building materials for Sustainable construction with emphasis on CO2 reduction. *Int. J. Environment and Sustainable Development* 10(4): 364–374.

Singh, R.K., Murty, H.R., Gupta, S.K. & Dikshit, A.K. 2007. Development of composite Sustainability performance index for steel industry. *Ecological Indicators* 7(3): 565–588.

UN-Habitat 2011a. Affordable Land and Housing in Africa. Volume 3. Africa. URL: http://www.unhabitat.org/pmss/listItemDetails.aspx?publicationID=3376 (accessed in 2019).

UN-Habitat 2011b. Ghana Housing Profile.URL: http://www.unhabitat.org/pmss/listItemDetails.aspx?publicationID=3258 (accessed in 2019).

Van Pelt, M.J.F. 1994. Environment and project appraisal: lessons from two cases. *The Annals of Regional Science* 28(1): 55–76.

Wong, J.K.W. & Li, H. 2008. Application of the analytic hierarchy process (AHP) in Multi-criteria analysis of the selection of intelligent building systems. *Building and Environment* 43: 108–125.

Xing S, Xu, Z, & Jun, G. 2008. Inventory analysis of LCA on steel and concrete Construction Office buildings. *Energy and Buildings* 40: 1188–1193.

CHAPTER 5

Geomorphic aspects of slope erosion in urban areas of Bouna in the Black Volta River Basin, Ivory Coast

S. Kambire
University Peleforo Gon Coulibaly, Korhogo, Ivory Coast

B. Kambire & N.P. Dangui
University Félix Houphouët Boigny, Abidjan, Ivory Coast

ABSTRACT: The City of Bouna (~60,000 inhabitants) is located in north-eastern Ivory Coast close to the Ghana border. Geologically, it consists of granitic bedrock. The overall relief is accentuated and the degree of urbanisation is (still) very low. However, the Gborontchara and Dassikélédougou districts of this town are well beyond 'classic' urban areas as they are particularly susceptible to hydric erosion. It is assumed and evidenced that the determining factors of erosion are linked to geomorphic features of slopes. Data obtained from a network of experimental runoff plots set up on typical slopes, indicate that surface erosion, runoff and risks of erosion are most intense on the scarps ('corniches') and on the steep lower foot slopes. Anti-erosive measurements should take geomorphology into account.

5.1 INTRODUCTION

Current climate studies predict an amplification of global warming that can already be observed in many regions (Meehl & Tebaldi 2004, IPCC 2007). This global climate change is characterised by high inter-annual climate variability (Giorgi 2006), an increase in extreme events such as intense rainfall, droughts and heat waves. Landscape erosion is defined as an excessive expression of extreme precipitation events (IPCC 2007).

Sea-level rise caused by climate change causes increased coastal erosion rates (Bicknell et al. 2009). Hydric erosion has become a global and multidimensional problem. It affects agricultural areas (gullying, mudflows flooding of fields and crops), urban infrastructure, and human and economic facilities (e.g. roads submerged by mud). Faced with the worsening observed over the past twenty years, a great deal of research has been undertaken on the processes and factors of water erosion. They have contributed to a better understanding of the mechanisms of soil erosion through field observations, modelling of phenomena and knowledge of physical mechanisms. They also help for a wider dissemination of this knowledge within regions to help local stakeholders to better understand and combat the risks associated with soil erosion (Le Bissonnais et al. 2002).

However, complexity of erosion processes and the large number of disciplines and the diversity of physical environments have to be taken into account. It is generally difficult to develop reliable global forecasts based on linear erosion models.

For several reasons, urbanisation has major effects on erosion processes. It is firstly the stripping and destruction of natural vegetation cover, modifying drainage patterns and compacting of soils by construction. Sealing of ground with an impermeable layer of asphalt or concrete accelerates rainfall runoff and surface wind speeds (Nir 1983).

The districts of Gborontchara and Dassikélédougou in the City of Bouna (~ 60,000 inhabitants) situated in north-eastern Ivory Coast, have very modest urban infrastructures: hard-surfaced roads are rare and pavements are rarely reinforced by lateritic clay and gravel. Construction of buildings is discontinuous and urban wastelands form an uninterrupted network. In addition, due to its position, the region of Bouna is located in a granitic terrain with rigid tabular mountains with ferricrete crusts on top (Avenard 1971). Gborontchara and Dassikélédougou districts are considered as the prototype of urban agglomerations showing a high sensitivity and vulnerability to water erosion.

It is was hypothesised that erosion phenomena in Gborontchara and Dassikélédougou, are mainly determined by topographic factors. The study was carried out in order to show both, the erosion phenomenon itself and the potential risks of erosion on the geomorphic scale in the environs of Gborontchara and Dassikélédougou districts in the City of Bouna.

5.2 STUDY SITE, DATA AND METHODS

5.2.1 Geographical location

The study area is located in the north-eastern part of Ivory Coast. Its geographical coordinates are 9°15 N and 3° W (Figure 1). The landscape settles on granitic bedrock and is characterised by isolated tabular mountains (Avenard 1971). The landscape is composed of a superposition of levelled surfaces that are individualised with respect to each other; the slopes connecting them are a minor geomorphic form (Rougerie 1964).

5.2.2 Data and collection techniques

The study of erosion processes in the districts of Gborontchara and Dassikélédougou has been carried out by focusing on slopes' characteristics. From a structural point of view, slopes often correspond to changes in the organisation of landscapes, and from a kinematic point of view, these slopes highlight the current organisation and dynamics of the morphodynamic environments.

A systematic mapping of all breakage and slope inflection lines visible on aerial photographs HAUTS-MONTS INC-CIV-96-015 to 016 from 24th February 1996 at a scale of 1:50,000 was undertaken. These breaks and slope inflections highlight the landscape segments. The objective was to recognize from the inclination the most common and 'typical' slopes in the area.

An IGN topographic map (1:50,000) was also available for the region. This topographic map was superimposed on the landscape segment map. This made it possible to obtain – with the maximum precision – the average slope angle between the contours that pass through the top and the base of a typical slopes, and the maximum average slope was defined by the nearest curves.

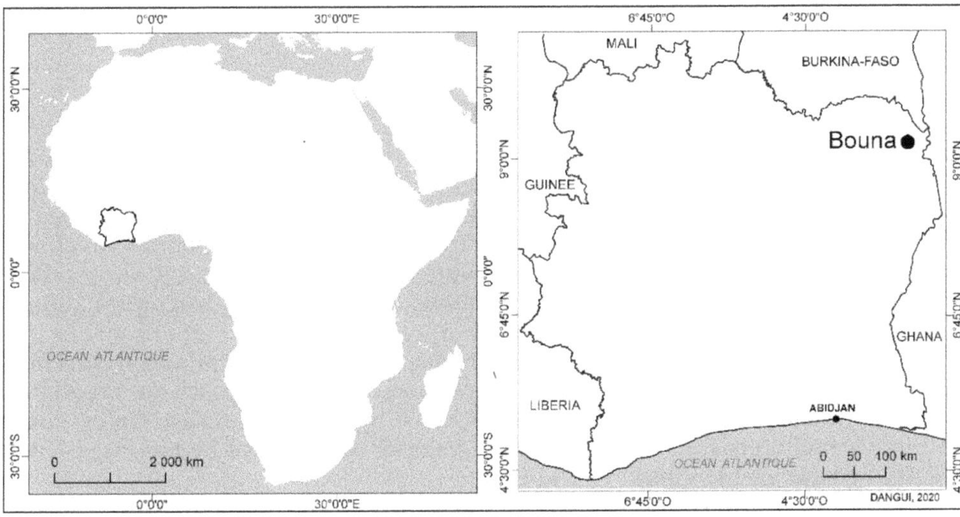

Figure 1. Location map of the study area (Cartography: N.P. Dangui 2020, Source: National Center for Remote Sensing and Geographic Information 2012, and National Institute of Statistics 2014).

The slope class generated is based on the Mayer (1990) classification. This classification was successfully used in several applications (N'dri et al. 2008). These authors differentiate four slope classes: 0-5% (slight slope); 5-15% (medium slope); 15-50% (steep slope), greater than 50% (very steep slope). In the field of study, low, medium and high slopes are represented (Figure 2).

In a second stage, and only in urban wastelands, experimental plots have been set on the slopes recognised. These apply to areas of about 100 square meters, in the shape of a rectangle elongated in the direction of the slope. A ditch, open upstream, delimits each plot while a gutter dug downstream ensures the recovery of the transported materials in bins. Each of these plots was subjected to runoff and erosion experiments on dry land under tropical rainfall conditions. The data presented correspond to three rainfall events of the same average intensity (approx. 60 mm/h of rain). They made it easier to compare the orders of magnitude relating to erodibility of land surfaces.

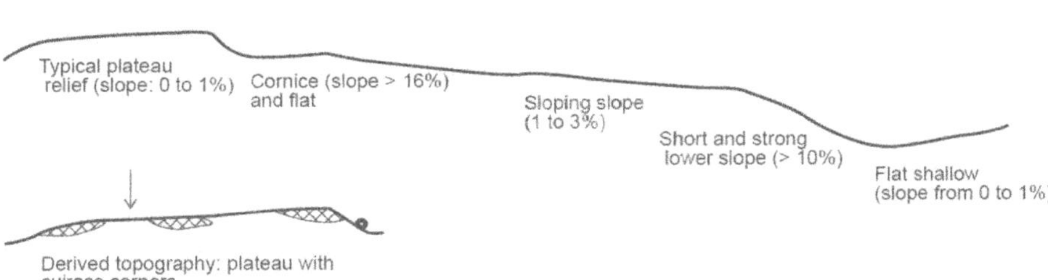

Figure 2. Typical topo-sequence of slopes in Gborontchara and Dassikélédougou (Draft: S. Kambire 2019).

5.3 RESULTS

5.3.1 Erodibility of urban sites

The topo-sequence in Figure 2 illustrates the typical landscape features and current morpho-dynamics of the natural environment in the Gboroncthara and Dassikélédougou districts. The relief corresponds to extended tabular mountains with ferricrete crusts on top. On the summit (slope from 0 to 1%) the ferricrete crust is discontinuous. A steep scarp ('corniche', > 16% inclined) with a lateritic crust (plintite) dominates a generally long slope. The more or less clear break and slope inflection lines shows on the slope a fairly differentiated profile. The hydrographic alluvial lowland is wide and flat (0 to 1% inclined).

In this framework, the data orders relating to the mechanical action of runoff water are different from one landscape segment to another. Areas of low erodibility are mostly found on the tabular mountain covered by a ferricrete crust. On dry land, a 60 mm/h rainfall caused erosion of 98 kg/ha after 42 min, 208 kg/ha after 57 min and 317 kg/ha after 121 min.

These areas of low degradation by water erosion are also found in lowlands. A 60 mm/h rainfall caused erosion of 53 kg/ha after 42 min, 175 kg/ha after 57 min and 287 kg/ha after 121 min.

The environments highly sensitive to erosion are located on the slopes. On the upper slopes, a 60 mm/h rainfall carried, on dry land, 323 kg/ha of soil after 42 min, 411 kg/ha after 57 min and 889 kg/ha after 121 minutes.

On the middle slope, 109 kg/ha of material was collected after 42 min, 268 kg/ha after 57 min and 601 kg/ha after 121 minutes.

The morphogenic degradation is very important on the scarps and at the foot of slopes. Under the same experimental conditions, on 'corniches', a loss of 381 kg/ha was recorded after 42 minutes of rainfall, 519 kg/ha after 57 minutes and 813 kg/ha after 121 minutes.

On the lower foot slopes, where erosion is most intense, land losses amounted to 401 kg/ha after 42 min of rain, 525 kg/ha after 57 min and 1027 kg/ha after 121 minutes.

5.3.2 Factors contributing to the vulnerability of land to water erosion

The long dry season in Bouna region led to an increase in the proportion of mechanical actions in the morphogenetic system. The heavy rainfalls at the beginning of the rainy season and on dry soils are notably aggressive to the geomorphic landscapes. The high slopes which are not protected by buildings, asphalt, channels for concentrating and draining rainwater against rainfalls' aggressiveness increase the intensity of water erosion (Table 1). From this point of view, morphogenetic processes in Gboronchira and Dassikélédougou districts have had quite favorable conditions.

If trees are planted in public spaces and high grasses of urban wastelands provide an effective protection by vegetation during the rainy season. However, this protection is unavailable in the dry season. It is during this short seasonal crisis in June/July-September/October, periods during which the region records its peak rainfall, when water erosion becomes more intense.

Soil texture and structure are also important factors in the morphogenetic degradation of the Gborontchara and Dassikélédougou sites. On the top of the tabular mountain with a ferricrete crust and in hydromorphic lowlands, erosion has little

Table 1. Classification of land vulnerability factors to water erosion in the districts of Gborontchara and Dassikélédougou.

	Erosion factor	Vulnerability to water erosion
Soil type	Hydromorphic soil	Low
	Armoured materials	
	Healthy rock	
	Sands, clays, silts of alluvial/colluvial origin	Very strong
	Typical ferrallitic soils on granite	
Slopes	0 - 1 %	Low
	2 -9 %	Average
	10- 50 %	Strong
	> 50 %	Very strong
Land use	Woody and high grasses	Low
	Urban building	Average
	Bare soil and unpaved street	Very strong

effect. In the first case, the resistance of the environment to water erosion is partly due to locally dismantled flush or sub-flush ferricrete crust. In the case of lowlands, these are the places where sediments are deposited (alluvia), torn off from the upper relief levels and transported by runoff water. The thick sandy alluvial deposits and colluvium then increase the quantity of water that can be absorbed by the soil and subsequently prevent it from flowing to the surface as erosive runoff. In areas where the soil is reticulated hydromorphic or clayey, gleyic and hydromorphic, the cohesion of the aggregates provides high structural stability.

But the value of the slope plays a major role. Relatively low (0 to 1% inclined) at the top of the tabular mountain and in the valleys, it slows down the flow velocity which is unable to concentrate linearly.

On the contrary, erosion is strong on the steepest slopes due to the impluvium effect caused by the relief. At the top and on the average slope (1 to 3%), it develops a diffuse or concentrated runoff. The force of showers and the spreading of the water, imposed by plant strains and dismantled ferricrete crust or rocks, cause the types of flow in anastomosed nets or in sheet erosion (denudation). In larger and bare spaces, runoff water exerts areolar ablation by dragging fine sands, clays and silts to spreading areas where they are abandoned.

On ledges and lower slopes with significantly steep slopes (>10%) or unprotected by ferricrete crusts, particularly on the urban roads, concentrated and linear runoff predominates. It represents the most aggressive and destructive form of erosion in the urban areas. It is more effective as it triggers other effective mechanical activities. Surface water buried in desiccation slits, open in the dry season, creates internal flows that generate landslides along the roads on the slopes.

The mechanical activities described still play a secondary role in erosion processes since it is geomorphology (topography) of the land that always determines the speed at which the surface flow will occur, which in turn determines the erosivity of the flow (Whisenant 2002, Blanco & Rattan 2010, Wainwright & Brazier 2011).

Buildings, due to the large quantity of water coming from their roofs, are also involved in surface erosion processes. They support linear gullying on loose soils, scouring and regressive gullying at the expense of ferricretes and on steep slopes. To this type of concentrated water movement are always added those related to human activity. In clayey-silt or finer sand, human settlements are the most effective erosion factors. The ridges developed for urban agriculture are the most typical elements of the peripheral urban landscape. After their abandonment, especially when they are made in the direction of the slope, they feed and accelerate the concentrated runoff because of fluvial channels dug between them

5.4 DISCUSSION

In north-eastern Ivory Coast, tabular mountains raise over the flat surface of the old West African Craton (Vennetier 1980). These tabular mountains show a ferricrete crust on top and appear relatively stable compared to the sedimentary regions of the Atlantic coast that are subject to strong erosion. Because the plintite plays a role identical to that of a resistant surface rock and structurally 'fortify' the landscape (Avenard 1971).

Like the regional landscape, the tabular mountains with ferricretes on top, the environs of Gborontchara and Dassikélédougou in the City of Bouna, are closely linked to long dry seasons which caused the pedogenetic induration of duricrusts typical for West African savannah environments.

Ferricretes result from the concentration and fixation of metal oxides and hydroxides released by hydrolysis (Cocque 1977). Under the A horizon of the surface of soil, depleted by the vertical migration of the most soluble substances, there is a B horizon followed by the underlying C horizon, which is currently being altered. It rests in greater depths (< 10 m) on unweathered rock. Former and recent evaporation triggers their desiccation. This accumulation, through internal migration, leads to the development of massif lateritic coatings (iron duricrusts) with or without enclaves of granitic rocks. The thickness and extent of this accumulation, depends on bioclimatic, geomorphic and lithological conditions. Granitic rocks rich in ferruginous elements (Fe/Al sesquioxides) intensify the process.

The slopes of incrusted tabular mountains in Gborontchira and Dassikélédougou districts present a topography with different susceptibility towards erosion. The surface of these urban 'bowè' undergoes disintegration of the ferricrete followed by recent decay during the rainy season. More decisive than dissolution of small basins cleaned by runoff subsidence of the ferricrete occurs that explains the accentuation of the relief. Erosion is most prevalent on marginal escarpments and on surrounding slopes. When heavy rainfall occurs, diffuse or concentrated runoff causes areolar ablation at the expense of ferricretes. It sweeps and transport plintite fragments, ferruginous pisolites and other fine material and spreads them at the foot slope. The subsequent dismantling of the scarps and of lower slopes, accelerated locally by the water outlet from internal leaching, leads to 'bloc landslides' and disintegration of the former crust.

The presence of a ferricrete does not hinder erosion, however, it significantly limits its effects. But the mechanical actions become much more intense when they occur on surfaces that are not protected by ferricrete crusts. The most 'catastrophic' of them are always linked to the intensification of natural aggressiveness due to the steepness of the slope (Blanco & Rattan 2010, Wainwright & Brazier 2011).

Water erosion is also known in Cameroon. Youngué-Fouateu et al. (2000) studied the topographical slope and structural forms and the morphogenic degradation of the Oyomabang site which caused a factor of local population's vulnerability. According to them the presence of hills and heterogenous lithological conditions (weathering of hard and soft rocks) are the main causes of urban land erosion.

5.5 CONCLUSION

Agents of erosion, especially runoff of surface waters are exerted on the weathering products of rocks (soils). The speed and energy of runoff allows the morphogenetic explanation of the landscape forms and the intensity of erosion. It depends largely on the inclination of slopes. In the districts of Gborontchara and Dassikélédougou in the City of Bouna the most intensive erosion phenomena and risks of erosion are caused by the intensification of natural aggressiveness due to the slopes' characteristics. Scarps and lower hill slopes are the most sensitive to water erosion. Institutional and technical measures to combat urban erosion must be taken into account, mainly on topography and morphodynamic processes.

REFERENCES

Avenard, J.M. 1971. Aspect de la géomorphologie. In Avenard, J.M., Eldin, M., Girard, G., Sircoulon, J., Touchebeuf, P., Guillaumet, J.L., Adjanohoun, E. & Perraud, A., *Le milieu naturel de la Côte d'Ivoire*: 11–68, Paris: ORSTOM 50.

Bicknell, J., Dodman, D. & Satterthwaite, D. 2009. *Adapting Cities to Climate Change: Understanding and Addressing the Development Challenges*. London: Earthscan.

Blanco, H. & Rattan, L. 2010. *"Water erosion". Principles of Soil Conservation and Management*. Berlin: Springer.

Cocque, R. 1977. *Géomorphologie*. Paris: Armand Colin.

Giorgi, F. 2006. Climate change hot-spots. *Geophysical Research Letters* (8): 110–115.

IPCC 2007. Global climate projections. In Meehl, G.A., Stocker, T.F., Collins, W.D., Friedlingstein, P., Gaye, T., Gregory, J.M., Kitoh, A., Knutti, R., Murphy, J.M., Noda, A., Raper, S.C.B., Watterson, I.G., Weaver, A.J. & Zhao, Z.C., *Climate change: the physical science basic. Contribution of working group I to the fourth Assessment Report of intergovernmental Panel on Climate Change*: 177–192. Cambridge: Cambridge University Press.

Le Bissonnais, Y., Thorette, J., Bardet, C. & Daroussin, J. 2002. *L'érosion hydrique des sols en France*. Orléans: INRA, IFEN.

Mayer, L. 1990. *Introduction to quantitative geomorphology*. Englewood Cliffs: Prentice Hall.

Meehl, G.A. & Tebaldi, C. 2004. More intense, more frequent, and longer lasting heat waves in the 21st century. *Science* 305: 994–997.

N'dri, B.E., N'Go, Y.A., Kouadio, H.N., Ouattara, A., Touré, B. & Biémi, J. 2008. Effect of soil slope and cover on runoff and rate soil loss from experimental plots in area of Attécoubé. *European Journal of Scientific Research* 21(3): 459–470.

Nir, D. 1983. *Man, a geomorphological agent: An introduction to anthropic geomorphology*. Dordrecht: Reidel Publishing Company.

Rougerie, G. 1964. *La Côte d'Ivoire*. Que sais-je? 1137: 1–128.

Rougerie, G. 1969. *Manuel de prospection pédologique. Initiations documentations techniques* 11: 1–128. Paris: ORSTOM.

Vennetier, P. 1980. *Les Atlas jeune Afrique, Côte d'Ivoire*. Paris: Édition Jeune Afrique.

Wainwright, J. & Brazier, R.E. 2011. Slope systems. In Thomas, D.S.G. (ed.), *Arid Zone Geomorphology: Process, Form and Change in Drylands*: 209–233. Chichester: Wiley & Sons Ltd.

Whisenant, S.G. 2002. Terrestrial systems. In Perrow, M.R. & Davy, A.J. (eds), *Handbook of Ecological Restoration: Principles of Restoration*: 83–105. Cambridge: Cambridge University Press.

Youngué-Fouateu, R., Baboulé, Z.B., Temgoua, E. & Bitom, D.L. 2000. Erosion hydrique en milieu urbain: le cas du site d'Oyomabang dans la ville de Yaounde. *Influence de l'homme sur l'érosion: bassins versants, élevage, milieux urbain et rural. Bull du réseau érosion* 20(2): 332–341.

CHAPTER 6

Impact of humans on tropical landscapes: The challenge of soil erosion for rural development in the Black Volta River Basin, Ivory Coast

S. Kambire & K.A. Alla
Université Peleforo Gon Coulibaly, Korhogo, Côte d'Ivoire

B. Kambire
Université Félix Houphouët Boigny, Abidjan, Côte d'Ivoire

ABSTRACT: In the Black Volta River catchment of northeastern Ivory Coast people practice traditional 'slash and burn' subsistence agriculture. This activity has negative agronomic consequences mainly caused by highly erodible soils. It was assumed that dynamic landscape changes are mainly controlled by natural processes linked to topography (morpho-pedological factors) and to vegetation cover. Direct measurements of erosion processes from a network of experimental plots on the slopes of the region's landscapes gave evidence that denudation by sheet erosion and gullying by torrents on steep slopes, poorly protected by vegetation, are mostly affected. Among recommended anti-erosion measures the 'stop strip' method seems to be the most appropriate for local farmers.

6.1 INTRODUCTION

Morphodynamic degradation of agricultural land can have several causes. Roose (2010) mentiones salinisation, carbonation, clogging of pores, compaction by trampling and motorisation, leaching of colloids or soluble elements in drainage waters, mineralisation of organic matter and 'skeletonisation' or enrichment of surface horizons with sand or gravel by selective erosion of finer particles. According to Blanco & Rattan (2010) water erosion and wind deflation are generally known as the two main causes of soil degradation. The combined factors are responsible for 84% of land degradation worldwide.

In France morphogenetic soil degradation by surface runoff has been considered specifically to regions with strong orogenic uplift (Alps, Pyrenees, Massif Central). Given the extent of damage mountain terrain restoration laws have been adopted. As a result, human induced erosion in mountain areas decreased significantly in the 20th century (Auzet 1987).

In Africa water erosion is a widespread problem that leads to a decline in agricultural productivity as food demand for growing populations continues to increase (Rattan & Stewart, 2019). After a first period of basic description of various processes of erosion (1940-1960), one began to quantify erosion at various scales from runoff plots (100 m^2) to micro watersheds of a few ha in size, up to large rivers' catchments thousands of square kilometres wide. In former Belgian Congo, geographers and agronomists from INEAC (*Institut d'Etudes Agronomiques du Congo*) observed degradation of soils when they were tilled, bare and exposed to the sun and to heavy rainfall. For francophone Africa, Fournier (1960) produced a first map of erosion hazards based on climatic and topographic indexes. He created the first network of runoff plots in half a dozen of countries: from Senegal to Madagascar with the aim to quantify runoff and erosion risks under various cropping systems (Fournier 1967). In Rhodesia (Zimbabwe), Hudson (1958, 1971) developed a large number of experimental plots at various scales to measure runoff and erosion under different vegetation cover, inclination and lengths of slopes, soil properties and tillage managements. Roose (1977, 1984) quantified soil erosion by water by many comprehensive empirical studies in West Africa (Roose 2008).

On etchplanated Gondwana surfaces with extended and shallow slopes (interflows, glacis), iron incrusted tabular mountains with low lying terrace landscapes of northern Ivory Coast, little attention has been paid to the erosion of slopes and on farmers' fields. However, Vogt (1953) already noted that "the problem of erosion appears (...) as an endemic problem" (cited in Auzet 1987). Geography has with rare exceptions neglected the topic of soil erosion of croplands in Africa. The first investigations specific to these regions were carried out by Richard (1983). He classified tropical soils according to their vulnerability to erosion by a system that was originally established by Roose (1977), subsequently criticised as being too general. Later on, it was focused on geomorphical mapping of the vulnerability of environments to erosion in the Korhogo area in Ivory Coast.

Northeastern Ivory Coast with the Black Volta (Figure 1) catchment remained a region where qualitative and quantitative research on soil erosion was almost absent. The lack of publications on erosion in a region does not necessarily mean that the problem does not exist, but often that it is not perceived as being of great interest. Indeed, "it does not seem possible to adopt universal solutions to erosion given the diversity of geomorphological, climatic, hydrological and local

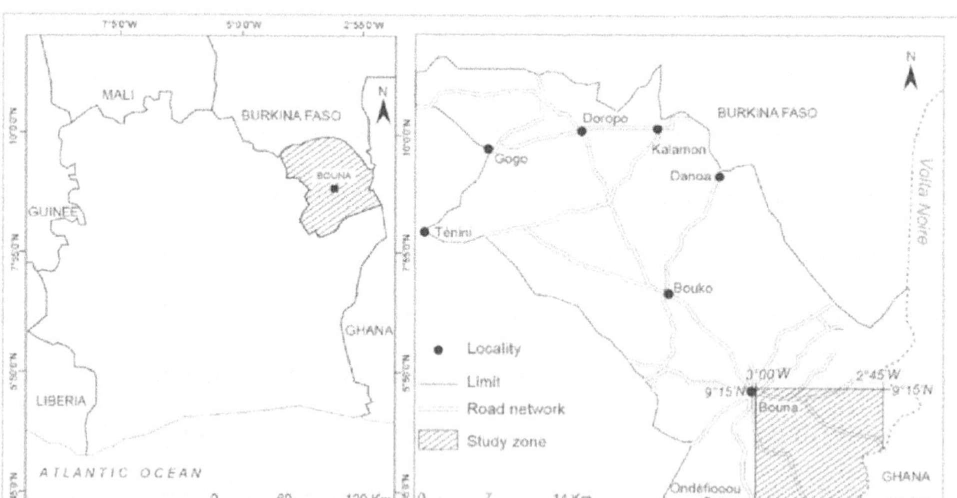

Source : BNETD/CCT, 2012 P. WGS 1984, Zone 30N Production : ALLA Kouadio A., 2019

Figure 1. Location map of the study area in Ivory Coast.

development situations" (Auzet 1987). The need to assess risks, to better understand constraints and to develop means of anti-erosion measures is not possible without taking into account the complexity of the local geomorphological and soil characteristics.

This study looks at the various conditions of vulnerability to erosion: the water erosion processes itself, and the solutions according to the farmers' logics in the Black Volta River region. The developed map of landscape susceptibility to erosion is therefore an essential tool (Figure 2). It provides an overall view of the threatened areas and locates the sectors requiring priority intervention from a sustainable soil management perspective.

In the different geographical areas of Ivory Coast, unsustainable agricultural practices are considered to be the single greatest contributor to the overall increase in erosion rates (National Research Council 2010). Tillage of agricultural land

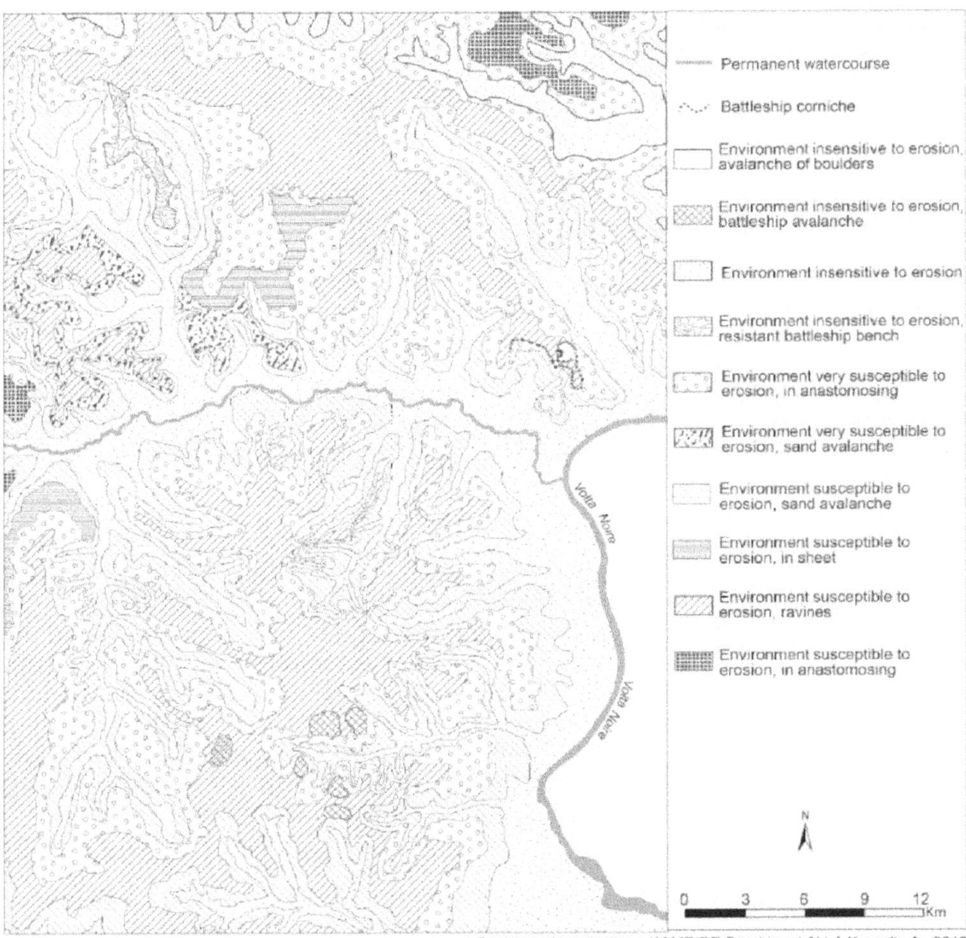

Figure 2. Map of landscape sensitivity and susceptibility to erosion in the area of the Black Volta River (compare Figure 1).

breaking up soil aggregates into finer particles is one of the main causes of accelerated soil erosion. The phenomenon has been worsened today with mechanised agriculture that allows deep plowing. It severely increases the amount of soil that is available for transport by water erosion. Due to mono-cropping, agriculture and land use on steep slopes, application of pesticides and chemical fertiliser displacing soil organisms that bind soil aggregates together, row-cropping and the use of surface irrigation are also involved in the process of accelerated soil erosion (Blanco & Rattan 2010).

Up to date, in the Black Volta River catchment agricultural mechanisation and irrigated agriculture are practically unknown. The local population is dependent on natural resources and practices 'slash and burn' subsistence agriculture with little use of fertilisers and/or pesticides. Vegetation is regularly burned for hunting of small animals (UNEP-GEF Volta Project 2011). Soil tillage is done exclusively with the hoe and mobilises superficial soil in form of compact decametric aggregates. Soil erosion in this area is often perceived as critical because of the immediate landscape and economic damage: loss of crops, flooding and discharge of sand downstream (UNEP-GEF Volta Project 2011).

The research question focuses on the determination of the role of landscape/slope interactions and vegetation-soil interrelationships and their importance to morphogenetic degradation of arable land. It is hypothesised that all environments are stable in their natural state of equilibrium as soon as a segment of the landscape's slope is poorly protected by vegetation. It is then probable and likely to observe a certain level of erosion regardless of the nature of the soil.

6.2 STUDY AREA, DATA AND METHODS

6.2.1 Geographical framework

The surveyed area is located in the northeast of Ivory Coast where the Lobi and the Koulango ethnic groups live. Its geographical coordinates are 9°-9°15′N and 2°45′-3°W (Figure 1).

From a geological and petrographic point of view it is a granitic country between the last elements of the Guinean ridge in the west and the Black Volta in the east. There are alternating geological structures that introduce a more differentiated relief (Avenard 1971).

All taxonomic landscape groups are represented. On schists, particularly along the border to Burkina Faso and the Black Volta River, rigid tabular mountains with a ferricretes on top, surrounded by smaller scarps, are the dominant geomorphic landscape elements. To the south-east, still along the Black Volta River, on thin phyllite outcrops passing locally to flysch-like and green rocks, incidentally migmatites and post-tectonic granites, it is the domain of extended slopes (interflows and glacis), tabular mountains with lateritic crusts, isolated crests with massive ferricretes and saprolited hilltops in shales. Rocky bars dominating glacis slopes from 100 to 200 m asl are structurally controlled striking in NW-SE direction. They overlook either the rigid surfaces of the glacis on schists, or the valleys incised by the Black Volta and its tributaries. Small isolated scarves of Precambrian Birrimian structures (schists, green rocks, dolerites, quartzites), whose direction does not conform to the classical orientation since it is NE-SW or NS, cause significant raising of massifs, hills, linear shapes of hillsides, rocky bars or crest lines and isolated inselbergs developed in melancratic metamorphic rocks. Typically, a geomorphic landscape succession of mounds with an iron lateritic crusts (plintite) on top – locally exceeding 500 m asl – can be determined.

Around Bouna, and in the north towards Doropo (Figure 1), as well as in the south towards Bondoukou, there are sandy or weathered tabular mountains surmounted by weakly plintited crusts; tabular mountains and mounds with a ferricrete crust weakly or partially dismantled; weathered hilltops on leucocratic rocks which settle on crystalline substratum (granites, gneiss, migmatites) at an altitude slightly higher than 300 m asl (Avenard 1971, BETPA 1978, Filleron 1984). A general feature is the flatness of the soils on the tabular mountains (Rougerie 1964). The savannah woodland on tropical ferruginous soils is considered as the natural vegetation. However, there are local islands of dry forests on shallow and tropical eutrophic brown soils (Adjanohoun & Guillaumet 1971).

All these tropical landscapes contribute to the rural economy of the Black Volta River basin: they correspond to the fields and food crops (corn, millet, sorghum, beans, yams) of the Lobi population (Chaléard 1996). But they are, more or less, suffering from land degradation through runoff, which tends to compromise the vital functions of the ecosystem and the economic future of thousands of people who depend on them.

6.2.2 Methodology

For the study of susceptibility of landscapes to erosion (landscape sensitivity) three methods were used. First, mapping of landscape segments was done. The technique consists of a systematic drawing of all slope break lines and slope inflection lines visible on aerial photographs HAUTS-MONTS INC-CIV-96-015 to 016 of February - 24th, 1996.

From a physiographic point of view, these slope breaks and slope inflections delimit topographic facets that are distinguished by their profile and plan view. For example, the amoeboid tabular hilltops facet contrasts with the facets of the steeper sloping slopes, whose lines are extended parallel to the surfaces defined by the lowlands.

Secondly, from a structural point of view, these slope breaks and slope inflexions highlight important changes in the organisation of environments. Thus, in the succession of landscape segments, the moment when the chain breaks, by the appearance or disappearance of a link, by the appearance or disappearance of a particular type of segment, necessarily coincides with another type of landscape, the latter being determined by different lithological conditions or soil materials.

Finally, from a kinematic point of view, these slope breaks and slope inflexions necessarily involve the current dynamics of the natural environment, those related to water balance, water sharing and concentration, surface erosion and accumulation, leaching and material transport. Thus, under the same vegetation and soil conditions, the almost tabular summit segments of the landscapes participate little in the overall kinematics. They appear to be relatively autonomous, and for a long time they retain the mark of soil-vegetation balance in relation to the lower slopes' segments, whose current dynamics are much more intense, trans-erosive or trans-accumulative depending on their slope value.

In a second step, the direct action of erosion processes is measured. The study of the overall kinematics of the environment becomes a preliminary work.

Several levels can be taken into account in the study of erosion. Priority has been given to the slope level. For each of the ortho-type slopes, i.e. with "normal" characteristics, the value of the slope is estimated. Slope classes are thus established.

On one or more slopes, corresponding to a given slope class, the shape and stage of erosion measured in the field is described.

Four classes of slopes, each capable of determining a stage of erosion, have therefore been identified. They are based on the classification by Mayer (1990). However, some slopes have one or more 'geon' located within them that may influence the erosivity of the precipitated water. Altogether,17 types of landscape environments, sufficient to identify the entire area, were thus subjected to runoff and erosion experiments under natural conditions of tropical rains and storms (Table 1).

The morphometry of the slopes was determined by measuring their angle (% or degrees). The superposition of the landscape segment map and the topographic map makes it possible to obtain, with the maximum precision, the average slope between the curves that pass through their vertex and their base, and the maximum average slope defined by the nearest curves.

The quantitative study of current erosion processes is carried out in the field. It consists of a network of experimental plots of a few hundred square meters area, in the shape of a rectangle, lengthened in the direction of the slope. A ditch, open upstream, delimits each plot, while a gutter dug downstream ensures the recovery of runoff water and transported materials in bins. The runoff and erosion experiments were carried out

Table 1. Types of environments studied and maximum of average slope.

Classes of slopes		Number of experimental plots	Characteristics of the studied environment
In %	In degrees and appreciation		
0 -1.5	0 - 1 slope from zero to very low	8	Ridge and ridge line, lithic ground Rounded gravel top Tabular sandy summit Tabular summit with ferricrete crust Tabular summit with weathered ground Lowland, cross-linked hydromorphic soil Shallow, hydromorphic gleyic soil Lowland and plain with alluvia and/or colluvia
1.5 - 4	1 – 2.30 low slope	4	High/Medium inclinated slope with gravel and/or indurated ground with ferricrete crust High/Medium inclinated slope with gravel and/or indurated soil with lateritic crust (plintite) Bottom of the connection side on sandy ground Bottom of slope, indurated soil with a lateritic crust (plintite)
7 - 16	4 – 9.30 moderate to steep slope	3	High inclinated slope, indurated soil with ferricrete crust Slope with sandy cover, Slope with clayey cover
> 16	> 9.3 very steep slope	2	Scarp ('corniche'), indurate soil with ferricrete crust Slops of hills and inselbergs, eutrophic soils

Source: Interpretation of 1996 aerial photographs.

under rainfall events of different duration but of the same intensity in order to compare the orders of magnitude on the slope categories selected.

The third step in the process of describing and quantifying erosion is followed by a survey of the components of the natural environment: the nature of the substrate and soil, the texture and structure of soil, and the state of vegetation. It is intended to estimate the influence of the structural state on the infiltration/runoff generation, the mechanical resistance of the soil surface to erosion, and the role of soil surface roughness and vegetation cover on retention and runoff. Between these new variables and the measured flow plot data, a rigorous report can be made and the results of this analysis can be projected on to the previous geomorphological variables (the nature and value of the slopes). It is the map, obtained at the end of this process, which allows to show the sensitivity of quasi natural environments to erosion.

Finally, an indication of traditional anti-erosion techniques mainly perceived by the Lobi people in the Black Volta River catchment is provided by a semi-directive interview with 207 stakeholders (heads of farmer's households) identified in the prospected area. The results of this survey made it possible, on the basis of cultural practices and local endogenous knowledge, to envision sustainable land management methods that take into account the internal logic of the farmers.

6.3 RESULTS

The map (Figure 2) shows most of the geomorphic dynamics and processes of the natural environment in the surveyed area.

6.3.1 Environments and landscapes not sensitive to erosion

They cover 7% of the landscapes surveyed. In these environments, erosion by storm water runoff is negligible. On dry land a rainfall event of 78 mm/h eroded 12 kg/ha of soil after 47 minutes, 38 kg/ha after 77 minutes, 73 kg/ha after 123 minutes and 109 kg/ha after 212 minutes.

This type of environment consists of different landscape segments. Their preferred domain corresponds to the tops of irregular tabular or convex hills (rocks with a ferricrete on top; hilltops with plintite). In the sectors of mountains, it is established on the tops of the tabular mountains with a ferricrete crust (Figure 3), on the tabular mountains and mounds with a lateritic crust or a plintite weakly or partially dismantled.

This immunity in surface erosion is due to practically zero slopes (0 to 1°) and the relatively dense natural or locally cultivated herbaceous and woody vegetation cover. The physical characteristics of the soil are a determining factor. At the bottom of alluvial flats and on hydrographic floors, thick alluvium and/or sandy colluvium increase the amount of water that can be absorbed by the soil and therefore prevented it from flowing to the surface in the form of erosive runoff. When the soil is reticulated hydromorphic or gleyic hydromorphic, its massive structure and cohesion of aggregates provide it with remarkable structural stability. Similarly, in indurated areas, the compaction of materials gives to the surface with a ferricrete crust a high mechanical resistance to erosion.

Moreover, the discontinuous and little dismantled ferricrete crust, the grass tufts and the roots of trees slow down runoff, which is carried out in more or less fine anastomosed nets and limits, in most cases, the erosion marks in extent and depth.

Figure 3. Part of a tabular mountain with a crest and ferricret crust on top (Photo: J. Runge, 2013) with sparse savannah woodland vegetation cover on pisolitic soil. This environment, armed with a lateritic crust is almost free of any trace of erosion. Some sand and gravel-like textures are deposited by sheet flow processes in the cavities of the ferricrete crust (plintite).

6.3.2 Environments and landscapes susceptible to erosion

They represent 54% of the study area and can be divided into two groups according to their behavior in the face of erosion (Figure 2). In their usual state of equilibrium, certain environments are stable with respect to erosion processes, especially when they are under natural vegetation. But they undergo soil erosion which becomes sensitive when the slopes have a marked convexity. Erosion becomes significantly stronger when these slopes are cultivated.

This type of environment (12% of the study area) has also been subjected to the experiment of runoff simulation. On dry soil, a rainfall of 78 mm/h of intensity eroded 127 kg/ha of soil after 47 minutes, 221 kg/ha after 77 minutes, 419 kg/ha after 123 minutes and 508 kg/ha after 212 minutes. It is located on a wide variety of landscapes. In the mountain areas, it occupies the entire top of pisolitic and saprolited tabular mountains with low, discontinuous, unclear scarps; gravel tabular mountains with low, continuous, unclear scarps; pisolitic and saprolited tabular mountains with irregular tops, unclear 'corniche' and very discontinuous scarps. In the sectors of regular tabular or convex hills, it occupies the entire pisolitic and weathered hillside.

All these landscapes have relatively gentle slopes (1° to 2.30°). The low erosion rate is therefore mainly linked to the behaviour of the surface materials. When they are lithic, the soils, well protected by stones and vegetation cover, they tend to be less prone to the erosive effect of rainfall. When eutrophic, they contain more clay and high levels of organic matter and tend to be more resistant to stripping because clay helps to bind particles (Mirsal 2008) while organic matter coagulates soil colloids and creates a stronger and more stable soil structure (Blanco & Rattan 2010).

Runoff occures in anastomosed nets with the possibility of collapse ('avalanches') of rock blocks and ferricretes collapsed by gravity breakage. In this case, no particular anti-erosion measures are recommended. Moreover, the Lobi people practice a culture under a tree-planted park. The roots of trees and plants hold back soil particles, preventing them from being eroded. In the same way, the slope and vegetation cover act to reduce the speed of flow and the kinetic energy of raindrops.

Other environments, in their normal state of equilibrium, are considerably affected by erosion. This type of environment is quite frequent (42% of the area). On dry soil, a loss of 307 kg/ha of soil was recorded under a 78 mm/h rainfall event after 47 minutes, 1068 kg/ha after 77 minutes, 1734 kg/ha after 123 minutes and 2009 kg/ha after 212 minutes.

Resurgence of erosion by concentration of runoff, with setting up of gullies, is localised on all the marked lower slopes (> 9.3° of slope) that characterize the summits of accentuated convex ridges; they are found on the tops of the monoclinal mountains (1° – 2.30°) (Figure 4).

Erosion is more often in the form of diffuse slick or anastomosing nets (Figure 4), without the formation of gullies, especially on Precambrian Birrimian substrates, with runoff rarely managing to concentrate at the expense of the massively structured soil. It then affects the entire inselberg edge foot plain (glacis); the foot slopes of the inselberg; the tabular weathered mountains with a well visible and continuous scarp as well as the connecting slopes of the tabular mountains. They result from the disturbed vegetation cover and, secondarily, from the low organic matter content of the soil.

The side effects of an improvement in tree coverage of the soil and herbaceous plants, already practiced by farmers, can be proved effective in many cases. Rational land use can, however, be advised on soil physics and soil retention. In concrete terms, this consists of the development, upstream of the slope, of a stony trap, capturing sediments that have been torn off at higher topographical levels, then a trap

Figure 4. Sandy soil on a planation surface close to the top of a monoclinal mountain (Photo: J. Runge, 2013). The low clay and organic matter content expose the soil to erosion. Sandy textures and leaf residues in the foreground give evidence of anastomosing runoff processes.

of forage grasses followed by a trap of dry cereal crops (sorghum, fonio, millet, beans). Finally, a plantation of fruit or fodder trees, yams or more moisture-intensive cereals such as maize are cultivated downstream of the hillside.

6.3.3 Environments and landscapes very susceptible to erosion

These environments represent 39% of the area mapped. In their usual state of equilibrium, they are susceptible to erosion due to runoff. Under the same experimental conditions, 805 kg/ha of eroded material was recorded after 47 minutes, 1429 kg/ha after 77 minutes, 1901 kg/ha after 123 minutes and 2023 kg/ha after 212 minutes.

This type of environment is mainly found in the inselberg sectors and concerns all the chains and hills (crests, whale domes and backs, rocky bars) where erosion is part of an almost natural morpho-pedological dynamics. Most of the time, the resumption of erosion is due to the strong topographic gradient, which exceeds 17°, and result in characteristic "staircase steps" (pedi-planation scarps). It leaves only the coarse elements on the surface; it is responsible for the removal and transport of the first loose horizons of the soil. The surface dynamic is called trans-erosive.

These environments and landscapes are also found in almost all the marked slopes downstream of the high mountains (slop values between 1-2.30°), hills and ridges (slope >9.3°), upstream and downstream of the extended glacis slopes (4-9.30°). In the latter cases, the superficial dynamics, generally trans-accumulative, can be trans-erosive. The value of the slopes (greater than 2.30°) and the poor protection of the soil surface at the beginning of the dry season are certainly in question. But morphogenic degradation is mainly linked to the nature of the geological bedrock. With a coarse sandy texture on granite, the soil is more sensitive to erosion and impoverishment than fine argillaceous soil with a massive structure of similar landscapes on Birrimian substrate (shale, green rocks and flysch-like rocks).

In this case soil erosion is part of a quasi-natural dynamic evolution and all these environments are highly unstable (Figures 5A & B). Most of these landscapes have not yet been cultivated, despite the pressure of increasing regional population densities and intra-regional migration of people. The issue of sustainable land management can, therefore, be considered as part of a more general management plan by trying to prevent late bush fires and to encourage shrub reforestation between contour lines. In cultivated areas, the farmers themselves protect the soil against erosion using traditional techniques of ridged mounds, oriented both perpendicular and parallel to the slope.

This type of environment can be cultivated by protecting the soil surface with drought-resistant herbaceous plants. Residues of crops can also be used (mulch). These residues, as well as the drought-resistant herbaceous cover, play a role in mitigating erosion by reducing the impact of rain drops that break down soil particles and of superficial runoff. The development of small basins, planted with fruit trees, on arid hillsides can also allow the sustainable establishment of more water-intensive commercial arboriculture such as the orange tree (*Citrus sinensis*), and the lemon tree (*Citrus limon*). Sediment and organic matter from runoff captured in the basins provides moisture and soil fertility that is beneficial to plants in the absence of irrigation. Alternatively, in a more intensive farming system, the organic content of the soil can be improved through mulching and manuring, so that straw and manure remain in place and enrich the soil. In the case of pioneer cultivation, one of the best anti-erosion techniques is the barrier strip method. It consists of alternating strips of

Figure 5. A and B. Soil and vegetation conditions in the Black Volta River catchment (Photo: J. Runge, 2013). Dry grassy soil cover (~50 cm large) protects against erosion on a glacis slope; accelerated soil erosion by runoff will occur by the steep slope on a neighboring inselberg in melanocratic rocks (5A); the total absence of surface plant cover in the dry season results in poor protection of the soil against runoff triggered from first thunderstorms at the onset of the rainy season (5B, fence post is ~150 cm large).

cultivated fields parallel to contour lines with narrower strips of natural permanent or long-lasting herbaceous and woody vegetation.

6.4 DISCUSSION

Some authors describe rainfall in the tropics as very degrading to the soil. Roose (2008) states that tropical rains are a hundred times more morphodynamically aggressive than temperate rains. Data measured on flow plots in the Black Volta River basin area showed that this is not always the case. Taking into account the experimental conditions (slope value, soil type and vegetation cover), the orders of magnitude obtained in the alluvial lowlands as well as on the tops of tabular mountains (from slope 0 to 1°) are not the same than those recorded in inselberg landscapes, chains and hills where the slope exceeds 16°. In the first case, soil losses are estimated from 12 to 109 kg/ha and in the second case they are between 805 and 2023 kg/ha under the same rain conditions (rainfall of 78 mm/h).

The topography of the land determines the speed at which the surface flow will occur, which in turn determines the erosivity of the flow. Similarly, steeper land is also more exposed to 'landslides' and other forms of gravitational erosion processes (Whisenant 2008, Blanco & Rattan 2010, Wainwright & Brazier 2011).

The influence of the topographical position at the bottom of the hills is sometimes more important than the value of the slope of the hillsides. Cosandey & Robinson (2000), like Roose et al. (1993), observe that Mediterranean ravines sometimes grow from the bottom of the valley to the top of the hill, because they take advantage of cumulative runoff or declivity at torrents bed (Roose 2008).

However, the effect of the length of the slope is minimal. Thus, the study by Boli et al. (1993) showed that on smooth sandy glacis in northern Cameroon by increasing the length (20 to 60 m) or surface area of a field (100 to 1000 m^2) runoff on slopes of 2% did not significantly increase erosion. In general, only plots with ravins loose more soil as the length increases (Roose 2008).

Like the slope structure, form and inclination, reduced vegetation cover is considered to be a very important factor in soil erosion through runoff. Seasonal variations in metabolism are of considerable importance, since "plant cover plays a very different role as a screen depending on the season" (Tricart & Cailleux 1965). The results obtained by ORSTOM, in a series of small experimental basins common to mesophilic forest areas in the western savannah regions in northern Ivory Coast (Ferkessédougou), indicate that the density of dry season vegetation cover spreads rainfall, and the runoff coefficient decreases. "Flow is also slowed down and the delay between showers and floods increases, indicating a braking effect by the vegetation cover, possibly also a higher hypodermic flow (Tricart & Cailleux 1965).

The low green cover is very efficient and acts like a stone mulch to disperse the rain drops and runoff energy.

In Africa, Hudson (1958) "made a nice demonstration of the impact of drop energy and their interception possibilities on a 4% slope with a stable Oxisol. A simple mosquito gauze at 15 cm over the soil surface was as efficient as a meadow to reduce erosion to 1% and runoff to 10% of what is observed on a bare runoff plot. The cover of the low canopy seems to be more important than the vegetation type and even more than roots (which are more efficient against linear erosion)" (cited in Roose 2008).

Bushfires have a significant effect on runoff from the first storms when vegetation is burned just before the rainy season (Roose 1996). As soon as wintering is over, the soil dries out, the leaves fall off, the grasses dry out and fall prey to fire. Then, the first morphodynamically intensive rains of the following rainy season are triggered on an unprotected soil.

"A real seasonal morphogenetic crisis corresponds to the beginning of the rains: the soil, desiccated and protected to a minimum, is particularly suitable for runoff" (Tricart & Cailleux 1965).

In the Black Volta River catchment area, the phenomenon is particularly noticeable on accentuated convex hilltops with weathered or gravel soil. It is also found on sandy monoclinal mountains, glacis, and sandy glacis with low clay content as well as saprolited tabular mountains. In fact, these landscapes correspond to the environments where yam (the main cash crop) and cereals (the main food crops, after the practice of burning by late fires) are planted.

The particularity of vegetation and plant cover in tropical Africa, however, is their ability to fastly regenerate themselves. This is clearly demonstrated by Adjanohoun & Guillaumet (1971) in the Konkodekro protected forest near Bouaké in central Ivory Coastin a pre-Guinean forest bioclimatic context. They show that early fires do not prevent reforestation of secondary formations. The integral fire protection of deeply modified areas, which evolve towards savannah ecosystems, leads to the reconstitution of a dense semi-evergreen forest. Similarly, in Ferkessédougou in northern Ivory Coast, under Sudanese bioclimate, the study of experimental plots indicates that the areas most affected by fires are distinguished by the abundance of *Imperata cylindrica*. But this herb, left to its own devices without any action that regenerates it (fire, cutting), is gradually destroyed over the years. The grouping it forms then becomes fleeting. In its place, woody vegetation develops towards a typical savannah woodland (Avenard 1971).

The issue of land conservation can be considered as part of a general management plan by trying to prevent late bushfires, and by encouraging longer fallows. Because three years of protected fallow was sufficient to increase bush vegetation and pedo-fauna activities, drastically increasing the infiltration capacity, the vegetation cover and limiting soil losses (Roose 1996, 2008). Fallow land is a structural practice in northeastern Ivory Coast, but with a shorter duration by Lobi people of 2 to 3 years maximum than in Koulango ethnic groups, over 8 years minimum (Chaleard 1996). These are integrated solutions that rely mainly on local farming practices. But of course, crop residue management, mulching, the use of drought-resistant herbs, the stop strip method to dissipate rainfall and runoff energy on hillsides are also part of the solution. The same applies to the spread of chemical fertilisers, and government must develop policies to maximize the availability of these expensive products. The aim is to disseminate natural phosphates on a large scale and to improve the transport of fertilisers.

Soil composition, moisture and compaction are also widely cited as factors in determining precipitation erosion (Blanco & Rattan 2010). According to Fournier (1967), in the absence of vegetation cover, the characteristics of surface horizons play a key role in the initiation and development of erosion, particularly structural stability and soil permeability. But the effect of the soil parameter is not always obvious without the influence of the slope. Thus, in the Black Volta River basin, most of hydrographic valleys and the tops of certain tabular mountains, with very slight slops (0 to 1°), are very sandy. However, it is not certain that these sandy areas are always sensitive to water erosion, because water can quickly infiltrate. In contrast, on hills, tropical eutrophic brown soils contain enough clay and organic matter to bind particles, coagulate colloids and give to the soil a more stable structure, i.e. a more resistance to erosion. Due to the steep slope, these areas have deep ravines which are advanced stages of morphogenic land degradation.

6.5 CONCLUSION

The mapping of the susceptibility or sensitivity of environments to water erosion in the Black Volta River catchment has made it possible to distinguish four classes of multifactorial vulnerability to the morphogenic degradation of land by runoff. Environments insensitive to erosion cover 7% of the study area while environments susceptible to erosion occupy 54% of the area. The class of the environments very susceptible to erosion covers 39% of the entire mapped area and concerns, in particular, the areas of hills, ranges and rocky bars. Their vulnerability is mainly due to hydro-morphometric parameters (slope value exceeding 9.30°, the significant curvature of the relief and the organisation of flows in sheets or ravines). The strong anthropic pressure on agricultural land at the expense of the vegetation cover is also an important issue.

The topographic slope is the determining and crucial factor in the multifactorial vulnerability to erosion in this region. The state of the vegetation ('geon factor') and the sandy or sandy-clay texture of many soils only contribute to the efficiency of the erosion processes.

The multifactorial vulnerability to water erosion obtained as a result of this integrated mapping of the natural environment could constitute a guidance document either for the restoration of land fertility and productivity or for rational land use in rural areas in the Black Volta catchment area.

The anti-erosion measures to be implemented are not only a technical problem; they must also and above all take into account the human context since the stakeholders' interests are not necessarily compatible with what is proposed from 'elsewhere'. The effectiveness of strategies is mainly based on farmers' own traditional knowledge, taking into account their experience, their ingenuity or, to go to the end, their internal logic. They must be based mainly on local practices and locally available materials. Thus, depending on the case, when the technique of cultivation under arborous park is not sufficient to control erosion, a combination of measures may be required.

REFERENCES

Adjanohoun, E. & Guillaumet, J.-L. 1971. La végétation de la Côte d'Ivoire. In Avenard, J.-M., Eldin, M., Girard, G., Sircoulon, J., Touchebeuf de L., P., Guillaumet, J.-L., Adjanohoun, E. & Perraud, A. *Le milieu naturel de la Côte d'Ivoire*: 161–263. Paris: ORSTOM (50).

Auzet, V. 1987. *L'érosion des sols cultivés en France sous l'action du ruissellement*. CEREG-UA 95. Strasbourg: CNRS.

Avenard, J.M. 1971. *Le milieu naturel de la Côte d'Ivoire*. Paris: ORSTOM 50.

BETPA, Bureau d'Etudes Techniques des Projets Agricoles, 1978. *Projet de développement rural intégré des savanes du Nord-Est: la région du Projet (le milieu physique & le milieu humain)*, Tome 2 & 3. Abidjan: 298 p.

Blanco, H. & Rattan, L. 2010. *Tillage erosion. Principles of Soil Conservation and Management*. Berlin: Springer.

Boli, Z., Roose, E., Bep A Ziem, B., Sanon, K. & Waechter, F. 1993. *Effets des techniques culturales sur le ruissellement, l'érosion et la production de coton et maïs sur un sol ferrugineux tropical sableux au Nord Cameroun* (Mbissiri 1991-1992). *Cah. ORSTOM Pédol.* 28 (2): 309–326.

Chaleard, J.L. 1996. *Temps des villes, temps des vivres, L'essor du vivrier marchand*. Paris: Karthala.

Cosandey, C. & Robinson, M. 2000. *Hydrologie Continentale*. Paris: A Colin.

Filleron, J.C. 1984. Régions morpho-pédologiques. In *l'Atlas du Nord-est de la Côte d'Ivoire*: 2-4. Abidjan: IGT-Université Abidjan.

Fournier, F. 1960. *Climat et érosion*. Paris: Presses Universitaires de France.

Fournier, F. 1967. Les facteurs de l'érosion du sol par l'eau: les recherches sur la conservation du sol en Afrique et Madagascar. *Sols Africains* 12(1): 6–51.

Hudson, N. 1958. Runoff and soil loss from arable land in Southern Rhodesia. *Int. Union for Conserv. of Nature*. 7th Technical session, theme 1C, Athens.

Hudson, N. 1971. Mesures de terrain de l'érosion et de l'écoulement des eaux de surface. *Bulletin pédologique de la FAO* 68, 153 p.

Mayer, L. 1990. *Introduction to quantitative geomorphology*. Englewood Cliffs (New Jersey): Édit. PRENTICE HALL.

Mirsal, I.A. 2008. *Soil degradation. Soil Pollution: Origin, Monitoring & Remediation*. Berlin: Springer.

National Research Council 2010. *Toward Sustainable Agricultural Systems in the 21st Century*. National Academies Press.

Rattan, L. & Stewart, B.A. 2019. *Soil degradation and restoration in Africa*. Florida: CRC Press Book.

Richard, J.F. 1983. Cartographie intégrée du milieu naturel (Tropiques humides). In *Actes du colloque sur les "Environnements et leur maîtrise humaine"*: 75–101. Kenzo Fujiwara.

Roose, E. 1977. *Vingt années de mesure de l'érosion en petites parcelles en Afrique de l'Ouest*. Travaux et documents. Paris: ORSTOM 78.

Roose, E. 1984. Causes et facteurs de l'érosion hydrique sous climat tropical, conséquences pour les méthodes antiérosives. *Machinisme Agric. Trop.* 87: 4–18.

Roose, E., Arabi, M., Brahamia, K., Chebbani, R., Mazour, M. & Morsli, B. 1993. Erosion en nappe et ruissellement en montagne méditerranéenne Algérienne: réduction des risques érosifs et intensification de la production agricole par la GCES. Synthèse des campagnes 1984-1995 sur un réseau de 50 parcelles d'érosion. *Cah. ORSTOM Pédol.* 28(2): 289–308.

Roose, E. 1996. Land husbandry: strategy and component. *Soil Bulletin* 70. Rome: FAO.

Roose, E. 2008. Soil Erosion, Conservation and Restoration: A few lessons from 50 years of research in Africa. In Dozzi, C. & Constantini, E. (eds.) *The soils of tomorrow: soil changing in changing world*: 159–180. Reiskirchen: *Catena*, (Advances in Geology) soil erosion and carbon dynamics. IRD Editions/CRC/Taylor and Francis.

Roose, E. 2010. La dégradation des terres et l'érosion des sols. In Roose, E., Sabir, M., Laouina, A., Benchakroun, F., Al Karkouri, J., Lauri, P. & Qarro, M. (eds.) *Gestion durable des eaux et des sols au Maroc: valorisation des techniques traditionnelles méditerranéennes*: 21–31. Marseille: IRD.

Rougerie, G. 1964. *La Côte d'Ivoire*. Coll. Que sais-je? N°137. Paris: PUF.

Tricart, J. & Cailleux, A. 1965. *Le modélé des régions chaudes (Forêts et savanes), Traité de Géomorphologie*, T.V. Paris: SEDES.

UNEP-GEF Volta Project 2011. *Analyse diagnostique transfrontalière du bassin versant de la Volta: Rapport thématique sur les écosystèmes du bassin, rapport final*, UNEP/GEF/Volta/RR., 102 p.

Vogt, J. 1953. Erosion des sols et technique de culture en climat tempéré maritime de transition (France et Allemagne). *Rev. Géomorphol. Dyn*: 157-183.

Wainwright, J. & Brazier, R.E. 2011. Slope systems. In Thomas, D.S.G. (ed.) *Arid Zone Geomorphology: Process, Form and Change in Drylands*: 209–233. Chichester: Wiley & Sons Ltd.

Whisenant, S.G. 2008.Terrestrial systems. In Perrow, M.R., Davy, A.J. *Handbook of Ecological Restoration: Principles of Restoration*: 83–105. Cambridge: Cambridge University Press.

CHAPTER 7

Stakeholders' perceptions and strategies to climate change resilience in Kara and Dapaong, Togo

E. Ouro Bitasse, Y. Moutoré & G. Dansoip
University of Kara, Kara, Togo

ABSTRACT: The issue of climate change requires a reconversion of disciplinary sectoral conceptions in order to better understand resilient and non-resilient factors. In this context the socio-cultural background that determines urban stakeholders' perceptions and practices was sociologically analysed. Individual interviews were conducted with the urban population of Kara and Dapaong in northern Togo. Socioprofessional categories such as gender, age and education gave evidence that socio-cultural perceptions of climate change are a function of the stakeholders' profession; they strongly determine their behaviour towards the environment. The cult of ancestors and the weight of the common past explain the strong and traditional dependence on firewood/charcoal and the uncontrolled management of waste. Urban populations also practice various deviances such as prostitution, homosexuality and different forms of religious practices that are interpreted as the anger of the deities which recently modify the climate with the aim to punish them. Therefore, it is revealed that each socio-professional group understands "climate change" as they conceive it in their own sectors of activities differently (e.g. depletion of rainfall for market gardeners or in the case of increased annual temperatures as a restriction for business activities). Environmental practices such as recycling of plastic bags are in line with the circular economy; they are observed in all of the screened urban societies.

7.1 INTRODUCTION

Climate change poses a threat to economy, especially for underdeveloped countries in Africa. With an average global warming reaching a level of 0.8 °C, these countries experience declines in agricultural productivity and vulnerability to epidemics. Limiting global warming below 2 °C will not even prevent the adverse effects of climate change (IPCC/GIEC 2014). Togo is already facing severe increase in temperature (heat), reduced and perturbed rainfall distribution, floods, droughts, strong winds and rising sea levels that exacerbate coastal erosion (République Togolaise 2015). Because of the damage they cause, changing climates and climate variability have a strong influence on the country's development and threaten to annihilate efforts for economic growth. Togo has been committed for several years to a proactive strategy for sustainable development and the fight against global warming. The main axes of intervention are: the fight against poor production practices in several economic sectors; the lack of demographic control, and the high incidence of poverty exacerbated by the negative impacts of climate change. These increase vulnerability of productive economic sectors and the pressure on natural resources (République Togolaise 2015).

This political will is manifested through the National Environmental Action Plan, the National Program for Environmental Management, the National Strategy for Sustainable Development (December 2011), the National Capacity Building Strategy for Environmental Management (October 2008), the National Program for Reducing Greenhouse Gas Emissions from Deforestation and Forest Degradation (REDD +, 2010-2050), the National Strategy for Disaster Risk Reduction in Togo (December 2009), the National Medium Term Priority Framework (NMTPF) for Togo (2010-2015), and the National Action Plan for the Management of Marine and Coastal Environmental Resources (République Togolaise 2017). Overpopulation of the region, economic activities, wealth, and towns are typical places of high energy consumption and consumption of natural resources. This exerts a strong anthropic pressure on the environment.

According to UN-Habitat (2011), the effects of urbanisation and climate change converge in a dangerous way as urban areas are the main contributors to climate change through the lifestyle and consumption of city dwellers: occupation of space (urban sprawl, urban construction), high energy consumption, population pressure on natural resources, waste, increased traffic and emission of greenhouse gases. According to the results of the 2010 General Population and Housing Census of Togo, the urban population represented 37.71% of the national population of 6,191,155 inhabitants. Estimates indicate that this proportion increases to 43.5% in 2020 (INSEED 2015b). The major challenge today is the establishment of a sustainable use of natural resources and the concern to preserve the environment. This is even more so, in the context of decreasing in relative air humidity, as we move from southern to northern regions in Togo. Accordingly, to the latitude division, Togo has three main climatic zones: The most southern sub-equatorial climatic zone extends from the coast to ~8° northern latitude; temperature varies in small amplitudes with an average annual rainfall between 1000 to 1400 mm. The growing period of plants is less than 240 days, divided into two rainy seasons: one from mid-March to late July and the other from early September to early November. In Togo, and especially in the Kara and Savannah Regions, climate constraints are one of the main factors limiting socio-economic development. In Mango in the northern Oti plain, there is a mean warming of 0.4 °C every decade. Dapaong seems to be more 'stable' with a single increase of 0.8 °C in the decade 1981-1990 (Figure 1), indicating a warming trend of climate compared to the previous decade (Badameli & Dubreuil 2015).

Climate change is generating various social problems that need to be elucidated in order to help better decision-making. Indeed, climate change in Togo would have

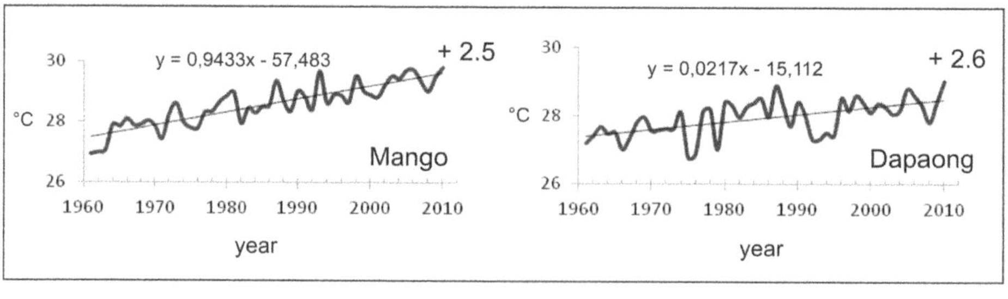

Figure 1. Interannual trends 1960-2011 of mean temperature at the meteorological synoptic stations of Mango and Dapaong in Togo (Badameli & Dubreuil 2015: 422, modified).

negative impacts on agriculture by compromising food security due either to drought characterised by declining rainfall and agricultural yields, or to heavy rainfall materialized by floods with the consequent devastation of fields and crops. According to the WHO (2014), climate change has become a major threat to public health. Changes in temperature and precipitation have severe effects on the development conditions of different disease vectors such as meningitis, respiratory diseases (coughs, bronchitis), cholera, dengue fever, yellow fever in urban areas and parasitic diseases such as malaria.

In addition, climate change would have harmful consequences on housing through the demolition of houses and on road and rail infrastructure as well as on hydraulic and energy systems. The interest of this research resides in its capacity to analyse the cultural elements which are resilient or not to climate change. The behavior of urban population in the different socio-professional categories is studied to understand how they perceive and understand "Climate Change", and how it influences their daily life. The study questions the cultural practices initiated by the citizens of Kara and Dapaong to reduce the effects of climate change phenomenon. To understand the reasons for the extent of climate change in recent years, it is attempted to analyse the practices of stakeholders in relation to the different social and cultural perceptions they have on climate change. It foremost questions the social representations of climate change that determine their practices, whether or not they are resilient to climate change. How do cultural practices impact the climate in Kara and Dapaong? The study is based on the assumption that the socio-cultural representations constructed by individuals are dependent on their profession (education) and determine their negative or positive actions on the climate system.

7.2 STUDY AREA AND METHODOLOGY

The towns of Kara (108,700 inhabitants) and Dapaong (66,100 inhabitants, estimation 2017) in northern Togo act as regional case studies. The urban population of both towns are considered to be vulnerable to climate change despite regional imbalances. The transdisciplinary research focuses on a multidimensional approach taking into account documentary and qualitative data on the basis of disparities according to gender and demographic indicators.

While the second methodological consideration refocuses the debate on social and cultural representations of climate change, it then makes it possible to construct the interdependence between population dynamics, urbanisation and the different manifestations of climate change. Thus, based on the manifestations of climate change identified in relation to social and cultural representations but also to the socio-professional category, the analysis of the practices of urban actors can be built around their typologies according to the socio-cultural and spatial context. The sample is drawn according to the location of climate change events by site selection. The people interviewed (farmers, breeders and civil servants) were directly met at their workplaces. Semi-structured interviews were carried out with the respondents. On the basis of a random draw according to socio-professional categories, 40 individuals were interviewed at the rate of 20 individuals in each of the two main towns (Kara and Dapaong). Individual interviews were also conducted with actors from civil society organisations, working on environmental issues and members of the neighbourhood development committees in both cities, to identify diverse perceptions of climate change in both towns.

The study first made use of a literature and data review: they comprise various study reports, policy documents related to climate change produced by the Togolese

State through the ministries and national bodies involved in natural disaster management. The objective is to document the manifestations of climate change.

To understand the drivers behind the attitudes and practices of residents in Kara and Dapaong towards their environment, in a context of resilience to the effects of climate change, the theories of cognitive vulnerability (Lammel 2015) and of urban resilience (Fournier & Zara 2015) have been applied.

Cognitive vulnerability is a cognitive state in which the subject has neither sufficient information/knowledge nor the information processing methods necessary for an optimal understanding of the phenomena with which he is confronted. This theory advocates that to understand the importance of the study of cognition and cultural context in individuals' ability to respond to climate change; it is necessary to be able to place the individual and their specific characteristics (age, gender, social status, etc.) at the center of a theoretical model of dynamic adaptation to change. This theory makes it possible to understand that, both Kara and Dapaong residents act positively or negatively facing the environment according to their cognitive state, which itself is culturally and socially constructed.

When it comes to adaptation to climate change, the term resilience is frequently used. In ecology, it refers to the ability of an ecosystem to recover its functions following major disturbances, whether natural or related to human activities. Resilience can, therefore, be defined as the ability of a system to "resist" after a severe disruption.

Introduced by the United Nations when it declared the 1990s "International decade for natural disaster reduction", the concept of resilience has since been increasingly used to build a new discourse on disaster management: resilience breaks with the idea of "zero risk" and implies a paradigm shift. Indeed, it is no longer a question of working to eliminate the disaster but of recognizing its inevitability, and preparing people by giving them the tools to reduce the impact and consequences of the coming disaster.

Applied to cities, this gives the concept of urban resilience which is defined by the Rockefeller Foundation as "the ability of individuals, communities, institutions, businesses and systems in urban areas to rebound/survive, adapt and grow despite chronic pressures and acute shocks" (cited in Fournier & Zara 2015). Exploring the theory of urban resilience is fruitful as it provides an account of the strategies deployed in both towns by residents, civil society organisations and state structures to reduce the risks associated with climate change.

7.3 RESULTS AND DISCUSSION

7.3.1 Socio-cultural representations of residents

7.3.1.1 *Definition/perception of climate change*

The problems posed by climate change today have brought light to regional disparities that make it possible to identify resilient behaviour in a context of globalisation. If the various international and national meetings make it possible to question all the actors with different status on the effects of climate change, the mobilisation is still disproportionate because of the social, political and economic context. To understand the practices that Togolese populations have been developing in recent years in response to the adverse effects of climate change, this research has focused on comparing their perceptions of the phenomenon that determine their way of life and their commitment to combat it. It is revealed that two factors determine the construction of representations, namely education level and occupation. Indeed, as proposed by the theory of cognitive vulnerability, understanding individual and collective behaviour in the context of climate change requires analysing the logic of actors based on their profile. Each actor's understanding of climate change depends not only on their professional and daily activities and the way in which climate change impacts them, but especially on their level of education. Most respondents understand climate change as:

> *"the transformation that our climate is undergoing [and that] extends the seasons in Togo (the seasons are no longer the same [or] causes an abnormal change in our seasons"* (comment by a student in Kara).

This representation reflects, in the Togolese context and particularly in the towns of Kara and Dapaong, an influence on the way of life of residents and their daily habits. This transformation of the climate imposes and modifies the periods of the usual tasks, such as washing clothes on the banks of the Kara River. In fact, market gardeners in particular and farmers in general define climate change as:

> *"the lack of water in rivers as a consequence of non-compliance with environmental rules"*

Climate change is manifested in recent years through the drying up of the Kara River and the Kporgou Dam at Dapaong. Unlike market gardeners and farmers, civil servants and related professions defines it in relation to the:

> *"current sunny climate compared to the old days; and unbearable heat waves"*

As the urban populations of Dapaong and Kara understand climate change in their own terms, their representations of causes and consequences are also built around cultural stereotypes.

7.3.1.2 *Causes and consequences*

The causes and consequences of climate change, although they may seem common, are nevertheless specific according to the social and spatial context but especially cultural. To understand the causes of climate change in Togo, it is therefore essential to question both upstream and downstream the cultural practices of which urban and rural populations represent themselves as determinants in the occurrence of new facts such as drying up, increase in temperature, radiation and flooding risk (Badameli & Dubreuil 2015). Indeed, for respondents, current phenomena resulting from climate change must be understood with the abandonment of traditions, which results in:

> *"the non-respect of traditions and the abandonment of rituals once performed by the forefathers to satisfy the gods. For example, formerly the rituals were performed each year at the Kara River to beg for nature's mercy by the gods. But today, this is no longer respected. It is normal for our ancestors to abandon us"* (Comment by a market gardener in Kara).

For the urban population of Kara, the linkage is mandatory between supernatural and human forces to create social but also natural balance. This is why the new phenomena that can be observed must be understood as the manifestations of God's action, which challenges all actors to the fundamentals of human existence. This vision confirms the cognitive vulnerability (Lammel 2015) of the urban population of the two case studies on the manifestations of climate change. This controversial perception explains and justifies the persistence of traditional habits and practices in the management of natural properties and resources. It is part of the cultural practices *en vogue* particularly in Kara where libations and religious sacrifices to ancestors are an essential principle for the functioning of communities; the cult of ancestors and the weight of the common past being the fundamental elements for the functional unity of society (Saint-Martin 2013). Although, precisely as R. Merton notes (cited in Saint-Martin 2013), the function conferred on social action may well prove to be an obstacle to the proper functioning of society and its sustainability, as in the context of climate change in Kara and in Dapaong.

In addition, beyond the weight of tradition, other respondents acknowledge that climate change is the consequence of human pressure through heavy deforestation that is not accompanied by reforestation in search of exploitable areas for agricultural activities and firewood, the use of non-degradable plastic bags or the general lack of awareness. For them, promoting the income-generating activities to increase people's incomes should be an effective response to limit deforestation. This is because it must be understood that the relatively high unemployment rate and, consequently, impoverishment, forces individuals in their struggle for survival to engage in harmful but essential practices in the absence of regulation and control in the field of initiatives.

For most of the respondents, the main actors in climate change are really the large production companies whose various activities are destructive on the environment. For example, there are factories such as the *Brasserie du Benin* and the iron processing plant in the free zone that: "pollute the city of Kara with the gases that their machines emit and especially the waste containing the chemicals that they emit at the end. It is the cars of these and other companies that emit intense fumes and intoxicate us. If not, how can we destroy?". These statements thus seem to reflect the escape of responsibilities from each other in the face of climate change. However, the market gardeners surveyed acknowledge that the use of pesticides in agriculture influences the occurrence of phenomena through, for example, erosion and the decline in agricultural productivity. Others acknowledge that poor waste management (household and industrial) is harmful to the environment:

> *"We have all once thrown a plastic bag or other waste into the street. These sachets often end up either in the gutters to slow the circulation of rainwater and create floods, or in the fields with its consequences on seeds and thus lower agricultural productivity".*

This testimony clearly shows the lack of awareness of the urban population on the harmful effects of the use of non-biodegradable plastic bags.

7.3.2 Practices engendered by social representations of the phenomenon. Residents' harmful practices in relation to climate change

7.3.2.1 *Strong dependence on timber and firewood (deforestation in Togo)*

High population growth (2.8% per year) coupled with rapid urbanisation in Togo (4%) and especially in the capital Lomé (6% per year) (INSEED 2015a; Gbenyon et al. 2006), have led to an increase in urban demand concentrated on wood energy, thus causing increased deforestation: "the increasing demand for charcoal in cities is now the main cause of one third of the deforestation observed in Togo" (MERF & MME 2008) apart from logging for house building and furnishings. However, charcoal production wastes a lot of wood; current carbonization processes have a weight efficiency of 15% only. It takes eight tons of dry wood, sometimes more, to produce one ton of charcoal (Girard 2002). Often charcoal is sufficiently used in urban areas, and it is clear that the gradual substitution of wood by charcoal accelerates deforestation in supply areas.

A study of the wood products market (Thiam 1991) showed an urban consumption rate of 76.6% of national consumption. Lomé is the largest consumer of charcoal, accounted for 36.40% (Baumer 1990). This same market study put forward consumption figures of 347 kg per habitants per year. Thus, with a density of 407 inhabitants/km² in the Maritime Region including Lomé, 96 inhabitants/km² (Savannah Region) and 66 inhabitants/km² (Kara Region) (INSEED 2015a), it is clear that population pressure on the natural resources - especially on vegetation cover -, becomes the most important driver of deforestation. In the same way, in his remarks during the interview, an NGO official involved in environmental protection in the Kara Region regretted the gradual loss of endogenous socio-cultural values and practices that used to protect the environment:

> *"Our socio-cultural practices, if they were done as our parents did, I believe they were practices that preserved our environment. For example, there are sacred forests in which people used to enter for ceremonies. In these forests, trees are not cut down, a mystery of their forests is kept, so that they had remained untouched. Currently, we have depraved everything. The forests into which the Akpéma were brought [young girls initiated in Kabyè country], were not open to everyone. But now with the current generation, the mysteries are lifted and the forests are destroyed for especially the needs in timber and wood energy of the city residents ".*

In addition, it also appears from the various interviews conducted in Kara and Dapaong that:

> *"Trades such as carpentry, masonry, sculpture, etc., have their place in climate change because of the need for timber, particularly for the construction of carpentry, slabs and furniture and works of art that are increasingly in demand by the urban population. In the context of global mobilisation to preserve the environment, overcoming the challenges related to the appropriation of alternative energy for cooking by Togolese urban populations and the control of deforestation remain an imperative for the State and civil society organisations in order to reduce the effects of climate change".*

7.3.2.2 *Anarchic management of household and factory waste (pollution)*

In terms of sanitation and hygiene the results of the QUIBB survey (Questionnaire des Indicateurs de Base du Bien-être, INSEED 2011) reveal that many households throw their garbage into nature (39.8%). The second mode of disposal is the wild dump (24.9%). The proportion of households using the healthy disposal method, i.e. authorized dumps and door-to-door collection, was only 27.9% in 2015. Depending on the place of residence, according to the same source as the use of the healthy household waste disposal method in urban areas is only 53%. Apart from pollution by urban households, linked to archaic methods of managing their household waste, it should also be noted that the few factories in Togolese cities contribute to pollution and emit greenhouse gases.

> *"The industries and companies that are emerging do not even take into account the environmental aspect. The BB brewery plant is a major polluter. It discharges waste water, discharges all its waste and also releases gases and bad smells [...] Plants such as BB, NSCT [*Nouvelle Société Cotonnière du Togo*] in Dapaong, emit greenhouse gases and pollute the environment. These plants are contributing in their own way to climate change".*

Adequate policy measures and sustained education to raise awareness, especially in urban areas, of the environmental challenges associated with the mismanagement of household waste are needed to stem threats to the environment.

7.3.2.3 *Inappropriate management of waste from computers and household appliances*

Waste of electronic and electrical equipment is generated by TV-sets, fax machines, modems, refrigerators, cameras, MP3/MP4 players, stoves, microwaves, radios, computers, digital tablets and end-of-life mobile phones. According to the QUIBB (INSEED 2011), communication equipment includes radio and TV-sets, mobile and fixed telephones, satellite dishes and microcomputers. In Togo as a whole, 77.4% of households own at least one communication asset. Taking into account the place of area of residence, the same source specifies that the proportion of households owning at least one communication property is 94.2% in urban areas and 64.4% in rural areas. This proportion is 76% in the Kara Region, and 62.7% in the Savannah Region. Nationally, 80.4% of households have at least one piece of furniture or household appliance. The proportion of households owning at least one piece of furniture or household appliance is 94.0% in urban areas and 69.9% in rural areas.

Household equipment is imported and distributed by an essentially informal network in new and used condition. Devices in new condition generally come from countries such as China and Dubai and are largely of lower quality (durability reduced on average to one year) and commonly referred to as “Chineseeries”. Those that come in a used state are commonly called “come” or “second hand” and are at the end of their life. This suggests that the increasing quantities of waste from these facilities in the urban solid waste circuit, which is subject to inappropriate management according to Gnane (2017):

> *“This electronic waste is managed by an essentially informal recycling system involving several actors using sorting, dismantling, land filling and even incineration practices… an ecologically inappropriate system, which poses a serious threat to the environment and public health with regard to the toxicity of certain components of this electronic waste”.*

Since towns and cities are recognized as the biggest consumers of electronic devices due to residents’ modern lifestyles it is remarkable that this inappropriate management of waste by urban populations from computer and household appliances undoubtedly constitutes a serious threat to public health and the environment.

7.3.3 Resilient practices of urban residents

7.3.3.1 *Ineffectual institutional solutions*

Faced with the challenges linked to the strong pressure of residents on firewood, the Togolese Government’s response through the control of traditional energies (wood and charcoal) and the development of alternative energies (renewable energies, gas, oil) falls within the framework of national strategies for the sound and sustainable management of natural resources and the environment in relation to the conventions on combating desertification, on biological diversity and on climate change in Togo. These strategies aim to reduce firewood consumption through an improved carbonization processes with diffusion of improved stoves as well as the wider domestic use of butane gas and the promotion of clean and renewable energy.

However, the major challenge in this area remains the strengthening of the institutional and regulatory framework for domestic energy and its effective implementation. This institutional and regulatory strengthening should make it possible to the establishment of an institutional and legal framework favourable to the development of the sector's activities and clearly delimiting the responsibility of the State and the various stakeholders. The permanent satisfaction of the population's basic needs in terms of clean energy; the efficient supply of energy to the country at lower cost; the assurance of efficient energy management by reducing losses and waste; the implementation of a plan to develop promising alternative energy sources for the production of electricity (in the case of renewable energy).

The institutional challenge lies in the need to federate the actions of both state and private sector actors, operating in the domestic energy sector, around a plan to popularize alternative energy for cooking with an effective communication strategy. This challenge is justified by the virtual absence of real actions on the ground to raise awareness of the need to adopt alternative energy for cooking. This is what emerges from the interviews with the managers in charge of the energy sector:

> *"The popularisation of alternative energy sources to firewood and charcoal requires the deployment of significant resources to intensify public awareness campaigns on the environmental, economic and other risks associated with the destruction of forests for cooking purposes and the need for them to opt for alternative energy sources".*

The promotion of improved stoves, renewable energy sources (e.g. butane and methane gas, hydropower, solar, wind, geothermics) is an alternative to population pressure on plant resources to protect the environment. To this end, the introduction in Togo of energy-saving cooking ovens, improved *atsu-tò* stoves in households and the *meule casamançaise* used by charcoal makers to produce coal with less primary wood, are energy innovations to be promoted further for greater efficiency in reducing pressure on forests through the uncontrolled felling of trees for charcoal production. Similarly, access to renewable energy through solar energy subsidies through the CIZO rural area electrification project launched by the Togolese government, whose initial phase consists in providing electricity to households, is one of the solutions for resilience to the effects of climate change.

7.3.3.2 *Recycling plastics and other waste (circular economy for sustainable towns)*

The recycling of plastics from waste produced by Togolese urban populations through the actions of certain civil society actors is proving to be an innovative strategy for the management of solid urban waste. This strategy is still in an embryonic phase by the promotion of circular economy in Togo, which is a kind of loop economy, thus eliminating the notion of "waste". Its objective is to produce goods and services while strongly limiting the consumption and waste of raw materials by nonrenewable energy sources.

Thus, in Kara and Dapaong, as in other cities of the country, one can recognize the emergence of certain NGOs collecting plastics from water consumption in sachets, commonly known in Togo as "Pure water", in order to reuse them for the manufacture of bags that are sold to the consumers on the markets. This strategy is very beneficial, especially since it helps to employ workers in the production chain of these new products while reducing environmental pollution.

7.4 CONCLUSION

Based on theories of cognitive vulnerability and urban resilience, aimed to show the perceptions of urban residents in two major towns of Togo (Dapaong and Kara), the study intended to expose with regard to social representations (stakeholders), practices which have a negative impact on the environment. Furthermore, it examined some innovative and resilient strategies that address the ongoing effects of climate change on these urban populations.

Two methods have been used: the documentary analysis and the in-depth individual interviews with forty individuals (20 men and 20 women) in the two towns. Among actors interviewed were resource persons from civil society organisations working on environmental issues, as well as urban market gardeners and members of the Neighborhood Development Committees of both cities, the aim being to take into account the possible diversity of urban social representations on attitudes and practices related to the phenomenon of climate change.

The analysis of documentary and qualitative data made it possible to identify the objective and subjective knowledge held by urban residents about causes and consequences of climate change as well as practices harmful to the environment and those adaptive to climate change.

With regard to residents' knowledge, it appeared that the understanding that actors have about climate change depends, not only on their professional and daily activities, the way in which climate change impacts them, but also, and above all, on their literacy and educational level. Thus, although from an objective point of view, all interviewees perceive climate change as "a transformation of the climate" by comparing it to one time period to another. Market gardeners define it as a change in the seasons, a scarcity of water in the rivers and dams they use for their market gardening while civil servants define it in terms of an intensification of "sunshine" and the current "heat" as compared to the past.

As far as perceptions on causes and consequences of climate change are concerned, although they may seem common, there are specificities according to the social, cultural and spatial contexts. The less educated attribute climate change to "*the non-respect of traditions and the abandonment of rituals once performed by forefathers to satisfy the gods*"; the linking being mandatory between supernatural and human forces to create social but also natural balance. For these actors, climate change is understood as the manifestations of God's wrath. This perception confirms the theory of cognitive vulnerability (Lammel 2015) among populations in the two selected towns as to manifestations of climate change.

Concerning harmful practices exerted on environment, the high dependence of residents on firewood and charcoal has been identified as an essential factor in deforestation in Togo. The uncontrolled management of household and factory waste, coupled with inadequate management of waste from computer and household appliances by residents, have been revealed as a source of environmental pollution and an aggravating factor of climate and environmental change.

With regard to the resilience of residents to the effects of climate change two aspects have been identified: Firstly, there are some mixed institutional approaches. In fact, the response of the Government of Togo through the control of traditional energies and the development of renewable energies is an adaptive strategy that needs to be reinforced by sustained actions to raise awareness among populations, especially in urban areas to reduce anthropogenic effects on nature. Secondly, the recycling of plastics from the waste produced by urban populations in Togo by NGOs (e.g. *Sainte Thérèse de l'Enfant Jésus,* Figure 2) for the manufacture of handbags, is an innovative practice in line with the circular economy for sustainable cities.

Figure 2. Bags produced by the association *Sainte Thérèse de l'Enfant Jésus*, based on the plastic sachets of "Pure Water" (made-in-togo.com 2019). https://made-in-togo.com/post/les-produits-a-base-des-sachets-de-pure-water-127 (accessed on January 15th, 2020).

After having shown the social representations of residents in Togo on climate change as well as harmful practices and those that are resilient to the phenomenon, it is paramount to seek institutional and local obstacles for more action towards towns and countryside for an effective resilience strategy, especially for developing countries.

REFERENCES

Badameli, A. & Dubreuil, V. 2015. Diagnostic du changement climatique au Togo à travers l'évolution de la température entre 1961 et 2010, XXVIIIe Colloque de l'Association Internationale de Climatologie, Liège, pp. 421–426.

Baumer, M. 1990. *Agroforesterie et désertification*, World Agroforestry Centre-Technical Centre for Agricultural and Rural Cooperation (ICRAF-CTA), Lomé, pp. 17–18.

Fournier, H. & Zara, H. 2015. Exemple d'une adaptation climatique: la ville de Medellin (Colombie). Actes pratiques et ingénierie immobilière. *LexisNexis*: 46, pp. 1–46..

Gbenyon, K., Kadoumta, M., Pana, E., Diasso, S., Nagou, M., M'gboouna, K. & Akpaka, K. 2006. *Femmes, environnement et développement au Togo: Pistes pour une politique en matière de gaz butane et d'énergie traditionnelle*. Lomé, pp. 1–33.

Girard, P. 2002. Quel futur pour la production et l'utilisation du charbon de bois en Afrique ? Document FAO, Rome (Italy), pp.30–34.

Gnane, N.M. 2017. Gestion des déchets électroniques et enjeux environnementaux au Togo. *ÉCHANGES, Revue africaine de philosophie, Lettres et Sciences Humaines*. Faculté des Sciences de l'Homme et de la Société (FSHS) de l'Université de Lomé, Sciences Humaines, 9, pp. 484–497.

INSEED, Institut National de la Statistique et des Études Économiques et Démographiques, 2011. *Enquête questionnaire unifié des indicateurs de base du bien-être 2011*. Lomé, pp. 1–25.
INSEED, Institut National de la Statistique et des Études Économiques et Démographiques, 2015a. *Enquête questionnaire unifié des indicateurs de base du bien-être 2015*. Lomé, pp. 1–47.
INSEED. Institut National de la Statistique et des Études Économiques et Démographiques, 2015b. *Perspectives démographiques du Togo 2011–2031*. Lomé, pp. 1–25.
IPCC/GIEC, Groupe Intergouvernemental d'Experts sur l'évolution du Climat 2014. Changements climatiques 2014, rapport de synthèse, Rapport du IPCC/GIEC sur l'évolution du climat. Genève, pp. 6–21.
Lammel, A. 2015. *Changement climatique: de la perception à l'action,* Les notes de la Fondation de l'Ecologie Politique, 5, pp. 1–3.
MERF & MMEE. 2008. *Programme d'appui à la maîtrise des énergies traditionnelles et de promotion des énergies renouvelables au Togo*, Lomé.
Republique Togolaise. 2015. *Troisième communication nationale sur les changements climatiques*, Lomé, pp. 62–108.
Republique Togolaise. 2017. *Plan National d'Adaptation aux Changements Climatiques du Togo*, Lomé, pp. 1–57.
Saint-Martin, A. 2013. *La sociologie de Robert K. Merton*. La Découverte, Paris, pp. 1–126.
Thiam, A.I. 1991. *Etude de marché des produits ligneux au Togo*, Lomé, pp. 1–225.
UN-HABITAT. 2011. *State of the world's cities 2010/2011, Bridging the urban divide, Nairobi, Kenya*, London, UK, pp. 1–244.
WHO, World Health Organisation 2014. *Quantitative risk assessment of the effects of climate change on selected causes of death, 2030s and 2050s*. Geneva: WHO, pp. 18–19.

CHAPTER 8

Traditional agroforestry in northern Benin and Togo and its contribution to climate resilience

J.B.K. Vodounou
University of Parakou, Parakou, Benin

ABSTRACT: Traditional agroforestry with *Vitellaria paradoxa* prevalents in northern Benin and Togo. It contributes to a favorable microclimate and helps to increase agricultural yields. Today, this ancestral practice is highlighted against the background of efficient use of natural resources and its contribution to climate resilience. Data and field studies of 62 plots (20 m x 20 m) allowed to calculate the Shannon species richness index, the equilibrium of Pielou, and to determine the specific diversity of areas with agroforestry. Calculations of density, basal area, distribution by circumference class and height were used to analyse trees' structure. It was found that areas dominated by *Vitellaria paradoxa* are not really diversified. The dendrometric structures of adult stems have a positive asymmetric distribution with the shape parameters (C = 1.26 m, C = 2.10 m and C = 2.57 m) using Weibull distribution. Coverage rate of the species distribution is 75%. However, traditional agroforestry based on *Vitellaria paradoxa* is one of the methods of soil protection and climate resilience. It helps to reduce erosion by surface runoff and promotes the near to ground convection process by evaporation.

8.1 INTRODUCTION

Agroforestry is the association of trees and crops or animals on the same plot (Anderson & Sinclair 1993). The traditional form prevailing in sub-Saharan Africa in general and particularly in Benin and Togo uses indigenous species such as *Vitellaria paradoxa* and *Parkia biglobosa* associated with crops or livestock. Nowadays, this traditional practice is rediscovered because it allows better use of resources, greater biological diversity and the creation of a favorable microclimate, and leads to the increase of agricultural yields (Vodounou 2016, Veldkamp 1994, Rao et al. 1998). The tree creates through its root system conditions in the deep layers of the soil that promotes water supply and nutrients to surface crops. Agroforestry techniques make it possible to arrange trees in such a way that they optimise crop production and return as little as possible to competition with them (Daoudi 2013, Kang & Reynolds 1989). However, the traditional practice leaves the roots as 'naturally grown' so that one does not observe any alignment like in plantations. This practice aimed at harvesting the fruits of *Vitellaria paradoxa* and *Parkia biglobosa* for food usage can be considered as a factor promoting climate change resilience. Trees provide diversity of species and habitats which are beneficial to crop and pollinator insects. In addition, trees have the capacity to absorb CO_2 and store carbon (Szott & Palm 1996, Zaltman & Stiff 1973). Agroforestry thus becomes an essential component of the agroecological landscape zones of northern Benin and Togo (Figure 1), which should be strengthened as a formal agricultural practice for the protection of biodiversity. As the ongoing climate change debate

Figure 1. Geographical location of the study area.

demonstrates it seems necessary, alongside mitigation actions, to initiate adaptation measures in rural areas.

The study area in West Africa is located between 8°55′ and 12°05′ northern latitude, and 0° and 4° eastern longitude. It comprises four departments of northern Benin and three regions of northern Togo with a total surface of 114,727 km². An annual average of 1100 mm rainfall, distributed over a five months rainy season (May to September) and a seven months long dry season (October to April) is typical. Geologically, the area of interest consists of levelled crystalline Precambrian basement rocks with quartzitic mountain ranges of the Pan-African orogeny (Atacora). The four departments in northern Benin are Alibori, Atacora, Borgou and Donga subdivided into 27 municipalities. The three Regions in northern Togo are Central, Kara and Savannah; they are subdivided into 16 prefectures (Figure 1).

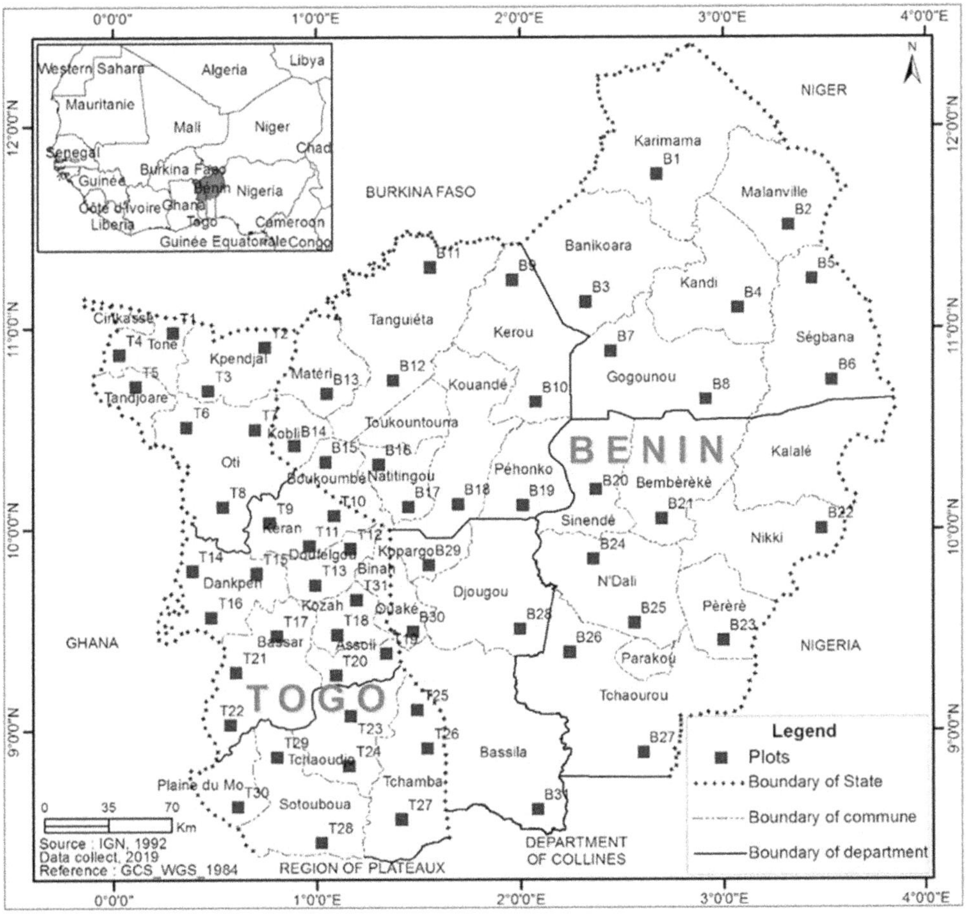

Figure 2. Spatial distribution of plots.

8.2 METHODS

Most of the research that have dealt with the formal and conventional dimension of agroforestry, none has taken into account the dimension of traditional peasants' agroforestry. Fieldwork was conducted in two phases: a preliminary field survey, followed by a ground truthing camping with data collection. Data was collected on 62 study plots (Figure 2, Table 3). Several parameters including the Shannon specific richness index, and the equilibrium of Pielou were collected to obtain information on the specific plant diversity in close to nature 'landscape parks'.

The study plots installed were designated by B (Benin) and T (Togo): 31 plots were located in northern Benin and 31 in northern Togo. Test surfaces covered 20 m x 20 m each for collecting dendrometric parameters. Calculations of density, basal area, distribution by circumference class and height of trees were used to analyse the different structure of the plots. Data collected through

Table 1. Specific diversity parameters of the studied parks.

	R		H		E	
Index and parameters	m	e	m	e	m	e
Vitellaria paradoxa	12.13	2.87	1.15	0.66	0.63	0.22

R = Specific wealth, H = Shannon index, E = Pielou equitability index, m = mean, e = standard deviation

random sampling of 204 key stakeholders allowed to document local perceptions of main factors that influence the dynamics of 'landscape parks'.

Spatial mapping of landscapes dominated by *Vitellaria paradoxa* was documented by the help of remote sensing data (LANDSAT-8 OLI/TIRS, supervised classification supported by verification in the field). To cover the study area, nine scenes of LANDSAT-8 OLI/TIRS 2019 were downloaded from the US Geological Survey Earth Explorer, www.earthexplorer.usgs.gov (Table 4).

Good quality satellite data covered the vegetation period (foliage time of *Vitellaria paradoxa*) between April and November. During fieldwork identification and collecting of coordinates of identified *Vitellaria paradoxa* specimen was carried out. Projection of geographic coordinates on a LANDSAT-8 mosaic of Band 4 (0.64-0.673 μm wavelength, 30 m spatial pixel resolution) allowed to identify from the supervised classification (Kappa index 91%) and overall accuracy of the tree mapping of is 95% (with a variance of 0.000019). The values of the pixels associated with the chlorophyll reflectance of *Vitellaria paradoxa* have been visualised by an ArcGIS thematic map (Figure 5) illustrating the distribution of *Vitellaria paradoxa* in northern Benin and Togo.

8.3 RESULTS AND DISCUSSION

Certain indexes such as specific diversity, structural characteristics (horizontal), circumference classes, height-circumference ratio, and spatial distribution of species as well as its impact on soil and perception of stakeholders on agroforestry are considered.

8.3.1 Specific diversity of 'landscape parks' dominated by *Vitellaria paradoxa*

The floristic composition of the 62 plots consists of 841 species belonging to 52 genera grouped into 27 families, including 341 individuals of *Vitellaria paradoxa*.

Table 2. Horizontal structure parameter (m = mean, e = standard deviation).

	Density		Earth surface		Average diameter	
Parameters	m	e	m	e	m	e
'Landscape parks' *Vitellaria paradoxa*	57.5	35.71	5	2.31	36.59	10.18

Table 3. Geographic coordinates of test plots (see Figure 2).

Benin			Togo		
Northern Latitude	Eastern Longitude	Name	Northern Latitude	Eastern Longitude	Name
11,761313	2,674835	B1	10,975074	0,294477	T1
11,508851	3,31681	B2	10,902942	0,748909	T2
11,126551	2,321388	B3	10,686546	0,467594	T3
11,097698	3,071561	B4	10,866876	0,027589	T4
11,241962	3,432221	B5	10,708185	0,106934	T5
10,737038	3,533206	B6	10,506216	0,359396	T6
10,881302	2,444012	B7	10,491789	0,698417	T7
10,643266	2,91287	B8	10,109489	0,539726	T8
11,234749	1,960727	B9	10,030144	0,770549	T9
10,62884	2,076139	B10	10,06621	1,08793	T10
11,299668	1,556788	B11	9,914733	0,965305	T11
10,737038	1,376458	B12	9,900306	1,167275	T12
10,672119	1,051864	B13	9,719976	0,994158	T13
10,412444	0,893173	B14	9,792108	0,388249	T14
10,333099	1,04465	B15	9,777682	0,70563	T15
10,318672	1,304326	B16	9,561286	0,482021	T16
10,109489	1,44859	B17	9,467514	0,806615	T17
10,123916	1,693839	B18	9,474727	1,102356	T18
10,116703	2,01122	B19	9,380956	1,340392	T19
10,196048	2,37188	B20	9,272758	1,095143	T20
10,051784	2,696474	B21	9,287184	0,604645	T21
10,001291	3,482714	B22	9,027509	0,575792	T22
9,445875	2,999429	B23	9,070788	1,167275	T23
9,849814	2,357454	B24	8,825539	1,160062	T24
9,532433	2,559423	B25	9,099641	1,491869	T25
9,388169	2,242042	B26	8,912097	1,542362	T26
8,890458	2,602703	B27	8,55865	1,412524	T27
9,50358	1,996793	B28	8,443239	1,023011	T28
9,820961	1,549575	B29	8,868818	0,806615	T29
9,489154	1,47023	B30	8,623569	0,611858	T30
8,609143	2,083352	B31	9,647844	1,196128	T31

Shannon's diversity and Pielou's equitability index give evidence that *Vitellaria paradoxa* 'landscape parks' are not very diversified (Table 1). They reflect the strong anthropic pressure on 'landscape parks' dominated by *Vitellaria paradoxa.*

8.3.2 Structural characterisation

8.3.2.1 *Horizontal structure*

The results of the horizontal structure are obtained through density, basal area and mean diameter (Table 2).

Table 4. LANDSAT-8 images used in the study.

	2018 images Landsat 8 OLI				
N°	Recording date	Entity ID	Coordinates/WRS type	Path	Row
1	2018-11-30	LC81910522018334LGN00	Lat = 12.61338 and 12.60284 Long = 3.30201 and 5.39231	191	052
2	2018-11-30	LC81910532018334LGN00	Lat = 11.16750 and 11.16033 Long = 2.99267 and 5.07222	191	053
3	2018-04-30	LC81920522019120LGN00	Lat = 12.61333 and 12.61492 Long = 1.74971 and 3.84615	192	052
4	2018-10-20	LC81920532018293LGN00	Lat = 11.16609 and 11.16977 Long = 1.42941 and 3.51474	192	053
5	2003-10-20	LC81920542018293LGN00	Lat = 9.72157 and 9.72668 Long = 1.11952 and 3.19784	192	054
6	2018-06-24	LC81930522019175LGN00	Lat = 12.60958 and 12.62346 Long = 0.18725 and 2.29101	193	052
7	2018-11-12	LC81930532018316LGN00	Lat = 11.16473 and 11.17928 Long = -0.12725 and 1.96779	193	053
8	2018-05-07	LC81930542019127LGN00	Lat = 9.72043 and 9.73496 Long = -0.42415 and 1.65807	193	054
9	2018-06-15	LC81940522019166LGN00	Lat = 12.61649 and 12.59537 Long = -1.33363 and 0.74937	194	052

8.3.2.2 *Structure by circumference classes*

The identified species (n= 341) of *Vitellaria paradoxa* were subject to dendrometric measurements. The processing of these data made it possible by circumference groups to observe the individual size of studied species. They give evidence of a pronounced anthropogenic character of the environment (Figure 3).

The interpretation of Figure 3 results in that the dendrometric structures of adult stems of *Vitellaria paradoxa* have a positive asymmetric distribution with the shape parameters (C = 1.26 m, C = 2.10 m and C = 2.57 m) of the Weibull distribution. This is characteristic of monospecific stands with predominance of young or low circumference stems. The study area will evolve towards a certain climax vegetation state if *Vitellaria paradoxa* is no longer exploited. However, presence of stems greater than 310 cm testifies an overexploitation of species in the recent past (Figure 3).

8.3.2.3 *Height-circumference ratio*

Tree height and circumference showed a clear relationship as illustrated in Figure 4. It has been shown that coefficients of determination of the model proved that almost 92% of variation of the heights of the stems of *Vitellaria paradoxa* are explained by variations in individual circumferences.

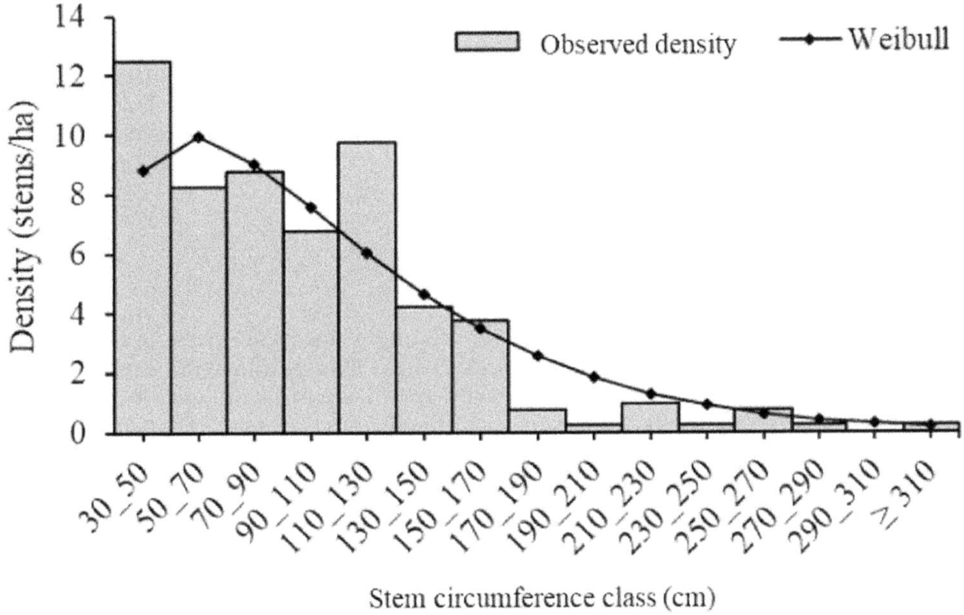

Figure 3. Structure by circumference classes of *Vitellaria paradoxa.*

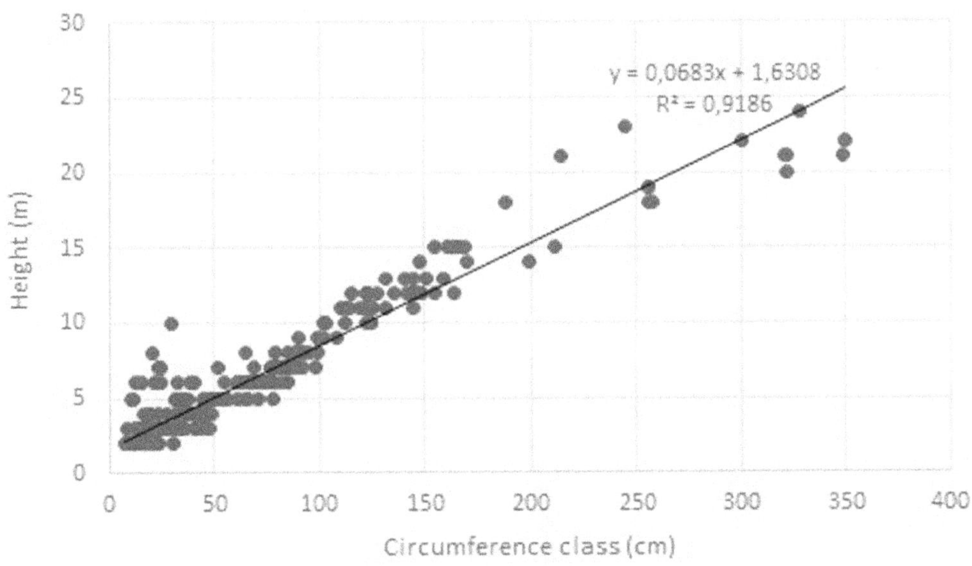

Figure 4. Height-circumference relationship of *Vitellaria paradoxa* in northern Benin and Togo.

8.3.2.4 *Regional distribution of Vitellaria paradoxa*

The species distribution map of *Vitellaria paradoxa* (grey halftones) was obtained from remote sensing data by evaluation of spatial and spectral features of image pixels. A good coverage (~75%) of individuals for northern Benin and Togo is recognisable (Figure 5). Values of multispectral pixels (16-bit) that indicated the occurrence of *Vitellaria paradoxa* were 8072; 8093; 8126; 8185; 8200; 8287; 8398; 8424; 8538; 8620; 8641; 8704; 8982 of LANDSAT-8.

8.3.2.5 *Soil erosion and soil fertility*

Traditional agroforestry based on *Vitellaria paradoxa* is one of the classic methods used in soil protection in rural areas. The presence of the species attenuates kinetic energy from precipitation (heavy rains) and superficial runoff Apart from direct field observations that evidenced the protective character of the plant, local stakeholders interviewed,

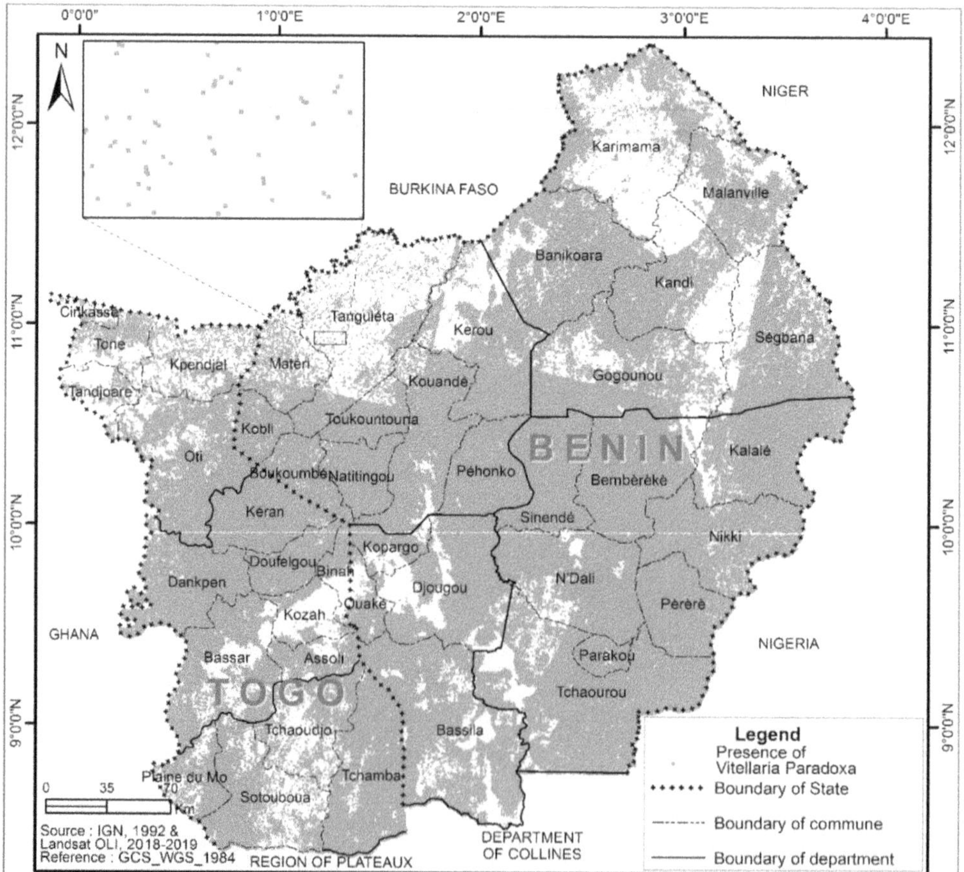

Figure 5. Regional distribution of *Vitellaria paradoxa* in northern Benin and in northern Togo (almost ~75% of the study area is dominated by this traditional agroforestry species).

confirmed that not only the protective character against soil erosion is of importance: the rotten leaves of *Vitellaria paradoxa* also serve as a natural mould and fertilizer for agriculture. Completely cleared sites, with any tree coverage, show more signs of erosion and soil leaching.

8.3.3 Adapting sustainable land management

The geological background and geomorphic landscape history of the study area favors shallow and leachable soils. Thus, conservation and cultivation practices such as reduced soil tillage and direct seeding are most important for sustainable agriculture. Direct seeding is a technique of minimally working the soil to maintain its structure and leaving crop residues on site to reduce erosion (Figures 6a, 6b). This technique is deployed by all farmers met and especially those who are in the group of Atacora chain.

8.4 DISCUSSION

Traditional agroforestry systems are applied by rural stakeholders for food production. The usual collection of *Vitellaria paradoxa* nuts is intended for the making of shea butter, mainly for cosmetics, and less for its dietary value. Over the years, this initial objective was gradually translated into the option of environmental protection and struggling against ecosystems' degradation as a consequence of mitigating effects of climate change. In these 'landscape parks' with numerous *Vitellaria paradoxa* individuals, they now serve more for land conservation and sustainable land management practices.

For example, concepts of soil improvement by trees in agroforestry systems were examined by Young (1989) and Buresh & Tian (1998). Soil improvement in agroforestry systems is linked to growth of nitrogen-trapping trees, or deep-rooting trees and shrubs that increase nitrogen availability through biological fixation, restore nutrients for plants in depth (especially in arid zones), and reinforce soil organic matter (Beer 1988, Rao et al. 1998).

Figure 6a, b. Savannah 'park landscape' dominated by *Vitellaria paradoxa* with strong anthropogenic impact by agriculture (left, 6a: harvested sorghum field with mulching and crop rotation; right, 6b: harvested field, before cultivated by a combination of maize and cowpea, currently used as an animal grazing space to collect dung) (Photos: J.B.K. Vodounou).

Research on agroforestry systems, particularly in Africa, initially focused on ways of maintaining soil fertility in annual cropping systems using bush legumes, such as *cajanus cajan, mucuna*, or in agroforestry parks (Charreau & Vidal 1965), or cultivation in band crops (Vodounou & Akpo 2015, Kang & Reynolds 1989). Furthermore, fallow fields have been ecologically improved by planting of trees.

In the current context of climate change, basic theoretical ideas of this study were that of the diffusion and adoption of innovations. Innovations here deal with traditional techniques used by people who today are proving themselves as protecting the environment. In traditional agriculture, which is practiced in the study area of northern Benin and Togo, natural resources are increasingly scarce, and innovation is a condition for sustainable growth. This is what Daoudi (2013) concluded in his study on technical innovation system in greenhouse market gardening sector towards Biskra (Algeria). Dissemination of this traditional agroforestry practice as an innovation will be of importance to technical contribution for adaptation to ongoing climate change in the study area. Understanding diffusion of innovations is an omnipresent topic in scientific literature on worldwide agriculture (Racine 2015). For Daoudi (2013), and in agroforestry, this concept is generally defined as being a new way of doing things, compared to an existing way of acting deemed to be less efficient, whereas in the case of this study the traditional practice obviously less efficient, reveals as an innovation to adopt. It is seen as a multidimensional, interactive and negotiated process, challenging a linear model of invention through research, transfer by extension services, and adoption by stakeholders (Leeuwis & Van Den Ban 2004, Ameur 2013). Moreover, according to Racine (2015), the search for profit maximization often involves improving techniques, adopting new practices, disseminating knowledge or even technical progress. Agroforestry work on technology adoption shows that farmers' decisions are influenced by multiple direct or indirect factors and the intrinsic characteristics of the technology (Vodounou & Onibon-Doubogan 2016, Ngondjeb et al. 2011). For a long time, researchers in particular fields of agricultural economics have recognized that adoptions are not instant events but the result of a decision process consisting in sequences of actions and decisions (Rogers 1995, Zaltman & Stiff 1973, Sunding & Zilberman 2001). Therefore, traditional agroforestry can be adopted and popularised as a relatively less expensive resilience model in a context of climate change.

8.5 CONCLUSION

Traditional agroforestry systems include several strategies in adapting to climate change. The traditional use of savannah trees such as *Vitellaria paradoxa*, although considered as an ineffective practice, because of its unconventional character remains today a solution to the phenomenon of climate determined land degradation. Indeed, many existing trees are intended for exploitation of fruits for shea butter production. However, over the years and in favor of widespread climatic distress, the practice of traditional agroforestry has become or can be considered as a strategy for adapting to current climatic conditions. In addition, soil fertility can be improved by *Vittelaria paradoxa*'s rotten leaves which provide a large supply of biomass (nekromatter) which is mineralised to soils. Thus, traditional agroforestry with integration of *Vitellaria paradoxa* has two main advantages: harvesting fruits for shea butter production, and also planting them for supporting peasants' agriculture, and to build up resilience against climate change.

REFERENCES

Ameur, F., Hamamouche, M.F., Kuper, M., & Benouniche, M. 2013. The domestication of a technical innovation: the diffusion of drip irrigation in two douars in Morocco. *Cahiers Agricultures* 22(4): 311–318.

Anderson, L.S. & Sinclair, F.L. 1993. Ecological interactions in agroforestry systems. *Agroforestry Abstracts* 54: 57–91.

Beer, J. 1988. Litter production and nutrient cycling in coffee (*Coffea arabica*) or cacao (*Theobroma cacao*) plantations with shade trees. *Agroforestry Systems* 7: 103–114.

Buresh, R.J. & Tian, G. 1998. Soil improvement by trees in sub-Saharan Africa. *Agroforestry Systems* 38: 51–76.

Charreau, C. & Vidal, P. 1965. Influence of *Acacia albida Del.* on the soil, mineral nutrition and yields of Pennisetum mils in Senegal. *Agronomie Tropicale* 20: 600–626.

Daoudi, A., 2013. *The technical innovation system in the Biskra greenhouse vegetable sector: dynamics and limits of a new model of production and circulation of knowledge.* Research Protocol: 28. Alger.

Kang, B.T. & Reynolds, L. (eds.) 1989. *Alley farming in the humid and sub humid Tropics.* Proceedings of an international workshop held at Ibadan, Nigeria, 10–14 March 1986. Ottawa: IDRC.

Leeuwis, C. & van den Ban, A. 2004. *Communication for Rural innovation: Rethinking Agricultural Extension.* Hoboken, New Jersey: Wiley-Blackwell.

Ngondjeb, Y., Nje, P. & Havar, M. 2011. Determinants of the adoption of water erosion control techniques in Cameroon's cotton zone. *Revue d'élevage et de médecine vétérinaire des pays tropicaux* 64(1): 9–19.

Racine, J.S. 2015. Mixed data kernel copulas. *Empirical Economics* 48(1): 37–59.

Rao, M.R., Nair, P.K.R., & Ong, C.K. 1998. Biophysical interactions in tropical agroforestry systems. *Agroforestry Systems* 38: 3–50.

Rogers, E.M. 1995. *Diffusion of innovations.* New York: Free Press.

Sunding, D. & Zilberman, D. 2001. The agricultural innovation process: research and technology adoption in a changing agricultural sector. *Handbook of agricultural economics* 1: 207–261.

Szott, L.T. & Palm, C.A. 1996. Nutrient stocks in managed and natural humid tropical fallows. *Plant and Soil* 186: 293–309.

Veldkamp, E. 1994. *Soil organic carbon dynamics in pastures established after deforestation in the humid tropics of Costa Rica.* Doctoral thesis: 117. Wageningen: Agricultural University Wageningen.

Vodounou, J.B.K. & Akpo, O. 2015. Agroecological analysis of the Okoriko watershed of Toui in the commune of Ouèssè in Benin. *International Journal of Innovation and scientific Reseach, ISSR Journal* 19(1): 103–114.

Vodounou, J.B.K. & Onibon Doubogan, Y. 2016. Farming and Climate Change Adaptation Strategies in Northern Benin. *Cybergéo, European Journal of Geography*: 01–27.

Vodounou, J.B.K. 2016. Climate change and agricultural production: Capitalization of cultural practices for food security in Benin. *International Journal of Innovation and scientific Reseach, ISSR Journals* 23(1): 78–97.

Young, A. 1989. *Agroforestry for soil conservation.* Wallingford: C.A.B. International/CIRAF.

Zaltman, G. & Stiff, R. 1973. Theories of diffusion in consumer behavior. In Ward, S. & Robertson, T.S. (eds.) *Technical Sources*: 416–681. Englewood Cliffs: Prentice Hall.

CHAPTER 9

Climate change and soil vulnerability under anthropogenic impact on the Dapaong-Bombouaka escarpments in northern Togo

L. Kankpénandja
University of Kara, Kara, Togo

D. Bawa, B. Afo, T.Y. Gnongbo & A.B. Blivi
University of Lomé, Lomé, Togo

ABSTRACT: The escarpments of Dapaong-Bombouaka, located in the Sudanese tropical climate area of northern Togo, and home to Gourma and Moba ethnic groups, have been impacted by a severe degradation of the physical landscape since 1980. This degradation manifested by both chemical and physical soil erosion is a major concern to the various stakeholders involved in regional rural development. The spatiotemporal evolution of climate and land use in connection to how they lead to soil erosion and fluvial systems' degradation is highlighted. Results show that land degradation is primarily linked to anthropogenic impact and also triggered by ongoing climate change.

9.1 INTRODUCTION

Land degradation has been a major concern for humanity for as long as various authors have pointed out such as Tricart (1978), Hudson (1981), Mietton (1988), Roose (1999), Bouzou Moussa (2000) and Morgan (2005). It has both on and offsite consequences. It degrades the living environment of urban and rural populations. It affects agricultural land, roads and recreational sites, and produces sediment flows that form nuisances downstream, including pollution of water and silting of rivers. It also damages the soil-absorbing complex and releases carbon that is carried by flowing waters or returns into the atmosphere. As a result, much research has been conducted to better understand the phenomenon and thereafter propose sustainable land management approaches (Lal 1976, Wischmeier & Smith 1960, McCormack & Young 1981, Zachar 1982, Hudson 1981, Renard et al. 1997 and Kankpénandja 2016). Significant advances have been made, but as the phenomenon of land degradation is complex and particularly worrying today due to continuously increasing and ongoing climate change. The tropical region in general and sub-Saharan Africa in particular, are those regions where degradation is likely to assume catastrophic proportions with severe socio-economic consequences.

The Dapaong-Bombouaka escarpments are located in the Sudan-type tropical climate zone between 10°30′ and 11°55′ northern latitude and 0°10′ western to 0°40′ eastern longitude (Figure 1). Home to Gourma and Moba ethnic groups, these escarpments have experienced an increase of population locally exceeding 300 inhabitants/km^2 since 1980. Because of the growing number of peasants,

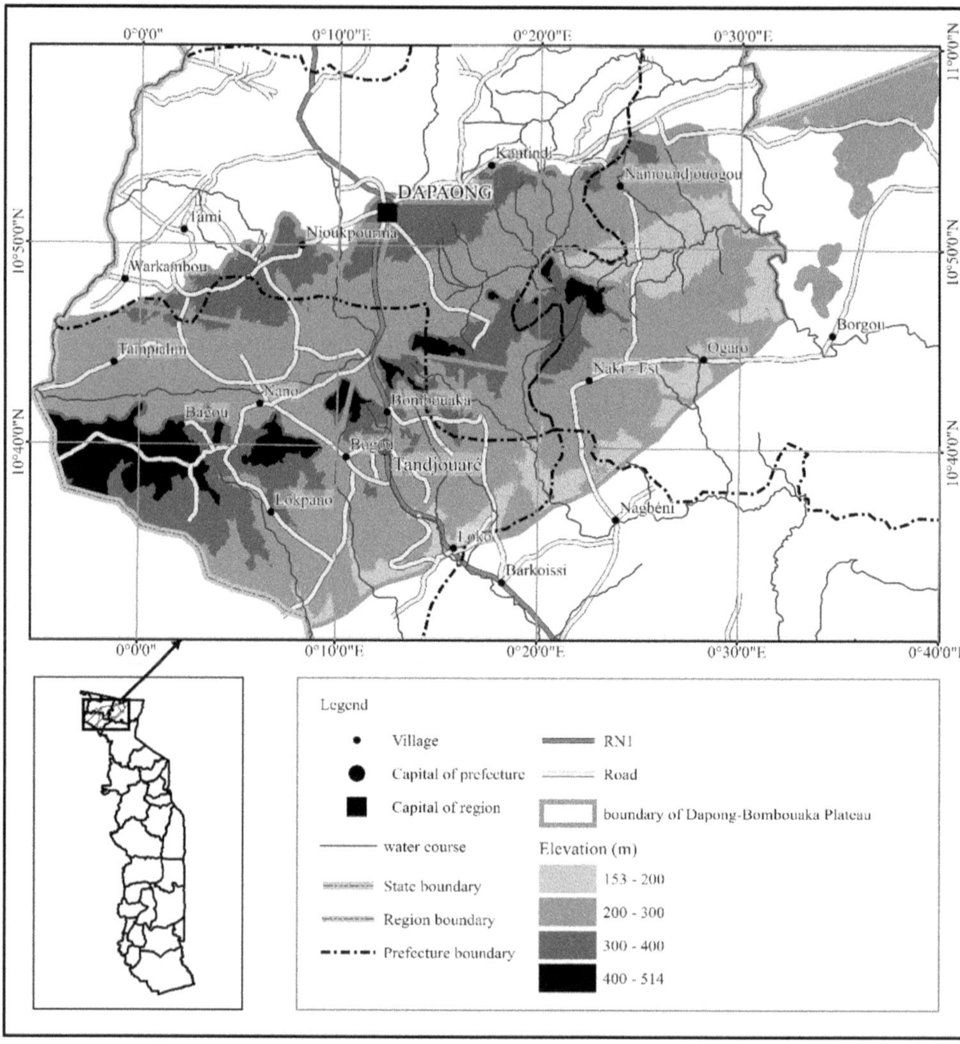

Figure 1. Location map of the study area (Source: IGN, Institut Géographique National, Paris 1989, Dapaong topographic map, scale 1:200,000).

a rapid extension of agricultural fields and an excessive cutting of firewood is recognisable. In consequence, landscapes are strongly degraded. The degradation of the physical environment has been noted by Addra et al. (1986), Brabant et al. (1996), Kankpénandja & Alassane (2012), Kankpénandja (2016), Kankpénandja et al. (2018). They have addressed land degradation in the Savannah Region without highlighting the particular role of climate change. This study focuses on those aspects that have not been deepened. It is assumed that ecosystems located on the Dapaong-Bombouaka escarpments are sensitive and vulnerable to their current pattern of land use and climate change. It aimed at assessing the vulnerability of landscapes to anthropogenic impact and climate change.

9.2 METHODS

9.2.1 Choice of study variables

Land vulnerability is measured by the degree of degradation under the influence of one or more factors. They are said to be vulnerable if the influence of the factors result in accelerated degradation. This degradation can be measured from the net rate of eroded land (E). It can also be evaluated by examining each of the various erosion factors expressed by E = RKLSCP, the universal land loss equation according to Wischmeier & Smith (1960), modified by Renard et al. (1997). In fact, according to the IPCC (2009, 2014) there is an interaction between land use and climate change. Climate change impacts on land degradation through drought, increased rainfall intensity, biomass production, rising sea level increasing coastal erosion, and decreasing soil organic matter. Since there are no inter-annual data on net erosion in the study area, the present study chose to examine the factors R (rainfall-runoff erosivity factor), K (soil erodibility factor) and C (cover-management factor) which are strongly influenced by humans and climate change.

9.2.2 Data collection.

The study of land vulnerability to anthropogenic forcing and climate change required the collection of different types of data, including morpho-pedological, climatic and land cover data.

9.2.2.1 *Geomorphological and pedological data*

Geomorphological and pedological data were collected through web sites, field and laboratory work. The basic terrain data (SRTM) that has been collected is a - 30 m-resolution Digital Elevation Model (DEM). These data were supplemented by some topographic surveys with GPS.

Topographic surveys in the field were coupled with a survey of surface materials. The description of surface materials has been carried using instructions by FAO (1990) and Poss (1996). Thus, the location, the depth of the base of layers, colour, content of clay, coarse elements, structure, porosity, cohesion, drainage class, and the local slope were determined for each soil profile. The laboratory analyses concerned essentially granulometry (texture) and the determination of the organic matter content.

9.2.2.2 *Climate data*

The baseline climate data collected are daily records, monthly and annual rainfall, temperature and evapotranspiration, for the period 1961 to 2018. They were collected at the General Directorate of National Meteorology (DGMN). The meteorological stations where the data were collected are Dapaong and Mango.

9.2.2.3 *Land cover data*

Land cover and land use data were derived from Landsat TM and ETM+ satellite images. Different types of secondary data, such as land use and vegetation indices have been derived from the remotely sensed data. The analysis of the inter-annual dynamics of the vegetation was based on the Normalized Difference Vegetation Index (NDVI).

9.2.3 Data processing

The data collected was processed using Microsoft Excel, SPSS and ArcGIS software. They have been normalized using the reduced centric anomaly method of Lamb given by the following formula:

$$x_i' = \frac{x_i - \bar{x}}{\sigma} \tag{1}$$

Where x_i = value of the variable i; $\bar{x}$ = the average of the series; σ = standard deviation.

The general trend and value dispersion parameters were calculated for the different types of data. The correlation and determination coefficients have also been calculated to evaluate the interdependence relationships between the different study variables. The central trend parameter used to characterize the average state of study variables is the arithmetic mean. This parameter is calculated according to the formula below.

$$\bar{x} = \frac{1}{n}\sum_{i=1}^{n} x_i \tag{2}$$

where x_i =observed value of rank i; n = number of observed values.

The parameter used to evaluate the dispersion of the values around the mean is the standard deviation (σ) given by the following formula:

$$\sigma = \sqrt{V} \tag{3}$$

where V = variance.

The relationship between different study variables, including dependent variables and explanatory variables was measured from the correlation coefficient and the coefficient of determination. The correlation coefficient is given by the following formula:

$$r = \frac{cov(x, y)}{\sigma_x \sigma_y} \tag{4}$$

Where cov (x, y) = covariance of the variables x and y; σx = standard deviation of the variable x; and σy = standard deviation of the variable y.

9.3 RESULTS

9.3.1 Indications of climate change

9.3.1.1 *Precipitation*

The data series (1961–2018) were divided into two sequences to analyse the evolution of rainfall during the WMO reference period (1961–1990) and subsequent changes (1991–2018). The analysis of climate tables of the first sequence (1961–1990) made it possible to detect two pluviometric phases. The first corresponds to the decade 1960–1970; it is characterised by a ten-year average greater than 1050 mm in the main stations of the region (Table 1). Positive anomalies are more frequently observed during this sequence (Figure 2). The second phase covers two decades

Table 1. Decadal rainfall average (mm) in the sub-watersheds of Oti (Source: MND data).

Stations	1961-1970	1971-1980	1981-1990	1991-2000	2001-2010	2011-2018	Average (1961-1990)	Average (1991-2018)	Average (1961-2018)
Mango	1147.9	1048.8	1048.0	1087.1	1027.6	1003.5	1081.5	1037.5	1060.3
Dapaong	1076.0	1008.4	918.0	1051.8	1129.1	1118.6	1000.8	1100.68	1041.8

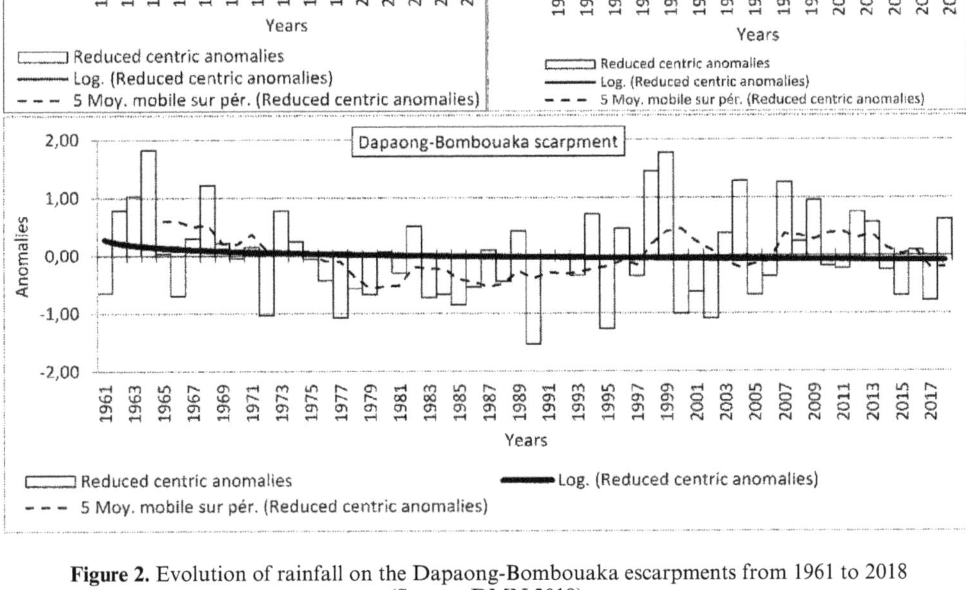

Figure 2. Evolution of rainfall on the Dapaong-Bombouaka escarpments from 1961 to 2018 (Source: DMN 2019).

between 1971 and 1990, marked by a significant deficit in rainfall that results in the frequency of negative anomalies. The ten-year averages of rainfall during this phase are lower than those of the 1960–1970 period. This evolution is reflected in a general downward trend in rainfall observed at all stations.

The second period (post-1990) is characterised by a rainfall pattern different from the first. Rainfall increased during the decade 1991–2000. The ten-year averages were greater than 1050 mm in each station. This upward trend continued for the Dapaong station but was reversed at the Mango station where rainfall was less. For the region as a whole, rainfall was slightly reduced during the period 1991–2018.

The analysis of results shows in general that the period 1961–2018 was characterised by a slight decrease in rainfall on the Dapaong-Bombouaka escarpments.

9.3.1.2 *Temperature*

Air temperatures are in general very high. Over the last 48 years, from 1971 to 2018, the average of the minimum temperature varied between 18 °C and 25 °C and those of the maxima between 37.8 °C and 39 °C. The absolute minima recorded are 17.1 °C for Dapaong and 15.5 °C for Mango, while the absolute maxima are 39.1 °C for Dapaong and 40.8 °C for Mango. The inter-annual evolution is characterised by an upward trend for both stations with a regional average ranging between 27.1 °C and 30.2 °C (Figure 3). Atmospheric warming is largely driven by regional and global scale factors, including greenhouse gases. However, local factors such as vegetation fires that precipitate the return of carbon stored in the biomass to the atmosphere, and agricultural activities that contribute to the accelerated degradation of soil organic matter and the CO_2 emissions may also contribute to an increase of air temperature.

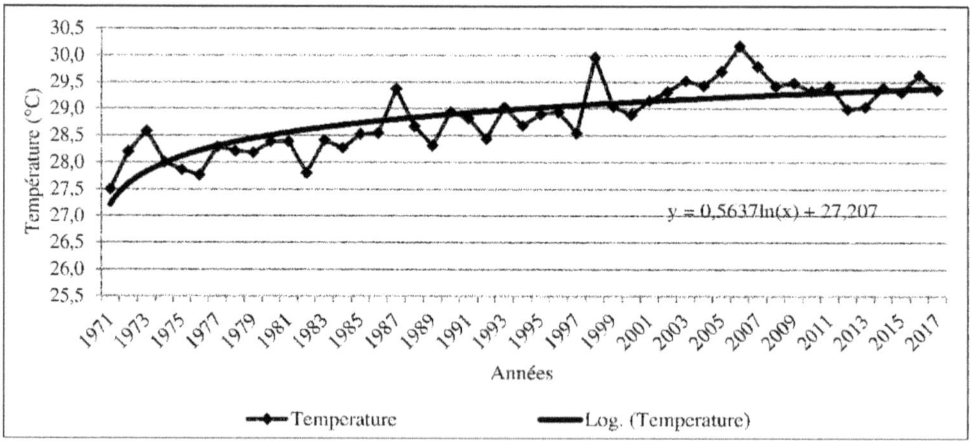

Figure 3. Evolution of temperature on the Dapaong-Bombouaka escarpments from 1970 to 2010 (Source: DMN data).

9.3.2 Land use changes

Due to the strong demographic pressure, land use has undergone a profound change during the period from 1980 to 2018 on the Dapaong-Bombouaka escarpments. The comparison of land use of the years 1980, 2000 and 2010 reveals a difference on the extension of different type of vegetation (Figure 4). In 1980, the areas cultivated represented less than 35.6%. They increased to 70.2% in 2000 and 80.4% in 2010. During the same period the non-cultivated areas have been sharply reduced from 70.4% to 19.6%. The have been also severely degraded. In fact, due to the strong pastoral pressure and excessive cutting of firewood, the formerly forested areas have been almost totally replaced by bush savannahs. These bush savannahs with bare land (rock outcrops), provide very poor protection of the ground. This land use dominated by cultivated areas results in the reduction of soil organic matter. Current agricultural practices accompanied by burning of crop residues or their use for other purposes hardly favour the enrichment of soils with organic matter.

9.3.3 Factors of soil erosion

9.3.3.1 *Cover-management factor*

The evolution of biomass on the Dapaong-Bombouaka escarpments during the period from 1981 to 2010 is characterised by significant annual variations (Figure 5a). Annual average of the NDVI varied between 0.24 and 0.40 during this period. Examination of the evolution of NDVI by types of land use clearly shows that biomass is

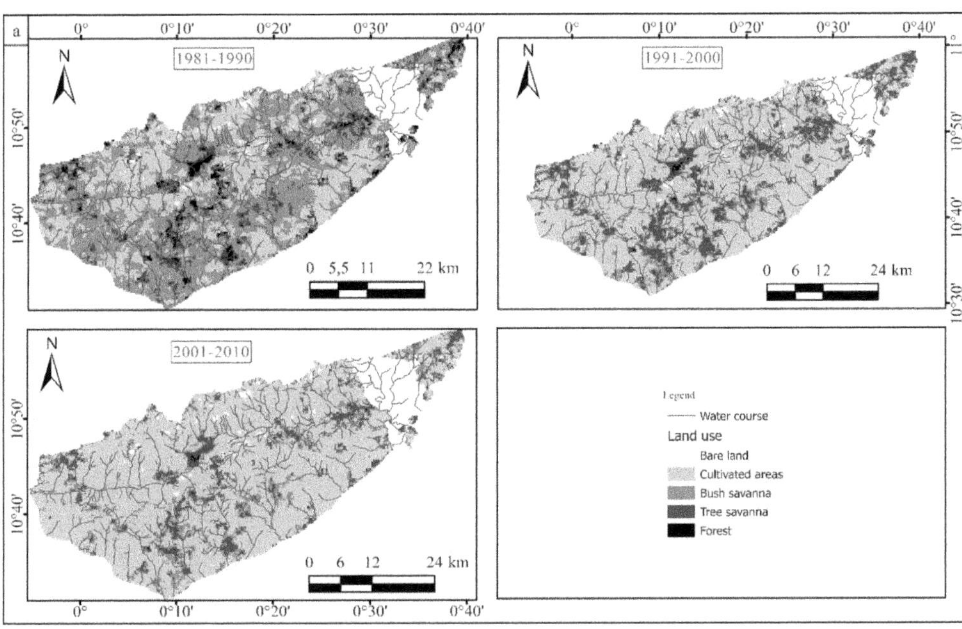

Figure 4. Spatial evolution of land use patterns on the Dapaong-Bombouaka escarpments from 1981 to 2010.

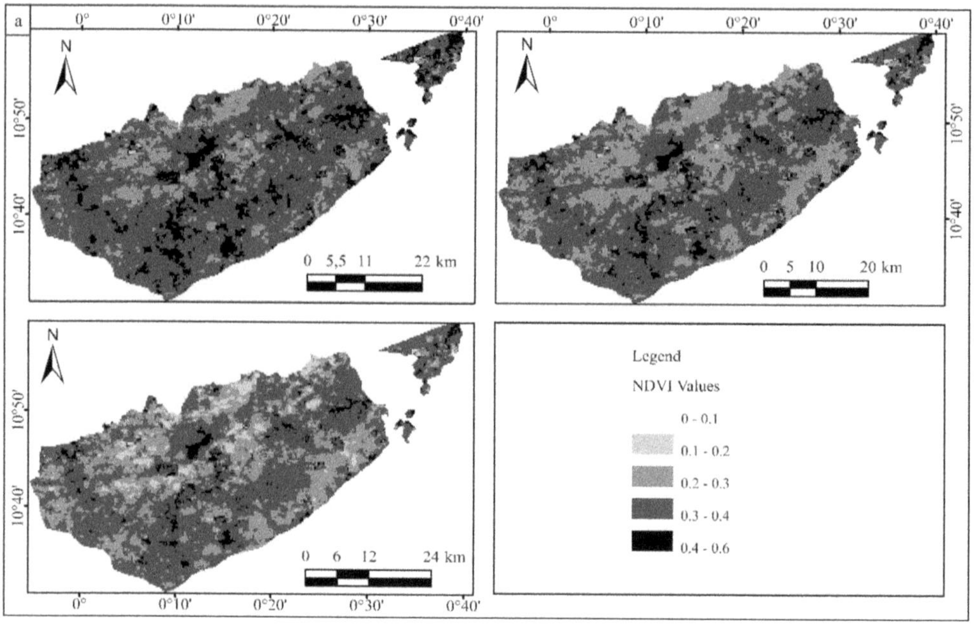

Figure 5a. Spatial evolution of NDVI values (Normalized Difference Vegetation Index) on the Dapaong-Bombouaka escarpments for time slices 1981-1990, 1991-2000, and 2001-2010 (compare Figure 4).

more important on natural vegetated areas than those on cultivated land. The NDVI ranged from 0.39 to 0.51 with an average of 0.45 for natural vegetation. On cultivated land, values varied between 0.21 and 0.43 with an average of 0.32. The NDVI is particularly very low on the surfaces with strong extension of outcrops where the annual average is 0.02 (Figure 5a).

The NDVI evolution curve is characterised by a downward trend resulting from more pronounced troughs especially after the 2010s (Figure 5b). This suggests less significant soil protection against both water and wind erosion.

The cover-management factor measured indirectly through NDVI integrates the influence of climate change and anthropogenic forcing. The coefficient of correlation between annual precipitation and NDVI is 76.3. The correlation is also very high with a rate of land use change of 56.8.

9.3.3.2 *Rainfall-runoff erosivity factor*

The rainfall-runoff erosivity determined from the Roose (1976, 1999) model shows strong variations with a general tendency to a slight decrease during the period from 1961 to 2018 (Figure 6). Erosivity values from the proposed Roose model for West Africa varied between a minimum of 404.2 MJ.mm/ha and a maximum of 664.5 MJ.mm/ha. The annual average over the period from 1961 to 2018 being 524.7 MJ.mm/ha. However, this trend can be misleading because Roose's formula is a simplified model. In fact, the region not only receives more and more intense rainfall, but also shows an irregular rainfall with torrential rains.

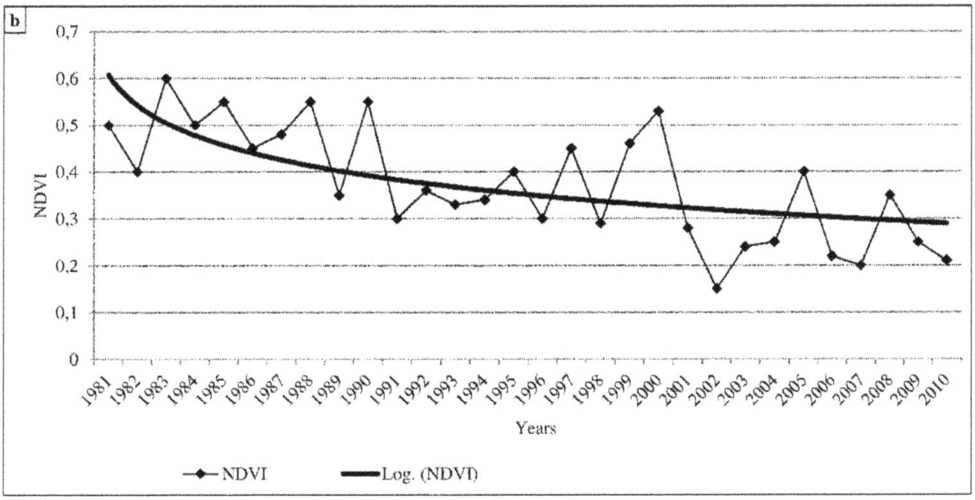

Figure 5b. NDVI values between 1981 and 2010 and logarithmic 'biomass' trend line for the Dapaong-Bombouaka escarpments (compare Figure 5a).

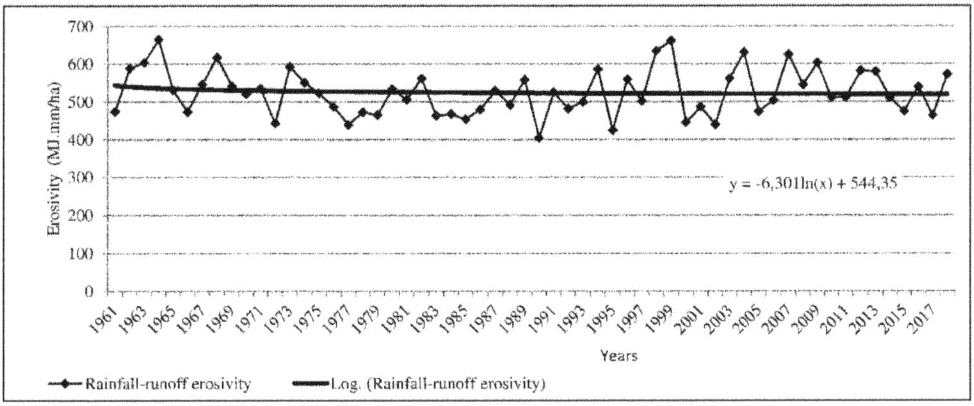

Figure 6. Evolution of annual average of rainfall-runoff erosivity (R) on the Dapaong-Bombouaka escarpments from 1981 to 2018.

9.3.3.3 *Soil erodibility*

Out of a total of 30 samples analysed, the organic matter level is 0.84% (Table 2). The highest rate of organic matter (2.1%) is observed in protected areas. The average rate of organic matter for protected land is 1.4%. The content of organic matter is moderately high on soils where fallow is practiced (1.3%). The minimum rate of 0.2% is observed on parcels regularly exploited without additional organic matter (mulch). On these soils, farmers use only chemical fertilization. The very low rate of organic matter on all cultivated areas reveals the vulnerability of land to anthropogenic forcing.

Table 2. Summary of statistics concerning organic matter in the superficial layers of soil on the Dapaong-Bombouaka escarpments.

Statistical parameters	Number of samples	Minimum (%)	Maximum (%)	Average (%)	Standard deviation
Values	30	0.2	2.1	0.84	0.38

9.4 DISCUSSION

The research hypothesis stated that the lands of the Dapaong-Bombouaka escarpments are vulnerable to their current use patterns and climate change. To verify this hypothesis, pedological, climatic and land-use data were collected, processed and analysed. Different lessons are learned on land vulnerability and its links to climate change and anthropogenic forcing.

9.4.1 Vulnerability of agricultural land

The Dapaong-Bombouaka escarpments are vulnerable in the current conditions which result in the activation of the erosion factors, notably the rainfall-runoff erosivity factor (R), the soil erodibility factor (K) and the cover-management factor (C). The average values obtained are 499.5 MJ.mm/ha.hr.year for R, 0.08 mt.hr/MJ.mm for K and 0.6 for C. They suggest a very high vulnerability of the land. These values are close to those obtained from other studies that have been carried out in the region. This is particularly the case of the studies carried out by Kankpénandja et al. (2012) who obtained values of R = 508 MJ.mm/ha.hr.year, K = 0.072 mt.hr/MJ. mm and C = 0.52 in the Boummonga watershed. Similarly, Kankpénandja (2016) obtained averages values of R = 478.1 MJ.mm/ha.hr.year, K = 0.07 mt.hr/MJ.mm and C = 0,50. According to Roose (1976, 1999), R varies between 100 MJ.mm/ha.hr.year at Sahel and 2000 MJ.mm/ha.hr.year in the humid tropical areas. He has found values ranging between 0.05 and 0.3 for K factor. Other authors have worked on land sensitivity in West Africa as well as in others regions of the world. These include Zachar (1982), Hudson (1981), Mietton (1988), Magrath & Arens (1989), Uri & Lewis (1998), Bouzou Moussa (2000) and Morgan (2005) who have shown the state of degradation of lands under the influence of different types of natural and anthropogenic factors.

9.4.2 Climate variability and land degradation

Analysis of meteorological data revealed that climate change on the Dapaong-Bombouaka escarpments over the last six decades is characterised by a downward trend in precipitation and rising air temperatures. The analysis revealed several sequences, including a sequence of excess precipitation from 1961 to 1970 and a deficit sequence from 1971 to 1990 for all stations. The evolution was then different according to the stations during the sequence of 1991 to 2018. Though the whole region recorded a slight decrease of the precipitations during this sequence, Dapaong, located in the north, records an increase in precipitation while Mango, in the south, is experiencing a downward trend. The evolution of the temperature remained uniform with a general tendency to increase for the two main stations during the period from 1971 to 2018. The question of the climatic evolution was treated by many authors who revealed this general tendency lowering of precipitations and rising temperature. These include Boko (1988), Charre (1988), Courel et al. (1992), Perard & Bokonon-Ganta (1993), Sarr (1995), Badaméli (1996), Klassou (1996), Vissin & Houssou (2004), Vissin et al. (2006) and Vissin (2007). All aspects of climate change and its impacts are assessed by the IPCC (1990) since 1988. In its first report of 1990, IPCC had concluded with certainty on an increase of the rate of the greenhouse gases which caused a rise of the average temperature of the earth surface. This report predicted an average temperature increase of 0.3 °C during the 21st century, with consequences on rainfall, hydrology and agriculture. Reports from later studies have shown that these predictions were well in line with reality. Indeed, according to the 5th report of the IPCC (2014), the average temperature of the Earth increased by 0.85 °C between 1880 and 2012, and for the first time, in 2015 the increase was 1 °C. This rise in temperatures could even reach 4.8 °C by 2100. Climate change is causing accelerated land degradation as noted by IPCC (2009, 2014).

9.4.3 Anthropogenic impact and land degradation

The assessment of the influence of anthropogenic impact on the Dapaong-Bombouaka escarpments was studied through the dynamics of land use. A precise mapping of the land use was made from Landsat TM and ETM+ images with a resolution of 30 m. It revealed since the 1980s by a strong extension of the exploited areas. It results in a weak protection of the land surface increasing soil erodibility. The question of the impact of anthropogenic factor on land in the Savannah Region has also been addressed by other such as Addra et al. (1986), Kankpénandja & Alassane (2012) and Kankpénandja (2016) who all revealed a strong anthropic pressure manifested by a rapid expansion of land used areas at the expense of natural vegetation. At the national scale, studies on land degradation status by Brabant et al. (1996) and land use by Afidegnon et al. (2003) confirm these findings. Brabant et al. (1996) attribute land degradation to various anthropogenic factors including unsuitable agricultural practices, overexploitation of vegetation, late and uncontrolled bush fires, overgrazing, commercial and industrial activities. The map of Afidegnon et al. (2003) reveals that, outside protected areas, vegetation in the north of Togo consists mainly of savannah "parks" degraded by agropastoral activities. Different studies carried out in the West African sub-region have also shown the role of anthropogenic impact on erosion. This is the case of Mietton (1988), Dipama (1996a, 1996b), Bandré (2000) and Bouzou Moussa (2000), evidenced anthropization in Burkina Faso and Niger resulting in the advancement of land degradation.

9.5 CONCLUSION

The concept of land vulnerability helped to understand the particular role of climate change and the influence of humans in the landscapes of the Dapaong-Bombouaka escarpments. The geo-ecological setting of the Dapaong-Bombouaka escarpments is sensitive and fragile because of the geomorphological and pedological conditions which are under strong anthropic pressure. Anthropic pressure is reflected in land use dynamics manifested by the strong extension of agro-pastoral areas. It results in land vulnerability and leads to a decline in biomass and to a reduction of organic matter in the soil. To these effects of ongoing anthropogenic pressure climate change processes are added. The study has shown that climate change is marked by decreasing precipitation and an upward trend in temperature. Declining and irregular rainfall increases water stress and reduces biomass production. The increase in temperature accelerates the mineralisation of organic matter. Climate change also results in an increased erosivity of rainfall.

REFERENCES

Addra, T.C., Fahem, K.A., De Jong, T. & Mank, T. 1986. *Atlas of Regional Development of Togo*. Lomé: UNDP-DTCD.

Afidegnon, D., Carayon, J. L., Fomard, F., Lacaze, D., Guelly, A.K., Kokou, K., Woegan, Y.A. & Batawila, K. 2003. *Map of the Vegetation of Togo*. Campus Project number 96-348. Paris: Terrestrial Ecology Laboratory (USP, France) and Botany and Plant Ecology Laboratory (UL, Togo).

Badaméli, K.M. 1996. *Climate variability and agricultural production in Togo*. Doctorate Thesis. Bordeaux: University of Bordeaux 3 Michel de Montaigne, 1–366.

Bandré, E. 2000. Vegetation cover dynamics in the Agricultural Supervision Unit (UEA) of Kié, Department of Solenzo (Burkina Faso). *Berichte des Sonderforschungsbereichs* 268(14): 505–520.

Boko, M. 1988. *Research on rainfall variability in Benin*. Climates and climatology, mixtures offered to P. Pagney, Univ. of Bourgogne, Dijon: 41–50.

Bouzou Moussa, I. 2000. Geomorphological responses to the problem of water and soil conservation in Niger. *Synthesis report presented for the diploma of habilitation to direct research*. Grenoble: University J. Fourier-Grenoble I, Institute of Alpine Geography.

Brabant, P., Darracq, S., Égué, K. & Simonneaux, V. 1996. State of Land degradation resulting from human activities in Togo. *French Institute for Scientific Research for Co-operative Development, Department of Environment and Agricultural Activity* – ORSTOM Paris, *Collection notice explicative* 112: 1–66.

Charre, J. 1988. Large regions of rainfall variability in West Africa south of the Sahara. *Satellite climate watch* 22: 21–25.

Courel, M.F., Chamard, P-C., Togola, M. & Jaqueminet, C. 1992. Recent hydro-rainfall evolution of the Inner Niger Delta (Mali). *Proceedings of the Colloquium of the A.I.C.* 5: 43–52.

Dipama, J.M. 1996a. *Impacts of the hydro-electric dam on the Kompienga watershed (Burkina Faso)*. Ph.D. Thesis. Bordeaux: University of Bordeaux 3 Michel de Montaigne, 1–392.

Dipama, J.M. 1996b. Analysis of vegetation cover dynamics by aerial photographs and LANDSAT images in Kompienga Province (Burkina Faso). *Cahiers of the Center for Studies and Research in Letters, Human and Social Sciences (CERLESHS), PUO* 23: 51–66.

FAO, Food and Agriculture Organization 1990. *Soil studies and surveys for irrigation*. Rome: FAO Soil Bulletin 42: 1–181.

Hudson, N.W. 1981. *Soil conservation*. 2nd ed. London: Batsford.

IPCC, Intergovernmental Panel on Climate Change 1990. *Climate Change, First Assessment Report, Scientific assessment*. Cambridge: Cambridge University Press.

IPCC, Intergovernmental Panel on Climate Change 2009. *Climate Change and Land, Chapter 4: Land Degradation*. https://www.ipcc.ch/site/assets/uploads/2019/08/2e.-Chapter-4_FINAL.pdf. (accessed 2019).

IPCC, 2014: Climate Change 2014: Synthesis Report. Contribution of Working Groups I, II and III to the Fifth Assessment Report of the Intergovernmental Panel on Climate Change [Core Writing Team, R.K. Pachauri and L.A. Meyer (eds.)]. IPCC, Geneva, Switzerland, 1–151.

Kankpénandja, L. & Alassane, A. 2012. Agropastoral activities and accelerated degradation of the upper Namiélé. Procee*dings of the 3rd Symposium of Sciences, Cultures and Technologies of UAC-Benin, Volume 1, Letters and Human Sciences*: 35–56.

Kankpénandja, L., Gnongbo, I.Y. & Tchalare, B. 2012. Erosion, risk of siltation and clogging of water reservoirs in the watershed of Sansargou, North-Togo and South-East of Burkina Faso. *Geographical Review of Leïdi Laboratory* 10: 312–326.

Kankpénandja, L. 2016. *Morphogenesis and management of sub-watersheds of Oti in Togo*. PhD thesis. Lomé: University of Lomé, 1–351.

Kankpénandja, L., Tchalare, B., Gnongbo, T.Y. & Blivi, A.B. 2018. Current morpho-sedimentary dynamics in the Gambara River valley in Northern Togo. *Àhh 5h 5, Journal of Geography of the Research Laboratory on Dynamics of Environments and Societies*, Special Issue: 28–44.

Klassou, K.S. 1996. *Recent climato-hydrological evolution and environmental consequences: the example of the Mono River catchment (Togo-Benin)*. PhD Thesis. Bordeaux: University of Bordeaux 3 Michel de Montaigne, 1–370.

Lal, R. 1976. *Soil erosion problems on an Alfisol in western Nigeria and their control*. Ibadan: *IITA Monograph* 1: 1–126.

Magrath, W.B. & Arens, P. 1989. The cost of soil erosion on Java: a natural resource accounting approach. *Environment Department Working Paper 18*. Washington D.C.: World Bank Policy Planning and Research Staff, World Bank.

McCormack, D.E. & Young, K.K. 1981. Technical and societal implications of soil loss tolerance. In Morgan, R.P.C. (ed.). *Soil conservation: problems and prospects*: 365–76. Chichester: Wiley.

Mietton, M. 1988. *Dynamics of the Lithosphere-Atmosphere interface in Burkina Faso, erosion in the savannah zone*. State doctoral thesis. Grenoble: University of Grenoble, 1–511.

Morgan, R.P.C. 2005. *Soil erosion and conservation*. 3rd ed. Oxford: Blackwell publishing.

Perard, J. & Bokonon-Ganta, E. 1993. Rain variability and socio-economic changes in Upper Borgou (North Benin). *Proceedings of the Symposium, A.I.C., Thessaloniki*: 509–518.

Poss, R. 1996. *Morphopedological study of North Togo*. Paris: Edition of O.R.S.T.O.M.

Renard, K.G., Foster, G.R., Weesies, G.A., McCool, D.K. & Yoder, D.C. 1997. Predicting Soil Erosion by Water: A Guide to Conservation Planning with the Revised Universal Soil Loss Equation (RUSLE). *Agricultural Handbook* 703. Washington: Government Printing Office.

Roose, E. 1976. Use of the universal soil loss equation to predict erosion in West Africa. *Soil Erosion: Prediction and Control, SCSA, Special Publication* 21: 60–74.

Roose, E. 1999. *Introduction to the conservative management of water, biomass and soil fertility (GCES)*. Rome: FAO Field Bulletin 70.

Sarr, B. 1995. *Climate and agriculture in tropical Africa: the case of rice cultivation in the developed areas of the Senegal River Basin*. Ph.D. thesis. Dijon: University of Bourgogne, 1–359.

Tricart, J. 1978. Applicable geomorphology. *Applicable geography collection*. Paris: Masson.

Uri, N.D. & Lewis, J.A. 1998. The dynamics of soil erosion in US agriculture. *Science of the Total Environment* 218: 45–58.

Vissin, E.W. & Houssou, C.S. 2004. Study of the variability of the water budget in the Sota basin (Benin, West Africa). *Proceedings of the 2nd scientific days of the National Universities of Benin*: 51–61.

Vissin, E.W., Houssou, C.S. & Houndenou, C. 2006. Rainfall and flow vulnerability in the Niger basin of the Niger River. *Environmental Sciences, LaRBE*: 75–92.

Vissin, E.W. 2007. *Impact of climatic variability and dynamics of surface states on the flows of the Benin basin of the Niger River*. PhD Thesis. Bourgogne: University of Bourgogne, 1–259.

Wischmeier, W.H. & Smith, D.D. 1960. A universal soil-loss estimating equation to guide conservation farm planning. *Proc. 7th Inter. Congress Soil Science Soc.* 1(7): 418–425.

Zachar, D. 1982. *Soil erosion*. Amsterdam: Elsevier.

CHAPTER 10

Impacts of climate change on plant adaptation and floristic composition in the Kpendjal Prefecture in northern Togo

M. Djangbedja
University of Lomé, Lomé, Togo

ABSTRACT: Climate change is rapidly modifying the spatial distribution of plants and causes changes in the floristic composition. In the Kpendjal Prefecture in northern Togo rainfall fluctuations are well evidenced, and this study aimes to characterize the variability in precipitation and its impact on vegetation cover. Diversity, structure, and spatial dynamics of plants in relation to climate variability are highlighted. The floristic inventory identified 184 species belonging to 106 genera and 44 plant families. Combretaceae and Mimosaceae are most abundant. The distribution of individuals by diameter and height class structure shows the low presence of individuals. According to the results of the stakeholders' surveys, 85% mention climate change, 25% demographic pressure, 15% 'carbonization', and 20% logging as main factors responsible for ecosystem degradation.

10.1 INTRODUCTION

In West Africa severe droughts with major famines between 1970 and 1980 highly affected the vegetation cover and destroyed woody species. In consequence biomass production was strongly reduced (Hiernaux et al. 2009). This situation led to an interest to explore the connection between climate dynamics and the state of vegetation cover (Justice et al. 1985, Le Barbé & Lebel 1997, Graef & Haigist 2001, Roerink et al. 2003). There is evidence that the spatiotemporal change in land cover became an important indicator that contribute to the evaluation of ecosystems' health. Indeed, intra-seasonal and inter-annual variations have strong impacts on the dynamics of vegetation.

Togo, like many other countries in West Africa, has been experiencing for some decades meteorological extreme events connected to climate change (Janicot 2012). These climatic 'disturbances' are increasingly frequent and lead to a significant change in vegetation composition. Various ecosystems from different environmental zones in Togo are nowadays characterised by deterioration of ecosystem services. Togo exhibits an annual rate of 4.5% of deforestation which is one of the highest worldwide (MERF 2011). Deforestation affects vegetation cover and composition, in particular in the Savannah Region. The northern part of Togo is not only threatened by desertification, but is also experiencing an unprecedented anthropic pressure. An intense exploitation of forests (a carbon sink!) and the search of firewood is responsible for this situation.

In face of this situation a growing interest in the conservation of forest resources is observed, but very few works have been completed on the present state of the vegetation types in this region (Lemoine 1976, Poss & Rossi, 1987, Kpongou 1994, Demakou 2009). Even though studies on vegetation types are documented, they remain scattered and fragmentary. Delpech (1993) has recently confirmed that plants are valuable bio-indicators to define the ecological characteristics of a region as they characterize the state of climate. Relationships between different climatic parameters and the vegetation types have been a matter of concern for many authors who acknowledge that the different aspects of vegetation types are in correlation with the integral conditions of the physical environment (Holdridge 1947, Birot 1965). It therefore appeared useful to study variations of climate parameters and to connect them to vegetation classes around the study area to assess the degree of disturbance of vegetation types. It is in this perspective to present the current state of vegetation types in the ecological context of climate change. This objective is pursued through the following research questions: What is the floristic diversity and the structure of the vegetation types in the east of the study region? What is the state of their spatiotemporal dynamics? What is the impact of climate change on these vegetation types? It was hypothesized that vegetation types in the east of the Savannah Region are rich and diverse; they present a structure with a predominance of young individuals and a specifically low diversity related to anthropic pressure and to climate variability; these vegetation types undergo a 'regression' as a result of rainfall variability.

10.2 STUDY AREA

The Kpendjal Prefecture lies in the northeastern part of the Savannah Region in Togo. It covers a surface of 1395 km^2 or 2,45% of the national territory. It is limited in the north by Burkina Faso, in the south by the Oti Prefecture, in the east by the Republic of Benin, and in the west by the Prefecture of Kpendjal. Geographically, the area is located between 10°37′-11°02′ northern latitude and 0°28′-0°54′ eastern longitude (Figure 1).

The climate is of Sudanian type with alternating of two seasons. Annual rainfall varies between 1000 and 1100 mm. It is characterised by a short rainy season (May/June to August/September) and a long dry season of six to seven months. The average annual temperature is high (28 °C). The relative humidity even in the rainy season is low (53-67%). Average insolation is 6.62 hours per day and the mean evapotranspiration is 1540 mm/year (Addra et al. 1984).

The relief of the Kpendjal Prefecture is composed of three great units: a peneplain located in the extreme north; a succession of two sandstone escarpments (Dapaong and Bombouaka, Runge 1990, see chapter 9 in this volume); and the vast alluvial plain of the Oti River (Petit & Addra 1978). Geologically three great structures can be distinguished into the Precambrian Birrimien basement, the Paleozoic sedimentary cover and the Oti River's alluvial plain.

Two major soil types prevail in Kpendjal: tropical ferruginous and hydromorphic soils. The Kpendjal River has shaped its flat valley into flat bedded strata of supra-tillitic mudstones (Kankpénandja 2002). The vegetation of the region is characterised by savannah, savannah woodland and dry forest. The population mainly includes peasants from the ethnic groups of the Moba and the Gourma, but there are also minorities from Nigeria, mainly active in trade, such as the Yorouba and the Haussa.

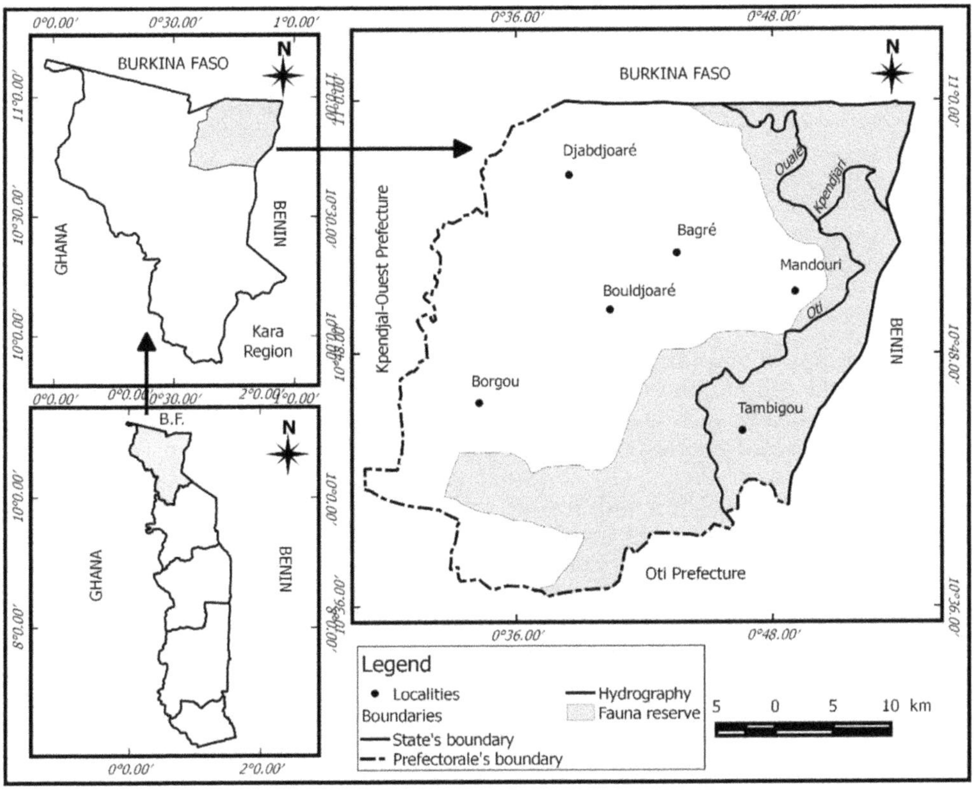

Figure 1. Location of the study area in northern Togo (Addra et al. 1984).

10.3 METHODS

Field transects were surveyed for the collection of floristic data. The proposed average length of the transects was 1000 m (Runge 1990). On each transect additional test areas 30 m x 30 m wide were selected, in open savannahs and in forest-savannah mosaics, based on the method of Braun-Blanquet (1932). Likewise, five square plots of 5 m x 5 m in size have been specified for a surveyed transect. Plant samples, data on topography (slope, downslope, upslope or summit), soil texture and general vegetation types were recorded. Soil texture was directly determined in the field by simple hand methods. In total, 41 spots at regular distances of 200 m were surveyed. In each spot, information about environmental parameters and dendrometric data (diameter at 1.30 m above ground, dbh) were collected. The height of trees was recorded for all the individuals with a dbh above or equal to 10 cm. The species not identified in the field were determined at the LABEV (Laboratory of Botanic and Vegetation Ecology) at the University of Lomé. The plant nomenclature was applied accordingly to Brunel et al. (1984). As for the characterisation of the vegetation types, the botanical system of Trochain (1957) was used.

Local rainfall data were collected to glean information on climate change. Half-conducted interviews were carried out with the five major groups of stakeholders. The general categories of people who were interviewed are listed in Table 1.

Ecological data processing of surveys was conducted using Excel spreadsheets. The Community Analysis Package (CPA 2000) software was used for distinguishing

Table 1. Proportion of categories of interviewed stakeholders.

Socio-professional categories	Proportions (%)
Farmers/Peasants	70
Livestock producers/herders	15
Charcoal merchants	10
Timber growers	5
Total	100

different groups of vegetation. The Detrended Correspondance Analysis (DCA) was applied for the ordination of data collected in the factorial plan, producing factorial maps and dendrograms. Data was analysed and interpreted by using the total number of collected species (Rs), the index of diversity of Shannon (*H*), and the equitability index of Pielou (*E*) (Frontier & Pichod-Viale 1995, Magurran 1988, Kent & Coker 1992). These indexes are most used in the study of the diversity of tropical vegetation. The total number of collected species (R) represents the total number of species in a given phytocenosis and in an area of a given sample. The diversity index of Shannon (*H*) is calculated based on the formula:

$$H = -\sum_{i=1}^{S} p_i \log_2(p_i) \tag{1}$$

whereby *H* is the diversity index of Shannon, p the proportion of elements in the total sampling belonging to the species i. $p_i = (n/No)$ represents the relative abundance of elements of the considered species; "n" is the number of elements of the considered species, and "*No*" is the total number of collected elements.

The formula used in the calculation of the equitability index of Pielou (*E*) is:

$$E = \frac{H}{\log_2 No} \tag{2}$$

whereby n is the number of elements of the considered species, *No* is the total number of collected elements. $\log_2 No$ is the theoretical value of the maximal diversity that may be reached.

The determined structural parameters are the basal area *G*, density of tree population and diameter classes. The basal area within each grouping is obtained based on the formula:

$$G = \Sigma \pi D^2/4 \tag{3}$$

with G = basal area in m^2/ha and D = diameter in m.

The density per grouping is given by the formula:

$$De = N/S \cdot 10{,}000 \tag{4}$$

with N = number of elements of the grouping; S = inventoried area. The density is extrapolated by hectare (Marchal & Rondeux 1995).

To assess the rainfall difference compared to the average it was calculated:

$$(X_i - \bar{X})/\bar{X} \quad (5)$$

whereby X_i is the value of annual rainfall for a year and $\bar{X}$ is the interannual average value of rainfall over the period studied.

The main biological types (life forms) and the phytogeographical affinity was identified referring to Raunkiaer (1934) and White (1986). The first is about life forms of trees (Phanerophytes) and Megaphanerophytes (MP), Hemicryptophytes (He), Chamaephytes (Ch), Geophytes (Ge), and Therophytes (Th). For phytogeographical areas and the following types were applied: Afro-American (AA), Pantropical (Pan), Paleotropical (Pal), and species in continental distribution regrouping such as Sudanese-Guinean, (SG), Afro-Tropical (AT), Afro-Malagasy (AM), African pluri-regional (PRA), Sudanese-Zambesian (SZ), Guinean-Congolese (GC), and species of fundamental Sudanese elements (S).

LANDSAT images of 1973, 1998, 2018 were classified to map land use dynamics in order to determine trends in vegetation cover. The processing of satellite images was done by using ENVI 4.7 image software with coloured image composition, digital classification and cartography of the classification results. Image processing enabled the generation of descriptive statistics on surfaces of different land use units for selected years. The comparison of exploitation maps of soil and the series of corresponding statistics allowed to highlight regional changes in vegetation cover.

10.4 RESULTS

10.4.1 Floristic diversity

The floristic inventories enabled to distinguish 184 species grouped into 44 families. The dominating botanical families are illustrated in Figure 2. It is evidenced that the Combretaceae (23%) and the Mimosaceae (19%) are the most abundant families in the study area.

10.4.2 Biological life forms and phytogeographical mapping

The biological type or the life form classification is dominated by Microphanerophytes (30%), Megaphanerophytes (22%), and Therophytes (15%) (Figure 3). The phytogeographical gross spectrum shows a dominance of Sudanese species (23%), Sudanese-Zambezian (22%), Afro-tropical pantropical (12%). The African American, Afro-Malagasy and cosmopolitan species are relatively less represented (3%) (Figure 4).

Legend for Figures 3 and 4: mp: Microphanerophytes; Mp: Mesophanerophytes; Th: Therophytes; G: Geophytes; np: Nanophanerophytes; MP: Megaphanerophytes; Ch: Chamaephytes, H: Hemicryptophytes.

S: Sudanian species; SZ: Sudanese-Zambezian sp.; AT: Afro-tropical sp.; Pan: Pantropical sp.; SG: Sudanese-Guinean sp.; GC: Guinean-Congolese sp.; Pal: Paleotropical sp.; PRA: African pluri-regional; AA: Afro-American sp.; AM: Afro-Malagasy sp.; Cos: Cosmopolitan sp.

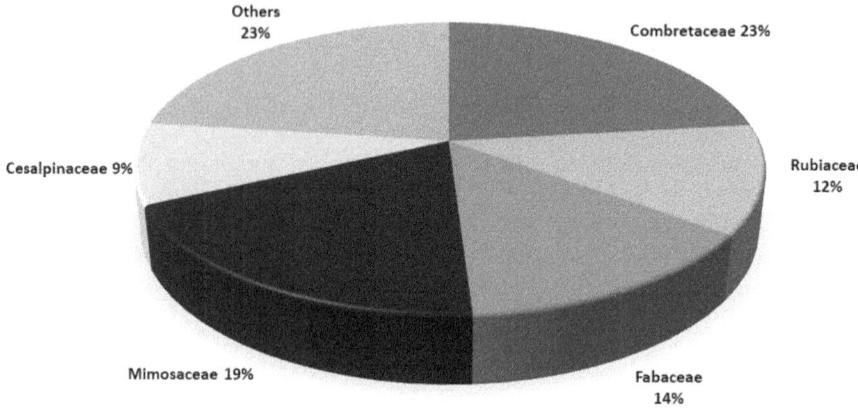

Figure 2. Spectra of major botanical families in the Kpendjal Prefecture in northern Togo (Sketch: L. Carstens 2020).

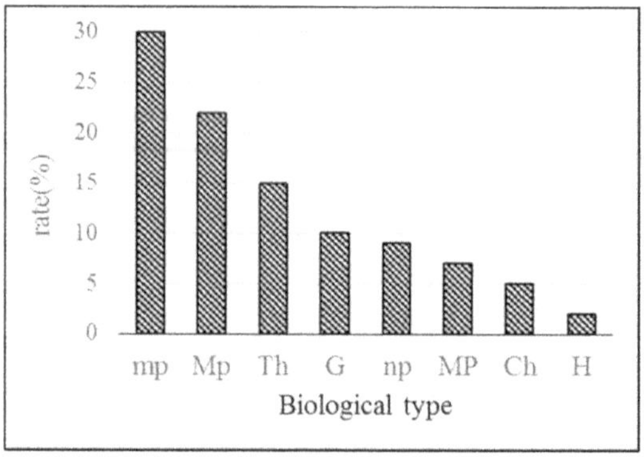

Figure 3. Biological life forms (for explanantion see page 127).

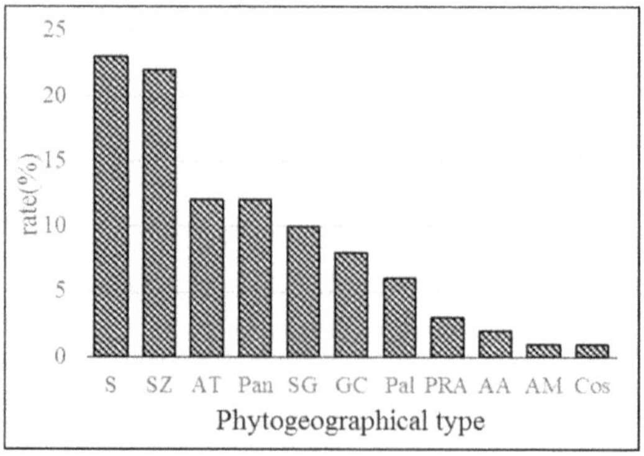

Figure 4. Phytogeographical types (for explanantion see page 127).

10.4.3 Typology of collected plant data

A rising hierarchical classification of the collected plant data according to Ward's method (Minimum variance, Ward 1963) was applied. It allowed to discriminate six predominant plant groups, Figure 5).

Gl Plant group with gallery forests (*Anogeissus leiocarpus*); G2 Plant group within bush savannah (*Acacia gourmaensis* on slopes and on ferrallitic hilltops); G3 Plant group with in 'Savannah Parks' (*Adansonia digitata, Borassus aethiopum*); G4 Plant group with gallery forests (*Acacia leata* on hydromorphic soils); G5 Plant group within bush savannah (dominated by *Detarium microcarpum*); G6 Plant group with tree savannah (savannah woodland predominated by *Lannea barteri*); R = sample

10.4.3.1 *Plant groupings*

Gallery forests with Anogeissus leiocarpus (G1)

The group G1 is located along water courses on deep alluvial soils. This plant group is characterised by a total number of 40 collected species on 11 sampling sites. Aside of *Anogeissus leiocarpus, Khaya senegalensis, Vitex doniana, Vitellaria paradoxa,* and *Diospyros mespiliformis* are common.

The diversity index of Shannon (*H*) is 3.85 with a Pielou (*E*) equitability index of 0.83 giving evidence to a rather homogenous species distribution. The basal surface represents 38.8 m^2/ha, the density of trees is 157 plants per hectare. The distribution of individuals by height classes presents a Gaussian distribution; regarding the distribution of height classes, individuals with 4-8 metres are more common (Figure 6).

Bush savannah with Acacia gourmaensis

This plant group (G2) often occurs on slopes and on ferrallitic hilltops. It has a total number of 16 collected individuals found in seven of the sampling sites. Other species identified were *Acacia senegal, Terminalia glausecens, Pseudocedrela kotskyi, Combretum glutinosum, Combretum nigricans, Bauhinia thonningii,* and *Grewia mollis.*

The diversity index of Shannon (*H*) is 2.5 with an equitability index of Pielou (*E*) of 0.76. The basal surface represents 32.66 m^2/ha, the density of trees is 57 plants

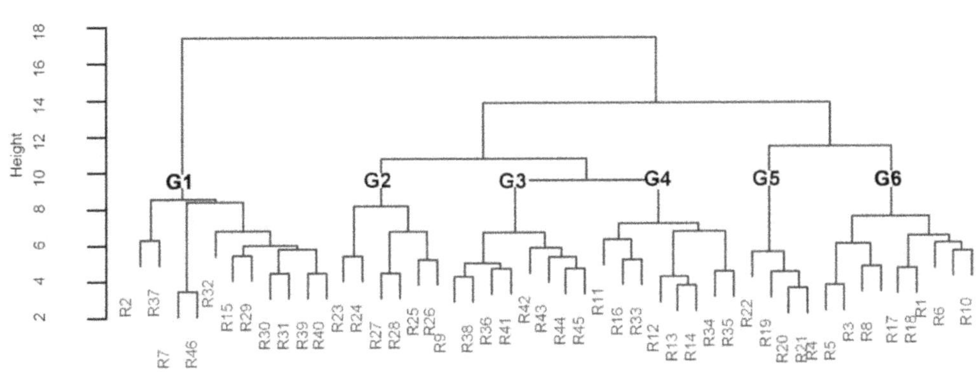

Figure 5. Aggregation of collected plant data from northern Togo according to Ward (1963).

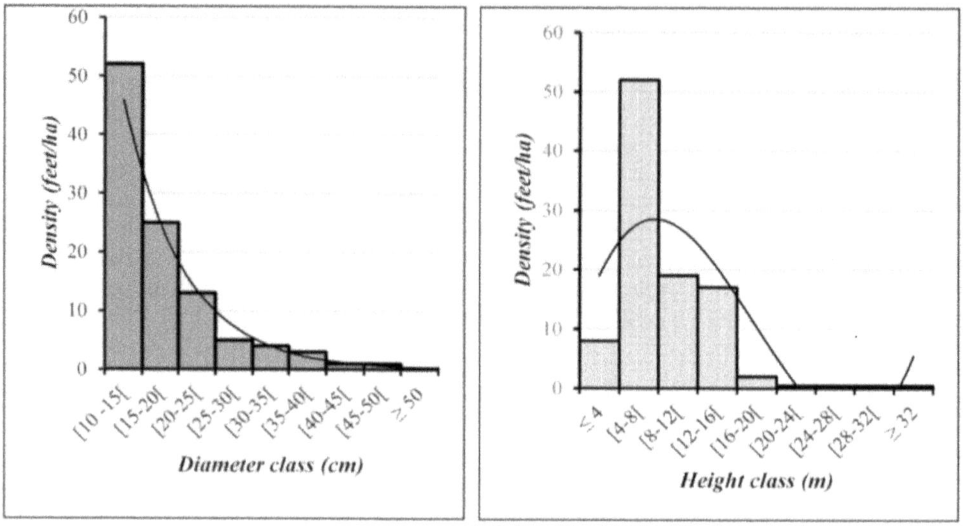

Figure 6. Demographical structure of gallery forests (G1) with *Anogeissus leiocarpus* in northern Togo (Source: fieldwork 2017).

per hectare. The distribution of individuals by height classes shows a trend in “L” (Figure 7, left) indicating a predominance of younger individuals. This distribution decreases regularly as one moves to Ward’s (1963) bigger classes. The Gaussian normal distribution is shown by the distribution of individuals by height classes. It indicates a preponderance of medium-sized woody vegetation (Figure 7, right).

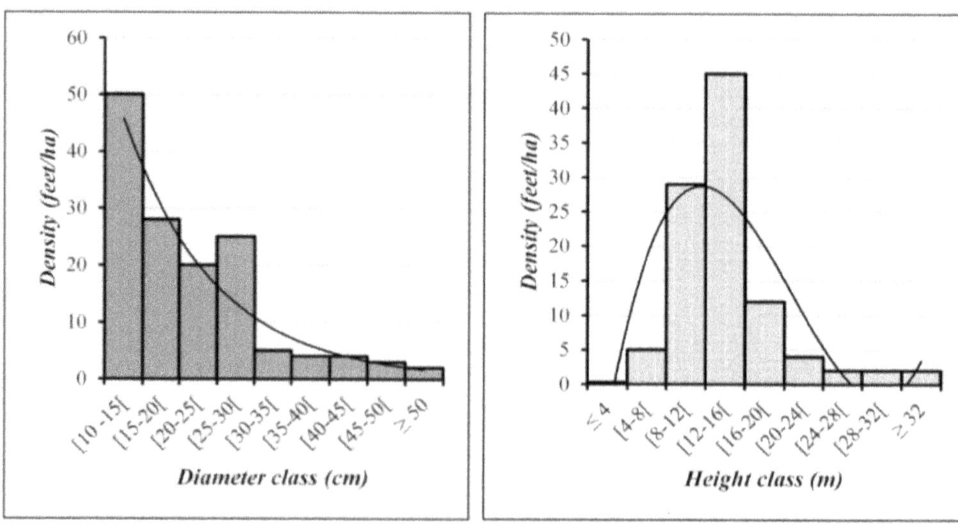

Figure 7. Demographical structure of bush savannah (G2) with *Acacia gourmaensis.* (Source: fieldwork 2017).

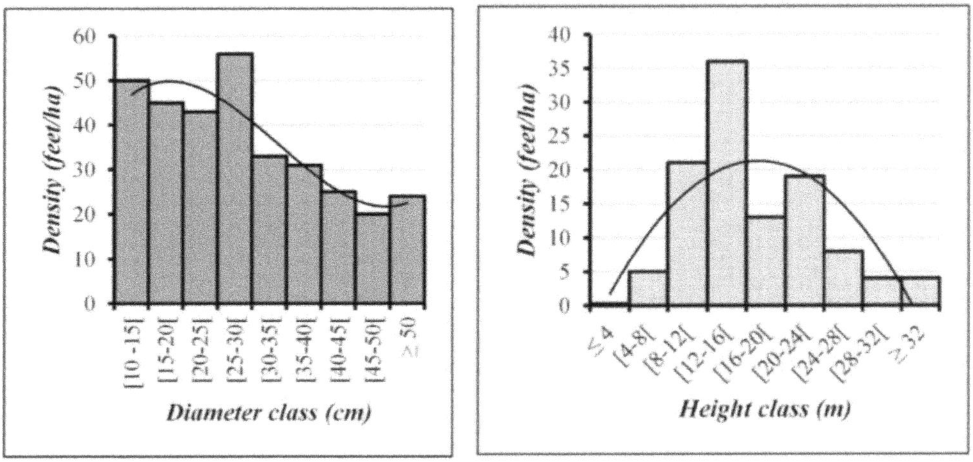

Figure 8. Demographical structure of 'Savannah Parks' (G3) with *Adansonia digitata* and *Borassus aethiopum* (Source: fieldwork 2017).

'Savannah Parks' with Adansonia digitata and *Borassus aethiopum*

These 'Savannah Parks' (G3) are commonly located on the foot slopes and in plains. The total number of collected species was 30 from seven sites surveyed. The floristic composition includes *Parkia clappertonia, Vitellaria paradoxa. Borassus aethiopum, Adansonia digitata, Mangifera indica, Vitex doniana, Tamarindus indica, Diospyros mespiliformis, Prosopis africana, Sterculia setigera, Ficus thonningii, Eucalyptus camaldulensis, Azadirachta indica*, and *Cassia siamea*.

The diversity index of Shannon (*H*) is 5.89 whereas the equitability index of Pielou (*E*) accounts for 0.97. The basal surface represents 14.20 m^2/ha, and the tree density is 175 individuals per hectare. The distribution of trees by classes of diameters and heights indicates a normal distribution. The latter reflects a predominance of medium aged individuals (Figure 8).

Gallery forests with Acacia leata

This plant group (G4) (Figure 9) mainly grows on hydromorphic soils in the study area. It is particularly rich with 42 taxons occurring in eight surveyed sites. *Acacia leata, Lannea barteri.* and *Anogeissus leiocarpus* are predominant.

The diversity index of Shannon (*H*) is 4.45 and the equitability index of Pielou (*E*) has a value of 0.98. The basal surface represents 18.69 m^2/ha, and the tree density is 141 plants per hectare. The distribution of individuals by diametre and height classes is illustrated in Figure 9. It shows that the gallery forests provide a distribution of an 'upside down S' leading to a heterogenous distribution. Some individuals with greater diametre exist but in relatively low proportion. However, individuals with small diametres whose class lies between 10 and 15 cm are rather lowly represented (19 plants per hectare). The distribution of individuals by height classes shows a Gaussian normal distribution with a high representation of individuals of the class of height 16-20 m.

Bush savannah with Detarium microcarpum

This plant group (G5) mainly occurs on sandy to clayey soils. It has a total number of 30 collected species within five of the surveyed areas. The diversity index of Shannon (*H*) is 3.1 and the equitability index of Pielou (*E*) shows a value of 0.74. The basal surface represents 12.73 m^2/ha, the distribution of trees by classes of diameters

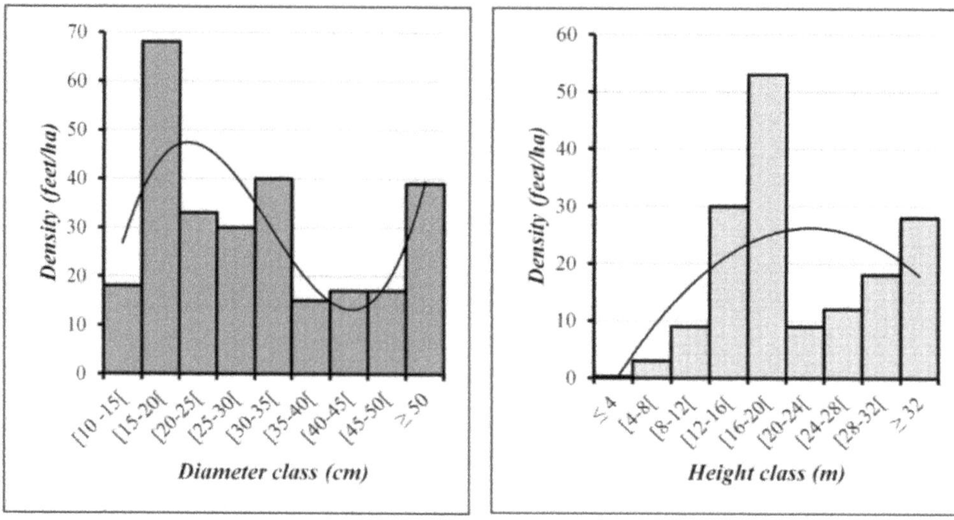

Figure 9. Demographical structure of trees from gallery forests (G4) with *Acacia leata* (Source: fieldwork 2017).

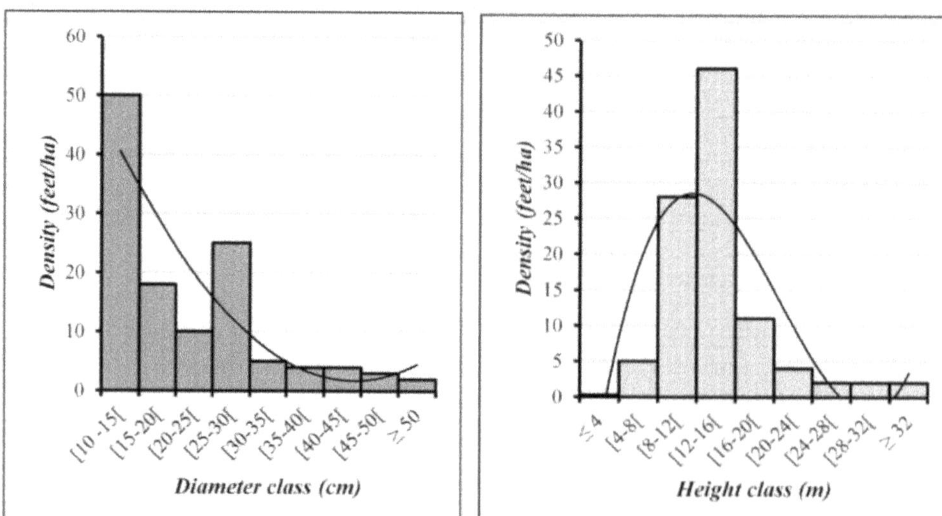

Figure 10. Demographical structure of bush savannah (G5) with *Detarium microcarpum* (Source: fieldwork 2017).

and heights is in normal distribution. Figure 10 shows a general distribution of individuals by diametre classes in «L». The distribution of individuals by height classes provides a trend in Gaussian normal distribution (Figure 10). The woody individuals of average height (12-16 m) are highly represented with 58 individuals per hectare.

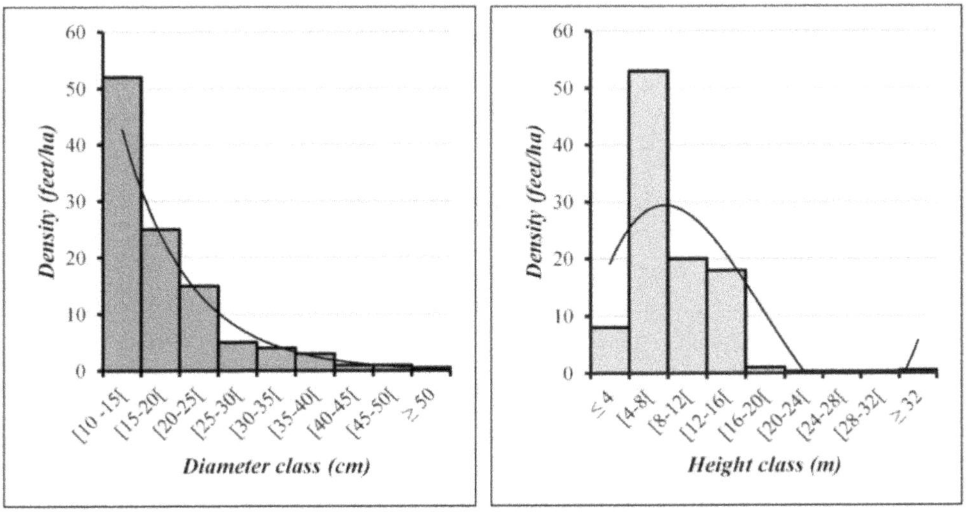

Figure 11. Demographical structure of tree savannah (G6, savannah woodland) with *Lannea barteri* (Source: fieldwork 2017).

Tree savannah (savannah woodland) with Lannea barteri

These plant group (G6) occurs on extended interflow slopes (French 'interfluves') with rock outcrops and on mid-slopes. The floristic composition is 29 species in eight surveyed sites. The index of diversity of Shannon (*H*) represents 3.120 ± 0.237, the equitability index of Pielou (*E*) is 0.99. The basal surface is 18.15 m^2/ha. The distribution of individuals by diametre classes provides a trend in «L» form interpreted as a predominance of young individuals. This distribution declines regularly as one looks at greater diametre classes. Height classes of individuals show a normal distribution (Figure 11); this gives evidence to the preponderance of an 'average' woody savannah.

10.4.3.2 *Variability of the rainfall regime*

Annual and average rainfall data and trends in precipitation for the meteorological station Dapaong are shown in Figure 12. The pluviometric record indicates a slight upward trend over the period 1961-2015 with strong rainfall variability. The five-year period 1961-1965 is marked by four surplus years with the exception of 1961 with a rainfall regression of (-9.74%) from the average. Over the 54-year period, 27 years were in humidity deficit with marked dry years in 1966 (-33.68%) and in 1978 (-31.68%) compared to the average rainfall. The 2005-2015 period was marked by an excess of rainfall with significantly marked surpluses in 2004 (+9.77%) and 2014 (+8.77%) compared to the average.

10.4.3.3 *Spatiotemporal dynamics of vegetation cover*

From 1973 to 1998 land cover and land use in the Kpendjal Prefecture underwent large spatiotemporal changes. In 1973, savannah vegetation represented 65.52% of the prefecture surface, followed by agricultural and fallow land with 22.4%. Up to 1998 the landscape slightly changed: however, vegetation was still dominated by

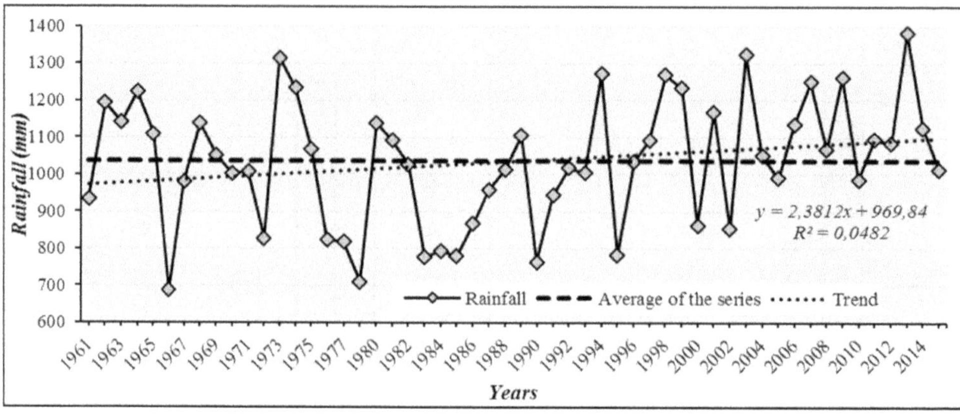

Figure 12. Fluctuation of annual rainfall at Dapaong between 1961 and 2015 (Source: Direction Générale de la Météorologie Nationale du Togo, DGMN, Lomé).

savannah (62.78%) and agricultural as well as fallow land (27.5%). In 2018, land cover and land use patterns considerably changed with a predominance of smallholders' agricultural land use mosaics (76.2%) (Figure 13).

10.4.4 Stakeholders' perception of local ecosystem degradation

According to the empirical results of sociological surveys, 85% of respondents mentioned climate change, 25% demographical pressure, 15% 'carbonization', and 20% forest exploitation as principal factors of landscape degradation. As far as climate change is concerned, it is essential to mention that smallholder farmers base their thinking and perception of climate variability on experienced climatic events such as abundance of rains, droughts, and floods. This perception of climate variability and change is based on the individual observation of punctual annual situations, and on a systematic follow-up of climate factors over a given interval of time. They perceive climate variability through the devastating damage to agricultural land. However, farmers' perceptions on rainfall variability indirectly confirm the results of measured rainfall trends over the last decades in the study area (Figure 12).

A review of the rainfall events shows a strong fluctuation in annual rainfall regime that has a top-down trend. In general the yearly trend of the rainfall system is in average downward over the whole period of 1985 to 2015. However, there are some nuances within the years when it rained heavily or moderately.

10.5 DISCUSSION

The study in northern Togo identified 184 plant species that belongs to 106 genera and 44 plant families. It gave evidence of a relatively important diversity (compare Demakou 2009, Dimobé et al. 2012). The predominant and omnipresent botanical families of Combretaceae and Mimosaceae are indicators of a generally semi-humid to semi-arid Sudanese type with low annual rainfall and year-round high temperatures (Thiombiano 2005, Abdourhamane et al. 2013).

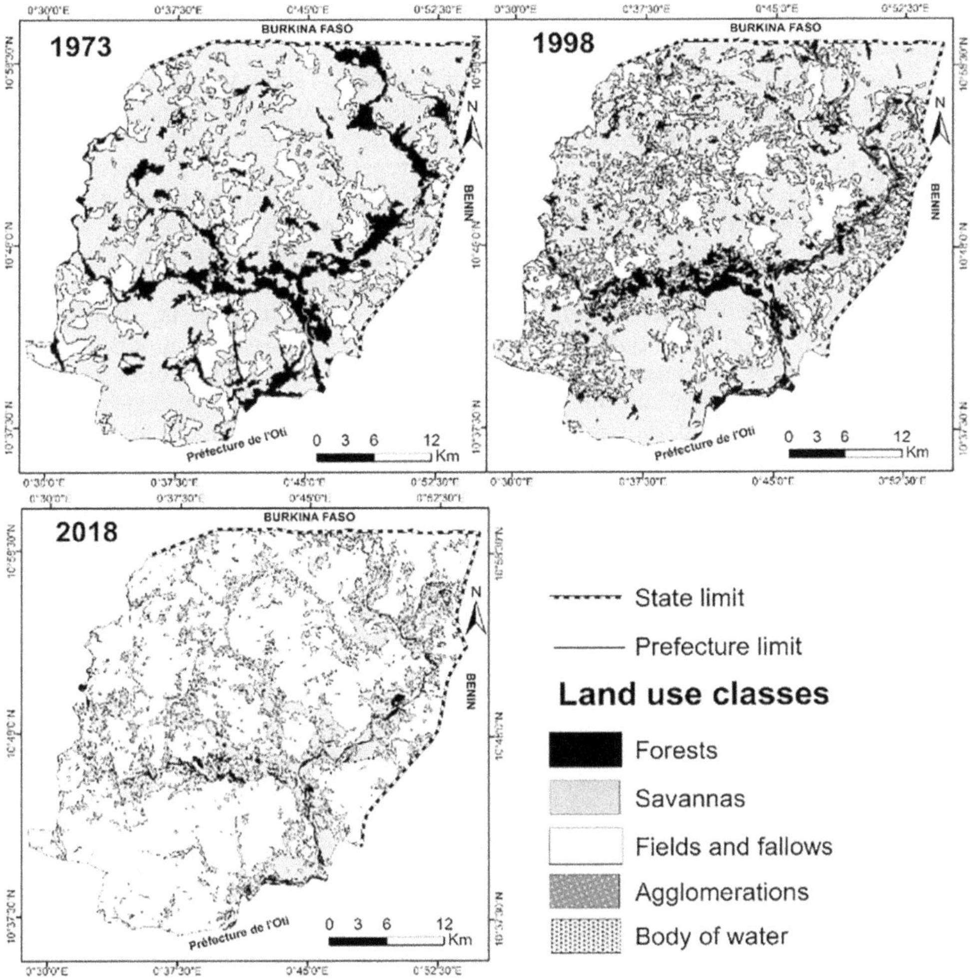

Figure 13. Land cover and land use dynamics in Kpendjal Prefecture between 1973-2018 according to LANDSAT satellite data.

The geoecological characteristics of different plant groups studied allowed to underline that the specific plant biodiversity is of importance and that the discrimination and local distribution of vegetation groups is also influenced by the geomorphic and edaphic factors. Soil texture and the topographic situation play a major role in the variations of environmental and structural parameters. The diversity index of Shannon (*H*) varied between 3.85 ± 0.4 and 3 ± 0.5. The lowest value is obtained in the vegetation groups established on sandy and clayey soils. On interflows and on watersheds the dominance of a small number of species was evidenced. The greatest diversity values were observed in the vegetation groups of down-slopes, on sandy and loamy to clayey soils. The same trend was found for the equitability index of Pielou (*E*). The density, the basal surface and the total number of collected plant species depend on the specific environmental site conditions. The lowest diversity values were observed in the plant groups growing on the mid-slopes and on shallow soils.

These results are consistent with those of Demakou (2009) who also highlighted the edaphic and geomorphic conditions in the discrimination of vegetation groups. The texture of the soil is the main environmental site factor that controls the floristic diversity, the structure of the distribution of the vegetation groups. The structure by class of diameter is dominated by the individuals with thin diameters, which underlined that ecosystems in northern Togo are deeply disrupted by humans.

The spatiotemporal dynamics showed a regression of savannah vegetation types followed by a strong increase of agricultural land use (fields, fallows and settlements). The regressive development of the former dry forest cover is coupled with the decrease in wood potential. Spatiotemporal change in land use modified structural parameters of vegetation. Soils with no vegetation cover are subject to runoff leading to landscape degradation. Other works accomplished in the region, Demakou (2009) and Djené (2015), confirm these results and allow to justify the regressive dynamics of vegetation types in the region.

10.6 CONCLUSION

This research enabled to identify 184 plant species distributed into 106 genera and 44 plant families in Kpendjal Prefecture in northern Togo. The demographic plant structure showed a high representativeness of small trees with thin diameters and a low distribution of greater, older trees. This was considered as a clear evidence of landscape overexploitation. The analysis of the determining factors of landscape degradation and vegetation loss showed that triggering aspects, such as climate variability and climate change are the most prominent ones.

REFERENCES

Abdourhamane, H., Morou, B., Rabiou, H., Mahamane, A. 2013. Caractéristiques floristiques, diversité et structure de la végétation ligneuse dans le Centre-Sud du Niger: cas du complexe des forêts classées de Dan kada Dodo-Dan Gado. *Int. J. Biol. Chem. Sci.* 7(3): 1048–1068.

Addra, T.C., Fahem, A.K., De Jong, T. & Mank, T. 1984. *Atlas du développement Régional du Togo*. Lomé: Minister du Plan et de l'Industrie.

Birot, P. 1965. *Geographie physique de la zone intertropicale (à l'exclusion des régions arides et semi-arides)*. Paris: C.D.U.

Braun-Blanquet, J. 1932. *Plants sociology*. New York, London: My Graw Hill.

Brunel, J.F., Hiepko, P. & Scholz, H. 1984. *Flore analytique du Togo. Phanérogames*. Eschborn, Berlin, GTZ.

Delpech, R. 1993. Deux clés essentielles pour la gestion raisonnée des prairies permanentes: typologie des peuplements et bioindicateurs. *Fourrages* 133: 3–21.

Demakou, Y. 2009. *Paysages végétaux, flore et dynamique des écosystèmes savanicoles du Nord-Togo (Région des Savanes)*. Doctoral thesis, Université de Lomé, Lomé.

Dimobé, K, Wala, K., Batawila, K., Dourma, M., Woegan, Y.A. & Akpagana, K. 2012. Analyse spatiale desdifférentes formes de pressions anthropiques dans la réserve de faune de l'Oti-Mandouri (Togo). *VertigO - la revue électronique en sciences de l'environnement* 14. URL: http://journals.openedition.org/vertigo/12423 [accessed 17/03/2020].

Djené, N. 2015. *Contribution à l'étude de la végétation xérophile du secteur nord-est de la Région des Savanes*, Frontier, S & Pichod-Viale, D (eds.). Écosystèmes: structure, fonctionnement, évolution.

Graef, F. & Haigist, J. 2001. Spatial and temporal rainfall variability in the Sahel and its effects on farmer's management strategies. *Journal of Arid Environment* 48: 221–231.

Hiernaux, P., Ayantunde, A.A., Kalilou, A., Mougin, E., Gérard, B., Paup, F., Grippa, M. & Diaby, B. 2009. Resilience and productivity trends of crops, fallows and rangelands Southwest Niger: impact of land use, management and climate changes. *Journal of Hydrology* 375(1-2): 65–77.

Holdridge, L.R. 1947. Determination of World Plant Formation from Simple Climate Data. *Science* 105: 367–368.

Janicot, S. 2012. Etat des recherches sur l'attribution de la variabilité décennale récente en Afrique de l'Ouest. *Cahier d'Outre-mer* 260: 163–477.

Justice, C.O., Townshend, J.G.R., Holben, B.L. & Tucker, C.J. 1985. Analysis of the phenology of global vegetation using meteorological data. *International Journal of Remote Sensing* 10: 1539–1561.

Kankpénandja, L. 2002. *Contribution à l'étude géomorphologique de la plaine alluviale du Kpendjal (Région des Savanes)*. Master thesis: Université de Lomé.

Kent, M. & Coker, P. 1992. *Vegetation description and analysis. A Practical Approach*. London: Behaven Press.

Kpongou, G.1994. *Rôle du reboisement dans la protection de l'environnement dans l'extrême Nord-Togo (Région des Savanes)*. Master thesis, Université de Bénin, Lomé.

Le Barbé, L. & Lebel, T. 1997. Rainfall climatology of the Hapex-Sahel region during the years 1950–1990. *J. Hydrol.* 188/189: 43–73.

Lemoine, G. 1976. Restauration forestière du Nord-Togo. Annotations sur le développement. *Rapport techn. PNUD-FAO*. Lomé: PNUD-FAO.

Magurran, A.E. 1988. *Measuring Biological Diversity*. Oxford: Blackwell Publishing.

Marchal, D. & Rondeux, J. 1995. Comment estimer la surface d'un peuplement? *Coll. Fiches techniques 3*. Gembloux: Faculté des Sciences Agronomiques de Gembloux.

MERF, Ministère de l'Environnement et des Ressources Forestières 2011. *Politique forestière du Togo*. Lomé: MERF.

Petit, M. & Addra, K.M. 1978. Géomorphologie des formations greseuses du Nord-Togo. *Annales de l'Université de Bénin V(1) série lettres*, 145–163.

Poss, R. & Rossi, G. 1987. Système de versants et évolution morphopédologique. *Z. Geomorph. N.F.* 31: 21–43.

Raunkiaer, C. 1934. *The life forms of plants and statistical plant geography*. London: Oxford University Press.

Roerink, G.J., Menenti, M. & Soebboer, Z.S.U. 2003. Assessment of climate impact on vegetation dynamics by using remote sensing. *Physics and Chemistry of the Earth* 28: 103–109.

Runge, J. 1990. Morphogenese und Morphodynamik in Nord-Togo (9°-11°N) unter dem Einfluß spätquartären Klimawandels. *Göttinger Geogr. Abh.* 90: 1–115.

Thiombiano, A. 2005. *Les Combretaceae du Burkina Faso: taxonomie, écologie, dynamique et régénération des espèces*. Doctoral thesis. Université de Ouagadougou, Ouagadougou.

Trochain, J.L. 1957. Accord Interafricain sur la définition des types de végétation de l'Afrique tropicale. *Bulletin de l'Institut d'Études Centrafricaines*, nouvelle série, Brazzaville (Congo): 55–93.

Ward, J. H., Jr. 1963. Hierarchical grouping to optimize an objective function. *Journal of the American Statistical Association* 58: 236–244.

White, F. 1986. *La végétation de l'Afrique: mémoire accompagnant la carte de végétation de l'Afrique*. Paris: ORSTOM, UNESCO.

CHAPTER 11

Climate change and agricultural organisations in Togo: Comparative analysis of cooperatives in Ogou Prefecture, Atakpamé, Plateau Region

K.M. Flevi & M. Nantob
University of Lomé, Lomé, Togo

ABSTRACT: Climate change and the Sustainable Development Goals (SDGs) are closely linked in the framework of international development efforts. The challenge of finding effective and efficient ways to mitigate and adapt to the effects of ongoing climate change is embedded in national development strategies, including stakeholders' needs in local communities. Located some twenty-five and seventy kilometers east of the town of Atakpamé in the Ogou Prefecture (Plateau Region, Togo), the Lonlon cooperative in Bétoè has implemented less costly and effective local mechanisms to combat the effects of climate change. However, socio-cultural heritage can be an obstacle to the implementation of adaptation strategies. It was evidenced that the local techniques for increasing resilience against climate change developed by the cooperative Lonlon had a favorable echo within the Bétoè community, unlike that of Okougnohou where socio-cultural burdens stifle Konto cooperative initiatives. In fact, the Lonlon cooperative in Bétoè succeeded in gradually establishing an approach furnished with tact and skills enabled the community to adopt technical field schools; composting for soil regeneration; introduction of new seeds, and use of improved stoves.

11.1 INTRODUCTION

In order to make development resources available to the current generation and accessible to future generations, all actions to improve living conditions are focused towards the achievement of the United Nations Sustainable Development Goals (SDGs). To achieve this ultimate goal, particular emphasis is placed on climate change and its challenges as evidenced by the succession of international climate agreements aimed at reducing the emission of greenhouse gases, notably the Kyoto Protocole in 1997; Copenhagen in 2009; and Paris in 2015 (De Perthuis & Jouvet 2015).

Today, it is clear that climate change is correlative to the occurrence of certain natural calamities or 'unusual' ecosystems requiring adaptation and mitigation efforts manifested by the emergence of new behaviors. After a study commissioned by the African Development Bank (AfDB 2013), it is shown on several axes that climate change leads to

> *"(I) the silting up of water courses and water bodies, (ii) the acceleration of land, water and ecosystem degradation, (iii) the erosion of biological and genetic diversity, (iv) the loss of cropland, the decrease in pasture areas and the halieutic potential with the corollary the reduction of agricultural, pastoral, silvicultural and piscicultural yields, and (v) the increase of the impoverishment of the populations"* (AfDB 2013: 4).

Recognising the vulnerability of societies to the threats of climate change, multisectoral interventions aim to strengthen the organisational capacity of populations, especially those living in rural areas, to increase their resilience and thereby contribute to the improvement of their living conditions.

However, it must be noted that despite the international summits and agreements to which Togo has subscribed, the effects of climate change continue to significantly affect all sectors of human life. It emerges from the foregoing that the attitudes and behavior of development actors are struggling to fit properly into international policies to fight climate change. Strategies for building community resilience to climate change threats are not prioritized and implemented in national development plans. Initiatives are being taken at the local level to get communities to put less pressure on the environment. Thus, several local structures, including Farmer Organisations (FO) serve as a channel for transmitting awareness messages on the protection of the environment and resilience of the population to cope with the effects of climate change. To this end, an "intern dynamics" (Balandier 1986: 17) has been developed in the Bétoè community (located east of Atakpamé, Plateau Region of Togo), which is impelling grassroots development, which consists of organising rural communities into small organisational structures (cooperatives, stakeholder groups) and providing them with the necessary services, technical and financial support to ensure the socio-economic development of their members in an appropriate environmental context. In this spirit, Belloncle (1982) highlights the importance of local structures by emphasizing that

> *"Village associations" constitute the "only possible basis for development at the service of African populations"* (Belloncle 1982: 9).

This article focuses on the cooperative Lonlon which has set up a local device for the protection of the environment including field school techniques, composting, promotion of ecological stoves in the locality, awareness on the need for reforestation unlike its counterpart Konto. Hence the questions to know: What are the impacts of the adaptation and mitigation strategies of the effects of climate change developed by the Lonlon and Konto cooperatives in the localities of Bétoè and Okougnohou? What are the obstacles to the success of adaptation and mitigation strategies for the effects of climate change in the communities of Bétoè and Okougnohou?

It is hypothesised that climate change adaptation and mitigation strategies implemented by the cooperative Lonlon contribute to improving the living conditions of the Bétoè community, whereas, under the effect of socio-cultural constraints, the said strategies initiated by the Konto cooperative in Okougnhou were not successful.

This paper aims to evaluate the impact of adaptation strategies to the effects of climate change developed by the cooperative Lonlon on the living conditions in the community of Bétoè. The presentation of the results obtained in the field are structured around the socio-cultural perceptions of climate change in the communities of Bétoè and Okougnohou, and the strategies of adaptation to the threats of climate change.

11.2 METHODOLOGY

Scientific research is based on an objectively defined approach that makes it possible to collect and process data from the field. For this research, the interactionist paradigm of Crozier & Friedberg (1977) is retained. It is based on three principles including the existence of a margin of maneuver for all the actors, the creation of different opportunities for all the actors, and finally the understanding of the functioning of the organisation through the reconstruction of the subjective logic of each actor confronted with the determinations of the global system or local systems of powers that Crozier & Friedberg (1977) calls "concrete systems of action" (quoted by Ferréol & Noreck 2003: 171). Methods used are documentary research, quantitative and qualitative approaches. Methodology is adopted and structured as follows: the research site; the target population and sample, and methods and tools for data collection.

11.3 STUDY AREA

The villages of Okougnohou and Bétoè located in the canton of Akparé and Ountivou in the Prefecture of Ogou (Plateau Region, east of Atakpamé), were used as case studies. In addition to the native population that are respectively the Ifè and the Adja-Èwé, one counts in these villages several ethnic groups in particular the Èwé, Ana, Kabyè; Kotokoli, and others. One meets here traditional religion, Christianity and Islam. The inhabitants live on traditional agriculture, processing of agricultural products and charcoal production. The first is located some 25 km east from Atakpamé, administrative town of the Plateau Region, the second village is situated at a distance of some 70 km east from Atakpamé.

Despite religious diversity, the way of life in these localities is strongly determined by traditional religion and its constraints, in particular the regularisation of life by totems and practices devoid of scientific bases.

11.3.1 Target population and sampling

By means of a reasoned sampling technique, 90 respondents including 38 and 52, respectively from the localities Bétoè and Okougnohou, were interviewed. Individual interviews were also organised with the village leaders and the technical partner represented by the NGO AVOCH (*Association of Volunteers of Christian and Humanitarian Works)* as facilitator.

11.3.2 Data collection

According to Grawitz (1999), the method leads to the reaching of truths. In addition to documentary research, which made it possible to build the theoretical base, the quantitative and qualitative methods respectively led to the design of a questionnaire and an interview guide. The quantitative data are processed using SPSS software.

11.4 SOCIOCULTURAL PERCEPTIONS OF CLIMATE CHANGE IN THE COMMUNITIES OF BÉTOÈ AND OKOUGNOHOU

Climate change is a phenomenon that results from human behavior vis-à-vis the natural environment. The exercise of the daily activities of the individual considerably affects the normal order of natural phenomena and consequently causes climatic anomalies leading to irregular seasons and global warming. Indeed, the scientific explanation of this phenomenon relates to anthropic works that manifest themselves in various ways.

In rural communities strongly determined by mechanical solidarity (Durkheim 1968) the perception of a phenomenon is derived from the socio-cultural heritage acquired through the process of socialisation which determines the attitudes, ideas and behaviors shared by the members of the community. Society, at the same time as the material results of these behaviors, manufactured objects. Beyond particularisms and societal diversity, the aim is to highlight the influence of institutions and customs on the personality (Ruth Benedict 1950, Margaret Mead 1963, Ralph Linton 1968, Abram Kardiner 1969; cited by Abeles 2012). The objective explanation of the phenomenon of climate change relating the actions of communities to the environment is not accepted and appropriate in the same way in these two localities (Bétoè and Lonlon) due to the attachment of a large part of the population to traditional values and stereotypes.

In the rural communities that were the focus of the study, the view or the ideas about climate change have a foundation in traditional heritage. This is to highlight the influence of culture which hinders the objective observation of the facts leading to a rational analysis in order to find appropriate solutions. In the same vein, a respondent relates:

> *At home here, people do not yet have the culture of development. They continue to believe that poverty is inevitable and that it is the gods who would like their fate to be so. For example, if it does not rain here, instead of being explained by our behavior in relation to the environment, we consult rather the oracles and we find spiritual causes. A second example: 25 members of the cooperative Konto are initiated and do not exercise any rural activity every Friday and Sunday, while the field work of the cooperative takes place on Fridays. In conclusion, tradition weighs all its weight on the emergence of peasant organisations in our village* (Source: December 2018, field interviews).

It is recapitulated that certain African cultures, because of their constraining nature, imitate a collective vision imposed on social agents, and place themselves at the antipodes of all innovations. Any change or initiative outside the socio-cultural heritage framework is reprimanded and marginalized. Based on the above, it is clear that culture is an obstacle to social emancipation and self-fulfillment in the Okougnohou community. Indeed, the causes of the phenomenon of climate change cannot be found in everyday practice if we don't delete prejudices and preconceptions based on tradition which is "one of the major obstacles to the development of African societies" (Amouzou 2009: 20).

Socio-cultural burdens have a psychological impact on the socio-economic emergence of African societies. In analysing the African context, Kabou (1991: 100) considers that development and its accessories are perceived as "the thing of the white", "an anti-African phenomenon". Through his postulate Meister 1977 (quoted by Bélloncle 1982: 18) points out that:

> *"Development can only be done by individuals. And since traditional structures oppose the emergence of "autonomous" individuals, the most urgent task is to break them down. Everything that goes in the direction of "disintegration" will be excellent, all that tends to confront them will be "negative".*

The author calls into question the socio-cultural bases that constitute, according to him, "the main obstacles to development". For this reason, he recommends a "deterioration of traditional structures and values" (Meister 1977: 42). The analysis of the negative influence of traditional practices on the socio-economic development of African communities has been the subject of scientific research on various aspects.

Indeed, some link the underdevelopment of Africa with socio-cultural perceptions. At the rank of the latter is Etounga-Manguelle (1991: 58) who states that:

> *"For us, the global cause, the unique cause, the one that is at the origin of all the deviations, is the African culture, characterised by its self-sufficiency, its passivity, its lack of ardor to go to meet the other cultures before the latter impose themselves on it and crush it, its incapacity once the harm is done, to evolve in contact with them without falling into an abject mimicry".*

For this author, it is necessary to redress the African culture through "a cultural adjustment program" (Etounga-Manguelle 1991: 1). That said, African culture must be sifted through careful analysis in order to prune out approximate and non-favorable values for socio-economic development and objective analysis of social or natural phenomena such as climate change. This review has the merit of placing itself above the plethora of cultures that make up Africa and instilling a culture of development in African societies especially in rural communities imbued with traditional prejudices and logic. The anchoring of sociocultural constraints in the perceptions of African societies is not conducive to the improvement of living conditions, especially in rural areas where precariousness is growing, as the World Bank (1995) argues, "in almost all the countries of the world, it is in rural areas that the poor are the most numerous and the most deprived" (cited by Bonnamour et al. 1996: 17). The situation of poverty in rural areas is perceived in terms of sociocultural constraints.

Thus, there are analysts who combine the precarious living conditions of African communities with cultural impacts. Speaking of the Togolese context, it must be said that

> *"Traditional communities are not immune to the effects of poverty. Sometimes regarded as a divine will by some, sometimes as a curse by others, poverty has no difficulty in developing in the population since it finds its roots in the heart of the culture of these peoples"* (Amouzou 2009: 55).

This highlights the poverty resulting from the weight of the culture of the Togolese communities which is manifested by certain practices and conceptions resistant to social change. In rural areas, the influence of culture is seen in many areas of social, political and economic life. With regard to the villages of Bétoè and Okougnohou, the sociocultural factor influences the objective taking of positions to facilitate the adoption of new behaviors to adapt to the threats of climate change. However, the level of influence of culture differs from one community to another. The perceptions of the inhabitants of the village of Bétoè about climate change are not the same if one speaks to the community of Okougnohou.

The information gathered in the field demonstrates in more than one way that the communities of Bétoè and Okougnohou are spatially and culturally close but develop different perceptions about the notion of climate change. For some, climate change is the result of the transgression of sociocultural norms. This conception gives rise to a system of interactions between the individual and his/her closest environment insofar as the action of the one causes the other to react differently and this permanently. As part of this dynamic, the respondents are all aware of the existence of interactions between the individual and his environment. However, it is the nature of the discourse on the causes of climate change that differentiates the position of the Okougnohou community over that of Betoè.

This work shows that global warming, environmental degradation and irregular seasons, resulting in low agricultural productivity, hence increasing poverty, are only a reflection of human behavior vis-à-vis nature. In addition, the effort to understand the attitudes of people to the environment is placed under the socio-cultural or spiritual angle at the expense of physical contact or pressure exerted by humans on nature. The comments made by a surveyor from Okunoghou's Konto group in these terms are quite suggestive:

> *"Here at home, nothing happens at random. The irregularity of recent rains is explained by the fact that people behave unworthily towards the ancestors and protective gods of the village. People deliberately break the prohibitions on the pretext that they are Christians and have nothing to do with tradition. When the village elders draw their attention to the totems, they refuse to obey but the consequences of their actions (the scarcity of rain and low productivity) spare no one. These consequences are manifested in other ways, in particular the devastation of agricultural produce by locusts and the appearance of diseases (smallpox, cholera, etc.)"* (Source: December 2018 field interview).

The analysis of this passage, like many others, gives a non-negligible place to the sociocultural element in the perception of climate change and the underlying causes of this phenomenon. Indeed, in communities strongly influenced by mechanical solidarity (Durkheim 1968), like Okougnohou, tradition is at the center of everyday life so that all reasoning revolves around it. This is the position of some authors who argue that culture is a necessary crutch for achieving inclusive and sustainable development. In other words, all development initiatives must blend into the sociocultural wanderings of the beneficiary community to ensure their chances of success. In Africa, social behaviors are determined by a community-specific cultural model that serves as a tool for assessing the level of social integration. That said, the sociocultural element is a very important, if not indispensable, factor in the analysis of social cohesion and change in a given community. All development actions in Africa, which do without the socio-cultural factors of the communities, risk losing the thread. Going in the same direction, Akibodé (1996) emphasizes that:

> *"The widely acknowledged failure of development policies in Africa today has created a new interest in culture. Recently occupied for secondary phenomena, cultural problems are now considered by a growing number of researchers as an essential dimension of social change, or even as the fundamental dimension of development"* (quoted by Amouzou 2009: 15).

The process of development, culture or the socio-cultural dimension is an essential link if not indispensable to social development. However, it is undeniably obvious to relativize because socio-cultural norms can obstruct an objective analysis of the facts as is the case with the perceptions collected in the village of Okougnohou about climate change. Unlike the interviewees from Okougnouhou village, those from Betoè village have developed a more rational approach.

In the Bétoè community, the information gathered is relatively focused on ancestral practices, and therefore more or less embraces modernism and its values. To this end, there are relatively objective and relevant remarks that highlight the physical action of humans on the environment in the context of the effects of climate change. Inspired by the comments collected in the field and especially those of a respondent who caught our attention:

> *"At Bétoè, we certainly worship our protective gods who help us when we invoke them. We also make the effort to respect their taboos. However, we manage to distinguish between things that belong to the gods and those that belong to our actions as a man, that is to say, a farmer living in agriculture. With regard to global warming and the scarcity of rainfall, we are aware that the cause comes from ourselves. If we continue to destroy the trees to make charcoal without planting and put bush fire to destroy forests and plantations, it is that we contribute to the degradation of the environment which in turn causes climate change"* (Source: December 2018 field interview).

It must be emphasised that the phenomenon of climate change, its impacts and approaches to solutions are well known to rural communities like Bétoè, which works to adopt new behaviors in order to increase its resilience. Moreover, the Bétoè community has a room for maneuver characterised by its ability to abandon the socio-cultural yoke in a dynamic of interactions to touch the real problems of climate change. Peer outreach is the preferred channel for passing messages on new provisions to adapt or mitigate the threats of climate change.

11.5 STRATEGIES TO MITIGATE OR ADAPT TO CLIMATE CHANGE THREATS

According to the Climate Change Convention, the phenomenon of climate change is defined as:

> *"climate changes that are attributed directly or indirectly to human activity that alter the composition of the global atmosphere and that add to natural variability of climate"* (Demaze 2009: 85).

Climate change is the result of natural and anthropogenic factors. It manifests itself in various forms depending on the geographical environment. The impacts of climate change on daily life at the beginning of the 21st century, are the subject not only of a dense literature, but also and especially of the organisation of international meetings in order to take resolutions in the direction of reducing emissions of greenhouse gases. Everywhere on the planet earth, the implications of climate change are particularly noticeable, the one month advance of the date of the harvest in France (Chuine et al. 2004, Seguin 2007) or the drift of giant icebergs without forgetting:

> *"Moderate increase in overall average temperature of 1.5 °C; the exposure of small islands to the rise in sea level or the destabilization of thawed ground in the far north, not to mention the increasing severity of extreme events (storms, droughts, heat waves, floods), well identified by reinsurance companies"* (Godard 2010: 287).

Indeed, the different characteristics of climate change are real in any ecosystem and in which resilience strategies are differently developed to make life possible. In Togo, as in many African states, human pressure on the environment is a daily practice that is observed through, among other things, river pollution, deforestation, wind and water erosion, the impacts of which are rising temperature, irregular rainfall and low agricultural productivity. In this logic, Sultan (2008) recalls that since 1960 there has been a general drop in rainfall, representing about 20% of the annual average across the continent. In Niamey, for example, it fell to 490 mm of rain of average per year from 1970 to 1990, against 650 mm from 1950 to 1970. Two strong impacts emerge: this decrease had brutal effects on the flows of the major rivers (the minimum runoff of the low water period has passed to 3 m^3/s of the Niger River at Niamey) and on the filling of reservoirs for agricultural or hydroelectric purposes. Secondly, associated with other factors, this drought also had repercussions on crop yields and on livestock. Thus, these impacts are perceptible on other human activities, notably fishing. The author underlines that there is a decrease of 50% of the fish caught in the interior delta of Niger in 10 years. Compared to livestock, the great drought of 1984 decimated more than 80% of the cattle herd in the Niger interior delta.

Every human society is dynamic and evolves in accordance with changes in its environment. Whether in urban or rural areas, strategies are developed to deal with the threats of climate change, or to adapt local actions to new constraints.

Engaged upstream, this effort is relayed by state structures, civil society organisations (CSOs) and local organisational structures that aim to maintain the healthy environment through daily acts of public interest. On the local level, this action is implemented and integrated into the operating plan of some agricultural organisations, like the Lonlon cooperative in Bétoè, which is trying to adapt its farming activities to the new climatic conditions.

The agricultural activities of the cooperative Lonlon and indirectly the community of Bétoè and its surroundings, have undergone profound changes ranging from the shift of seasonal planting deadlines, to the introduction of new agricultural varieties that adapt relatively to the threats of climate change as one interviewee points out:

> *"three years ago in Bétoè, everyone masters the seed period and the requirements linked to the selected varieties (corn; soya; bean) thanks to the techniques of field-schools"* (Source: December 2018, field interview).

The analysis of the interviews conducted in this locality reveals that the school-field practice has made it possible to restore degraded lands. This land degradation "reduces the productive capacity of cultivated soils" (FAO 2009, cited by Birame et al. 2014: 21) and consequently the purchasing power of the farmer, hence precariousness. To do this, composting techniques are developed that produce organic matter to restore the fertility of cultivated soils. This technique consists in mixing poultry, goat or sheep waste with feces recovered from ECOSAN toilets in order to provide nutrients necessary for the development of cultivated plants.

It should be added that the community management of natural resources is another technique implemented which favors a better conciliation between breeders and peasants in a climate of social cohesion quieted. Birame et al. (2014: 24) will argue that community-based natural resource management is "a consensual response to define rules for access and sustainable use of shared pastoral resources". The recurrence of disputes between nomadic herders and sedentary peasants is mitigated by the awareness campaigns organized by the members of the Lonlon cooperative, with the dual aim of getting the former to better supervise their herds and to respect the defined transhumance corridors by the State, and the latter to accept cohabitation with the shepherds and specially to avoid violence on them or their property.

Finally, the Lonlon cooperative is raising awareness of the need to protect forests through the use of improved wood stoves in order to significantly reduce deforestation and provide a sustainable solution to energy problems. In the same vein, it must be remembered that the analysis of the information gathered has shown that, unlike the Konto group of Okougnohou, the Lonlon cooperative of Bétoè has been able to put in place effective local strategies that respond to the threats of climate change in the Bétoè community through outreach. That said, these strategies only made sense in the new organisational context imposed by climate change, which creates a system by generating different opportunities for all actors. The members of the Lonlon cooperative took the opportunity offered by the context of climate change to gain notoriety in the Bétoè community. To understand how the social organisation works in the context of climate change, the following construction of the cooperative's subjective logic is necessary:

> *"Our vision is to find less costly local solutions to the threats of climate change in our locality"* (Source: December 2018 field interview).

Stakeholders are confronted with the determinations of the global system or local systems of power that Crozier & Friedberg (1977) calls "concrete action system" (quoted in Ferréol & Noreck 2003: 171).

During this confrontation, it transpires that the local system of power developed by the Lonlon cooperative was born out of the failings of the global system, particularly the socio-cultural heritage, whose philosophy is more or less favorable to social change. The methodological approach that served as the basis of analysis for this study made use not only of qualitative data but also of quantitative data.

Table 1. Distribution of respondents according to appreciation of local strategies to adapt to the threats of climate change (Source: December 2018 field data).

	Peasant Organisation					
	Konto		Lonlon		Total workforce	
Assessments of local adaptation strategies to climate change threats	Number	%	Number	%	Number	%
Very bad	2	3.84	0	0	2	2.22
Bad	30	57.69	0	0	30	33.33
Poor	15	28.86	0	0	15	16.67
Good	5	9.61	3	7.9	8	8.69
Quite well	0	0	20	52.63	20	22.22
Excellent	0	0	15	39.47	15	16.67
Total	52	100	38	100	90	100

The comparison of the field data collected in the two localities made it possible to obtain statistical data and to detect trends relating to the respondents' position regarding the performance achieved in terms of adaptation strategies to the threats of climate change. It is listed in Table 1 the assessment of the local strategies implemented by each agricultural organisation.

The scientific reading of the statistical data compiled (Table 1) reveals globally that the respondents are not satisfied with the local strategies of adaptation to the threats of climate change. This is also the position of the Okougnohou respondents, who believe that, in the face of changes in the climate and irregular rainfall, the Konto group has not taken effective adaptation action. However, adaptation actions to climate change threats developed by the cooperative Lonlon reflect the assessments of the interviewees in this locality. That said, the subjective logic put in place and materialized by the awareness on the protection of the environment on the one hand, and the use of improved stoves, new seeds and the system of composting in the regeneration of soils of other on the other hand, are progressively appropriate and integrated into the socio-cultural heritage so as to lead both to understand that man is the main actor of climate change. The set of climate change adaptation strategies introduced by the Lonlon cooperative in the Bétoè community is causing a change in mentality and attitudes.

The two agricultural cooperatives that were the subject of this research, only the Lonlon cooperative in Bétoè has developed mechanisms to respond to the threats of climate change.

11.6 DISCUSSION

The investigations conducted in the localities Bétoè and Okougnohou have highlighted the influence of sociocultural constraints on the social emancipation of the Lonlon cooperative and the Konto group. This result is more observable in the locality of Okougnohou where ritual practices prevent certain members of the Konto group to freely go about their business as testified by the words of a respondent:

> *"[...] 25 members of the group Konto are initiates and do not exercise any field activity every Friday and Sunday, while the field work of the grouping takes place on Fridays. In conclusion, the tradition weighs all its weight on the emergence of peasant organisations in our village"* (Source: December 2018 field interview).

By examining the weight of African culture on the socio-economic development of communities, it is obvious that this result goes beyond the limits of this study. Several previous works have highlighted the cultural dimension in efforts to analyse the causes of Africa's lagging development. The works of Kabou (1991) and Etounga-Manguelle (1991) have attempted to demonstrate, through relevant analyses, that African culture is a 'bottleneck' to the development of this continent insofar as development and its implications are perceived, like "the thing of the white" (1991: 100). If traditional practices are unfavorable to innovation in the Okougnohou village, what about the village of Bétoè?

Like many rural communities in Togo, life in the village of Bétoè is regulated by socio-cultural heritage. The members of the Lonlon cooperative have been able to integrate their innovations into the socio-cultural context without, however, upsetting the established social order.

The comparative research of the Lonlon and Konto cooperatives resulted in the strategies implemented to respond to the threats of climate change. In this regard, it should be noted that only the cooperative Lonlon has developed field school strategies, composting, community management of natural resources and finally the promotion of improved wood stoves. These strategies are guided by technical support from the NGO *Association of Volunteers of Christian and Humanitarian Works* (AVOCH) and the technical services of the State including the *Institute of Consulting and Technical Support* (ICAT) and the Institute. *Togolese Agricultural Research Institute* (ITRA). They aim to reduce emissions of greenhouse gases, and especially to adapt to the consequences of climate change. This study carries insufficiencies that give rise to a few limits.

The limits of this article relate first of all to the choice of the methodological approach that served as a basis for analysing the data of the investigations, notably the interactionist paradigm of Crozier & Friedberg (1977). This paradigm, in the name of the interrelationships between individuals, is on the side of Weber's comprehensive sociology which aims to reflect social realities from the rationality of individual actors. This sociology places more emphasis on the qualitative aspect of field data (Dubar 2010). Secondly, it must be remembered that the non-exhaustive nature of the survey did not allow us to obtain various points of view that could enrich the analyzes. The focus groups were not carried out in order to confront the positions of each other and to deepen the analyses. However, these limits do not undermine the approach taken nor do they prevent the extrapolation of the results obtained to the population as a whole.

11.7 CONCLUSION

The issue of climate change is proven and manifests itself in a variety of ways depending on the geographical environment and tropical ecosystems. In sub-Saharan Africa there is an increase in air temperature, the irregularity of rainfall giving rise to low agricultural productivity, leading to the disappearance of certain animal and plant species. To cope with these effects, which undoubtedly provoke the precariousness of the populations, a common vision is set up structured around the Sustainable Development Goals (SDGs) which advocate the socio-economic development flourished in a favorable environmental context. All development actors, regardless of their natures and areas of intervention, are required to take ownership of this global approach.

At the local level, efforts are being made in the climate change sector through local strategies promoted by agricultural organisations such as the Lonlon cooperative and the Konto group. But it is clear that the approaches to solutions to the threats of climate change proposed clash with socio-cultural realities especially in the village of Okougnohou. The present study on Lonlon and Konto cooperatives produced two main results: Field investigations evidenced that the socio-cultural heritage is a handicap to the implementation of climate change adaptation strategies, especially in the Okougnohou village. Despite the weight of tradition at Bétoè, the cooperative Lonlon has succeeded in gradually establishing a process furnished with tact and addresses that allowed the community to implement appropriate field school techniques; introducing composting methods for soil regeneration and new seeds, as well as the use of improved stoves for cooking.

In view of these results, it is appropriate to assert that the research hypothesis was confirmed that climate change adaptation and mitigation strategies implemented by the Lonlon cooperative contribute to the improvement of life in the Betoè community; what is not the case of the cooperative Konto of Okougnohou.

In some communities, development initiatives by local structures with the technical and financial support of development partners are trustful and have less difficulty in being appropriated despite the influence of traditional culture. In other communities, socio-cultural factors are opaque and stand in the way of any initiative whatever its nature. This being said, contrary to the "top and down" model (Théveniaut-Muller 1999) of development perceived as rigid and authoritarian, grassroots development is gradually contributing to the "social change" in rural areas perceived as:

> *"Any observable transformation in time, which affects, in a way that is not only temporary or ephemeral, the structure or functioning of the social organisation of a given collectivity and modifies the course of its history"* (Rocher 1969: 326).

REFERENCES

Abeles, M. 2012. *Culturalism*. URL: http://dvdweb.universalis.fr/v16/plusSurInternet=E952411SI2, accessed on 03/03/2019.

AfDB, African Development Bank 2013. *Integrated Program for Agricultural Development and Adaptation to Climate Change in the Niger Basin (PIDAACC/BN): Preliminary Studies of Multipurpose Infrastructure*. Tunis: AfDB.

Amouzou, E. 2009. *The influence of culture on development in Black Africa*. Paris: Editions l'Harmattan.

Balandier, G. 1986. *Senses and Powers*. Paris: Editions PUF, 334p.

Belloncle, G. 1982. *The Peasant Question in Black Africa*. Paris: Editions Karthala,114p.

Birame, D., Henry, M.., Bounama, D., Oumar, S., Ousmane, F. 2014. Towards climate-smart agriculture in Senegal: *A compendium of good adaptation and mitigation practices, on behalf of the science-policy dialogue platform on climate change, agriculture and food security in Senegal (C-CASA), S/C Directorate of Agriculture, Institution focal point, Dakar*: 180.

Bonnamour, J., Bret, B., Bruneau, M., Chavrey, J-P., Diry, J-P., Drain, N M., Landy, F., Renard, J., Raison, J-P., Rey, F. & Vaudois, J., 1996 Agricultures and campaigns in the world. *Agricultures and campaigns in the world*. Paris: Editions SEDES.

Crozier, M.& Friedberg, E. 1977. L'acteur et le système. Les contraintes de l'action collective. Editions Seil, Paris.

Chuine, I., Yiou, P., Viovy, N., Seguin, B., Daux, V. & Roy Ladurie, E. 2004. *Grape harvest dates and temperature variations in France since 1370*. Paris: Editions Nature.

De Perthuis, C. & Jouvet, P.-A. 2015. What financial strategy for a climate agreement in Paris 2015? Financial Economics Review 117: 31–48.

Demaze, M. 2009. International Conventions on the Environment: States of Ratifications and Commitments of Developed and Developing Countries. *Geographical Information* 73: 84–99.

Dubar, C. 2010. *Sociology-the big currents*. URL: http://assistance.universalis.fr/universalis.2012, accessed on/03/20/2020.

Durkheim, E. 1968. *Education and Socialism*. Paris: Edition PUF.

Etounga-Manguelle, D. 1991. *Does Africa need a cultural adjustment program?* Paris: Editions Nouvelles du Sud.

Ferréol, G. & Noreck, J.-P. 2003. *Introduction to Sociology*. Paris: Editions Armand Colin.

Godard, O. 2010. Adaptation to climate change file. This ambiguous adaptation to climate change. *Natures Sciences Societies* 18: 287–297.

Grawitz, M. 1999. *Lexique des sciences sociales*. Paris: Editions Dalloz.

Kabou, A. 1991. *Et si l'Afrique refusait le développement?* Paris: Editions L'Harmattan.

Meister, A. 1977. *Participation for Development*. Paris: Editions Ouvrières.

Rocher, G. 1969. *Introduction to Sociology*. Montreal: Editions HURTUBISE HMH Ltd.

Seguin, B. 2007. Global warming and its consequences for viticulture. *Conference communication Global warming: what are the likely impacts on vineyards? Dijon, University of Burgundy, March* 28–30.

Sultan, B. 2008. *20 study of variations and climate change in Africa*. URL: https://fr.m.wikipedia.org, accessed on 03/26/2020.

Theveniaut-Muller, M. 1999. *Local development, a political response to globalization*. Paris: Editions Desclée de Brouwer.

CHAPTER 12

Preliminary results of spatiotemporal mapping of vegetation and land use patterns in the Assoli Prefecture, Kara Region, Togo

I. Djeri, Z. Koumoï & T. Soussou
University of Kara, Kara, Togo

ABSTRACT: In Togo uncontrolled exploitation of forest resources for firewood and charcoal production as well as clearing of land for agricultural and industrial production has contributed between 1987-2015 to the progression of savannah woodland ecosystems and degradation of open dry forest (ODF) resources. In the Assoli Prefecture in the southern Kara Region the problem of land degradation is particularly striking. By applying remote sensing techniques with Landsat data (1987, 2001, 2015) unsupervised spatiotemporal mapping of land use patterns produced preliminary results: a significant area regression of open dry forests (ODF from 22% to 10%), and of forest-savannah mosaics (FSM from 49% to 32%), whereas agricultural fields, fallows, bare soils (FF) and urban areas (UA) showed over 30 years a strong increase in area (from 19% to 46% and from 10% to 12%, respectively). The main factors responsible for these changes are traditional shifting cultivation (slash and burn agriculture), charcoal and firewood production, increase in local population and maintenance and construction of roads.

12.1 INTRODUCTION

Like other sub-Saharan countries, Togo has been suffering for several decades a decrease in rainfall combined with increasing anthropogenic pressure on natural resources. Consequently, this has led to a severe degradation of arable land and forest ecosystems. According to Koumoï et al. (2013) and Badjana et al. (2014), degradation of natural resources in northern and central Togo is mainly due to unsustainable management practices such as overgrazing, wildfires, deforestation and urbanisation. These farming practices and population growth led to an overall degradation of the vegetation cover. Togolese forests show an annual average loss of 2.6% for mountain forests, and 3.7% for other dense forests. Uncontrolled exploitation of forest resources for firewood and charcoal production and clearing for cultivation of agricultural products have led to an increase in open savannah woodland ecosystems (World Bank Togo 2010).

In the Assoli Pefecture (Kara Region, northern Togo) studies on the evolution of the vegetation cover showed that plant cover underwent remarkable transformations between 1956 and 1996. Tchamiè & Bouraima (1997) proved on the Soudou-Dako Plateau in the Atakora mountain range that the forested area decreased from 44,812 hectares to 35,041 hectares between 1956 and 1976, a loss of forest from 46% to 37% of the total area. This corresponds to a deforestation rate of 9772 hectares over 20 years (489 hectares per year). According to Bouraima (1996), forest cover in

Assoli amounted in 1996 just 25,056 hectares (27%) against 68,696 hectares (73%) overgrown by savannahs, fallows and fields. During the period 1976-1996 Assoli Prefecture lost in average 1.6% of its forests per year, this corresponds to 560 ha/year.

This highlights the vulnerability of vegetation and plant cover due to peasant practices. Changes in land use are commonly known, but their magnitude differs strongly according to the type of pressure and the ecological and environmental conditions (Soulama 2014). This paper aims to present first and preliminary results of spatiotemporal mapping and evaluation of the dynamics of different types of land cover between 1987-2015 in the Assoli Prefecture in the southern Kara Region (Figure 1).

12.2 STUDY AREA

Assoli Prefecture is 937.5 km^2 large and one of the seven prefectures in the Kara Region. It is limited to the north by the Kozah Prefecture, to the east by the Republic of Benin, to the south by the Prefecture of Tchaoudjo (Central Region) and to the west by the Prefecture of Bassar (Figure 1). Bafilo, with ~30,000 inhabitants, is the administrative centre of the Assoli Prefecture, located between 9°12′ to 9°30′ northern latitude and 1° to1°24′ eastern longitude.

The relief of this district is composed of two main geomorphological units: north of the Dako-Bafilo-Soudou line (Figure 1) extends a plain and south of this line an uplifted peneplain surface. The altitudes vary according to whether one is on the Alédjo Plateau, in piedmont or in the Kara depression. Thus, the plateau of Alédjo in

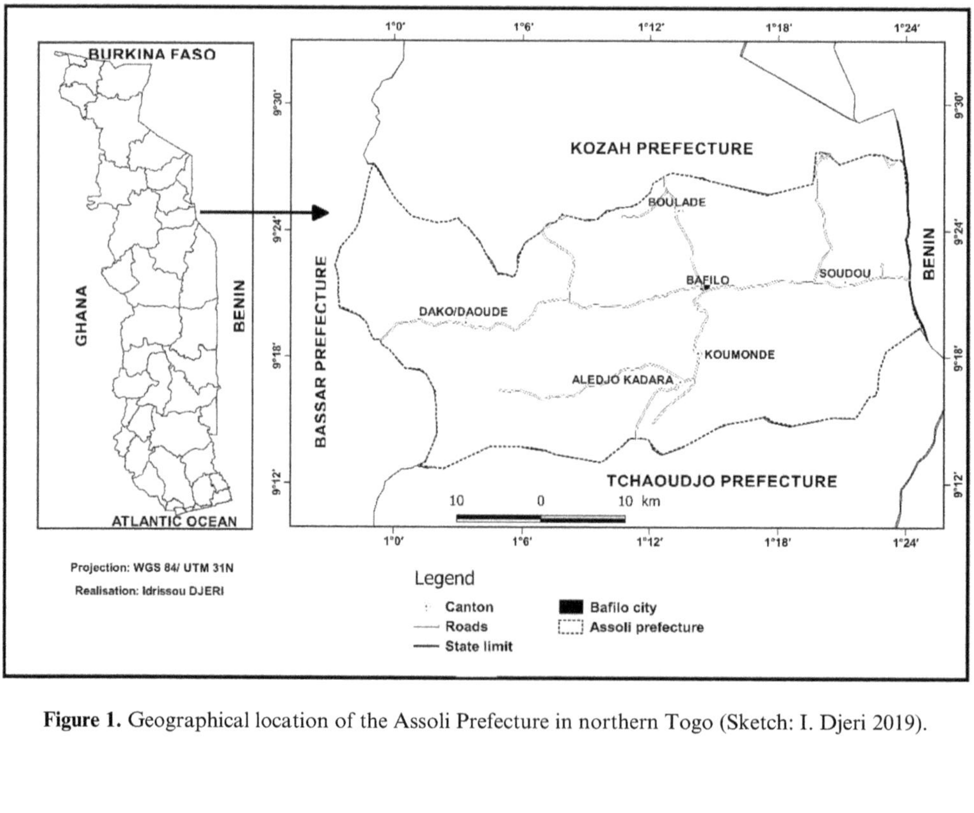

Figure 1. Geographical location of the Assoli Prefecture in northern Togo (Sketch: I. Djeri 2019).

the south culminates at the village of Alédjo Kadara at 850 m asl. (Djeri 2014). Oriented with an East-West strike direction, the Tchaoudjo Massif belongs to the Soudou-Dako chain; it has a tabular form with a few linear fluvial incisions (Alfa 2007). All streams take their source on the plateau which forms the watershed between northern (Volta-Oti basin) and southern Togo (Mono basin). These rivers show reduced (or any) runoff during the dry season, especially those of the Kara plain.

The tropical Sudanian-type climate is characterised by the alternation of a long dry season from November-May and a rainy season for the rest of the year. The configuration of the relief and the severity of the climate determines major vegetation types. Assoli Prefecture is covered with open and dry forest (ODF) and forest-savannah mosaics (FSM). They are growing on different soils, but there are more or less homogeneous entities allowing to classify them into three groups (DRDAT/Kara 2009). In Assoli Prefecture, the typical soils are lithic, tropical ferruginous, and ferralitic soils. These soils are exploited by the dominant ethnical group Tem. The Tem are the main indigenous ethnical group, and they live in harmony with the Kabyè, Lamba, Bassar and Peulh who migrated from neighboring districts to Assoli to develop agriculture. The coming together of these social and ethnic groups explains the rapid growth of the overall population in Assoli Prefecture. The main activities are food crops production accompanied by logging, production of charcoal, firewood and construction wood.

12.3 METHODS

Land use and vegetation cover were classified into four groups: Open Dry Forest (ODF), Forest-Savanna Mosaic (FSM), Field and Fallow (FF) and Urban Area (UA, including bare soil). The spatiotemporal analysis of three Landsat images of 1987 (TM), 2001 (ETM+) and 2015 (OLI/TIRS) and field surveys confirmed the proposed (preliminary) classification scheme of four land use/vegetation cover units by unsupervised image processing for 1987, 2001 and 2015 in the Assoli Prefecture.

12.4 RESULTS

12.4.1 Spatiotemporal analysis of land use and vegetation cover (1987-2015)

Patterns of land use and vegetation types in Assoli Prefecture obviously showed a certain and significant surface dynamic over the period from 1987 to 2015 (Figures 2a, 2b, 2c). Overall, clear/open and dry forest and forest-savannah mosaics decreased in size while fields and fallows and soils and urban areas increased (Figure 3).

Unsupervised spatiotemporal image interpretation over 28 years resulted in the following observations concerning vegetation cover and land use in the study area: open dry forest has lost 9662 ha in area, an overall regression rate of 12%. The most sensitive period was between 2001 and 2015, when the open dry forest lost 7639 ha in area, a decline of 9%. The forest-savannah mosaic also showed a regressive evolution of a 17% rate, which represents 13,696 ha. This vegetation cover suffered a sharp deterioration in the first period from 1987 to 2001 (Table 1).

The fields and fallows and the bare soils as well as urban areas showed an increase in surface. It should be noted that the area of fields and fallows has evolved more from 14,909 ha in 1987 to 37,159 ha in 2015, an increase rate of 28%. Bare soils surface and urban areas increased from 8149 ha to 9366 ha, at a rate of 2% (Table 1).

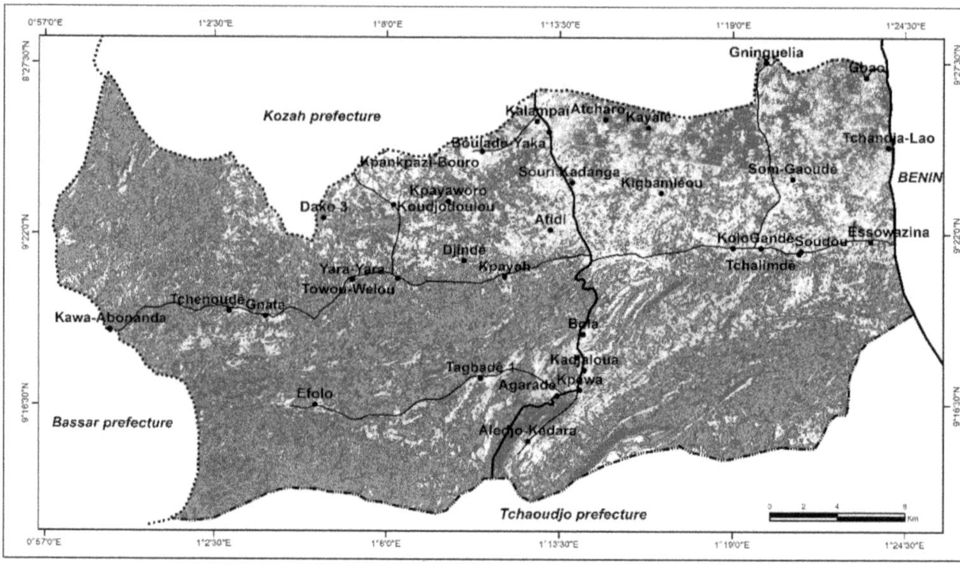

Figure 2a. Land use and vegetation cover in the Assoli Prefecture in 1987: more than two-thirds of the area are covered by open dry forests (ODF) and forest-savannah mosaics (FSM) (see more map details in Figure 2c).

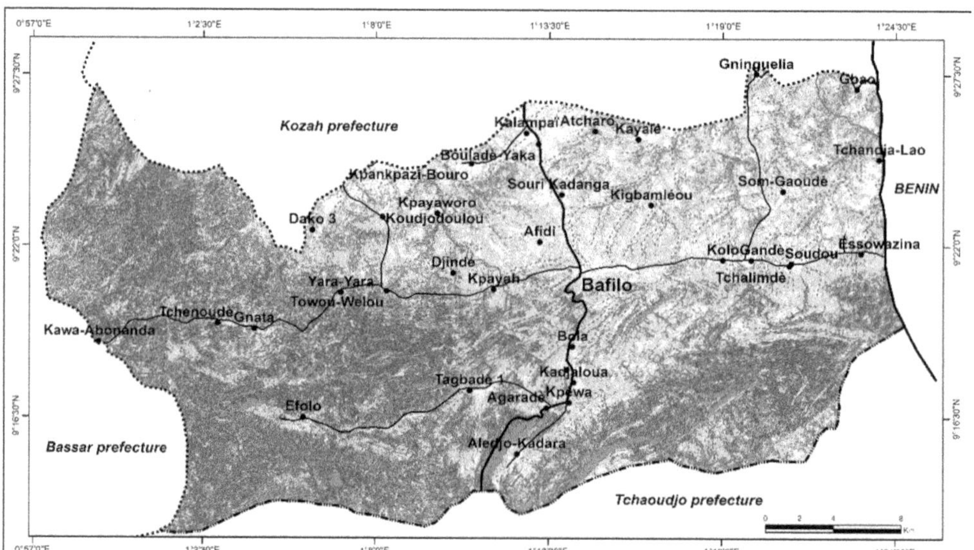

Figure 2b. Land use and vegetation cover in the Assoli Prefecture in 2001: natural to semi-natural dry forest and savannah ecosystems have been significantly reduced by growing population and increased agriculture (see more map details in Figure 2c).

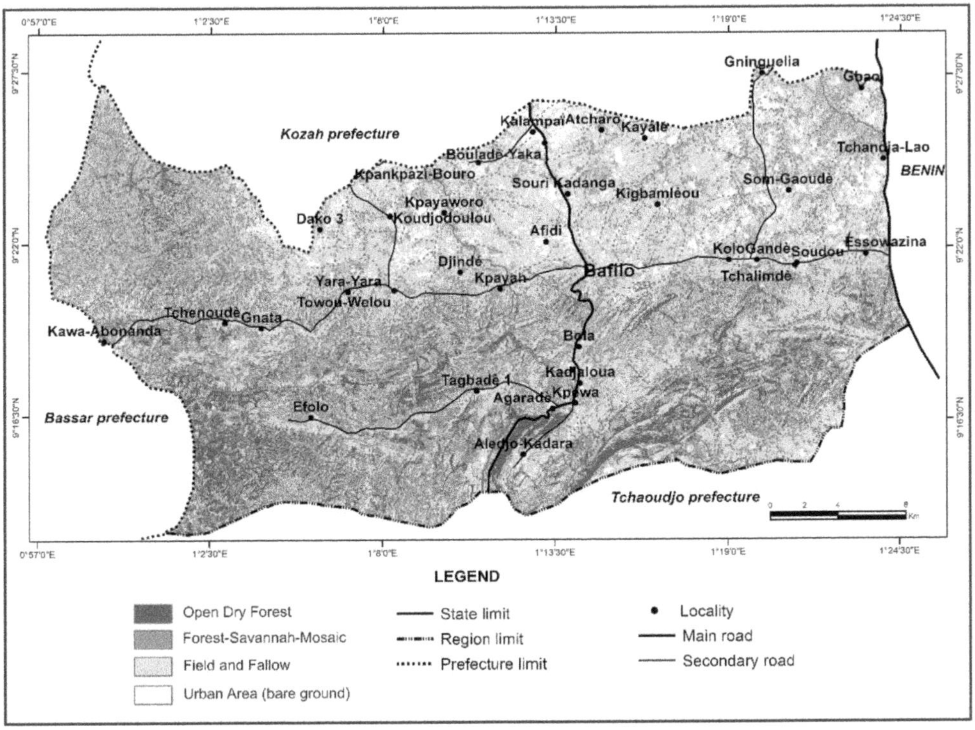

Figure 2c. Land use and vegetation cover in the Assoli Prefecture in 2015: over 50% of the area are finally used by traditional slash and burn agriculture. Open dry forests have been strongly degraded because of bushfires (hunting), charcoal and firewood production.

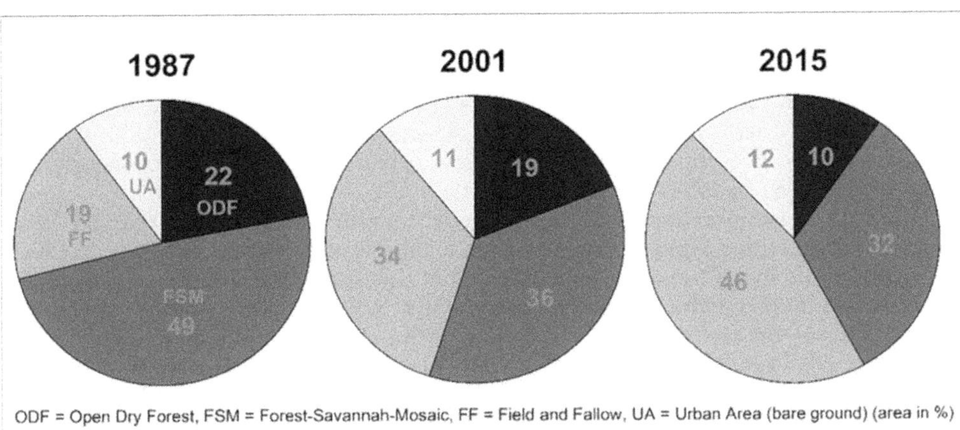

Figure 3. Comparative development and dynamics of land use and vegetation cover (areas in percent) in Assoli Prefecture for time slices 1987, 2001 and 2015 according to Landsat satellite images (compare Figures 2a, 2b, 2c, Table 1, sketch: J. Runge).

Table 1. Dynamics of land cover units (Processing of Landsat images 1987, 2001 and 2015).

	Area (ha)			Area variation	
Land cover units	1987	2001	2015	(ha)	%
Open dry forest	17,562	15,539	7900	-9662	-12
Forest-savannah mosaic	39,637	28,762	25,941	-13,696	-17
Fields and fallows	14,909	27,523	37,159	+22,250	+28
Bare soils and urban areas	8149	8486	9366	+1217	+2
Total	80,257	80,310	80,366		

The types of land use and vegetation cover identified in 1987 to 2015 in the Assoli Prefecture are open and dry forests, forest-savannah mosaics, fields and fallows, bare soils and urban areas. Open and dry forest has a regressive evolution while fields and fallows, bare soils and urban areas have increased during the observation period.

12.4.2 Assessment of the evolution of land use units and vegetation cover from 1987 to 2015

In 1987 open dry forest occupied 22% of the total area of Assoli (Figure 3). Open dry forests (ODF) covered to a large extent the Alédjo plateau, south of the Dako-Bafilo-Soudou line (Figure 2a). Forest-savannah mosaics (FSM) or 'savannah woodland' in the plain, north of the Dako-Bafilo-Soudou line, and in the extreme south of the study area occupied 49%. In contrast, in 1987 human-influenced areas, such as agricultural fields and fallows (FF) came up to 19% of the area. Only 10% of land use was assigned to bare soils and sealed surfaces of urban areas (UA). The latter two land cover types were found not only throughout the plain north of Bafilo, but also on the Alédjo plateau (Figure 2a).

In 2001 the land and vegetation cover of Assoli Prefecture already showed strong modifications (Figure 2b, 3): open dry forest which still represented 22% in 1987 (Figure 2a, 3) was now found with 19% of the total area of Assoli Prefecture. Forest-savannah mosaics (FSM) occupied 49% in 1987, and 'shrinked' to a surface share of 36%. Fields and fallows (FF), which covered 19% in 1987, strongly increased up to 34% in 2001. Bare soils and urban areas (UA) increased slightly (from 10% in 1987 to 11% in 2001).

In 2015, vegetation and land cover (Figure 2c) were clearly dominated and 'overstamped' by increasing human activities. Figure 3 illustrates that open dry forest (ODF) was reduced to only 10% in area, unlike the initial situation (1987) where it still occupied 22% (19% in 2001) of the Assoli Prefecture. Only a few isolated 'shreds' of this dry forest remained on the Alédjo plateau (Figure 2c). Forest-savannah mosaics (FSM) occupied in 2015 still 32% against 36% in 2001 (and 49% in 1987, Figure 3). Fields and fallows (FF) have evolved by 34% in 2001 to 46% in 2015. Bare soils and urban areas (UA) only slightly increased up to 12% of the area, an increase of 2% towards the initial situation of 1987 (Figure 2a, 3). The latter land use class dominated the plain north of Bafilo, and the southeastern part of the Alédjo plateau.

The assessment of land cover units' evolution over almost 30 years gave evidence that no 'real' natural vegetation cover has progressed since 1987 in the Assoli Prefecture. An ongoing push back of natural and quasi-natural

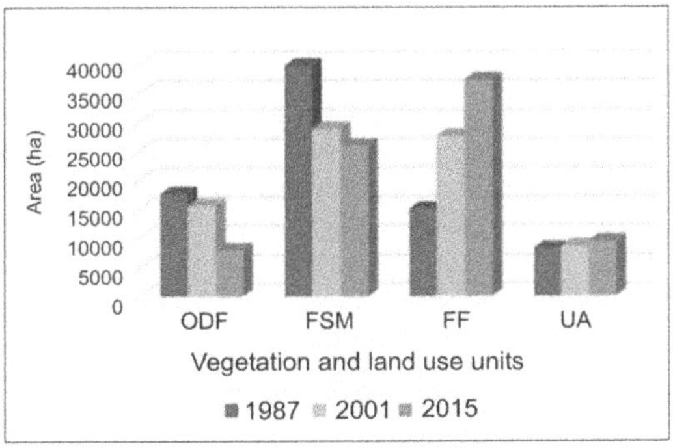

Figure 4. Assessment of the evolution of vegetation and land use units from 1987 to 2015 according to Landsat satellite images (for abbreviations see Figure 3).

ecosystems is obvious. Figure 4 expresses the changes of vegetation and land cover in absolute numbers (hectares). Degradation and size reduction of open dry forest (ODF) and forest-savannah mosaic (FSM) is well recognisable. Natural vegetation cover decreased from 17,562 ha in 1987 to 7900 in 2015 whereas human induced land cover (FF) proves a steady growth from 14,909 ha in 1987 to 37,159 ha in 2015 (Figure 4).

12.5 DISCUSSION

The findings have revealed a regressive evolution of vegetation and land cover in the Assoli prefecture. It demonstrates the important role of humans on the dynamics of land use.

The results highlighted that open dry forest and forest-savannah mosaics (semi-natural vegetation types) experienced a decline in area in favor of anthropogenic land use. Several authors have achieved similar results including Soussou (2009) in the Litimé plain, Takou et al. (2012) in the western Plateau Region of Togo; Koumoï et al. (2013) in central Togo, Kpedenou et al. (2016a) in southeastern Togo, Kpedenou et al. (2016b) in the Yoto prefecture, Nourou et al. (2016) in communes of Banikoara and Karimama in Benin, and Sidi et al. (2018) in the limestone quarries in south-eastern Togo.

In Togo, the degradation of natural vegetation cover in favor of anthropogenic vegetation cover is accentuated by humans, particularly logging and shifting cultivation, correlated with strong population growth. Population of Assoli district strongly increased by 19,044 inhabitants between 1981 and 2010, a growth of 37% in 29 years (RGPH4-Togo 2010). This led aside of the extension and increase of traditionally cultivated land, to more bare surfaces and a general extension of sealed urban areas, but also to a most probably irreversible degradation of ecosystems.

Logging causes the degradation of closed forest cover. Thus, Tchamiè (1998), World Bank Togo (2010), Koumoï et al. (2013, 2017) and Soussou (2013) showed that local use of certain species as timber (*Milicia excelsa, Pterocarpus erinaceus, Khaya senegalensis*), building wood (*Khaya senegalensis, Pterocarpus erinaceus*) and for the manufacture of charcoal (*Prosopis africana, Khaya senegalensis, Anogeissus leiocarpus, Burkea africana*) supports the reduction and destruction of natural forests and woodlands.

Agriculture is a big consumer of land resources. The increase of agricultural land in the Assoli Prefecture for cultivation of cassava, maize, yam and tomato leads to a destruction of the natural plant cover and causes land degradation by erosion. Authors who have reached similar results are Yonkeu et al. (1995) in the Bourzanga Basin; Bouraima (1996) in the Assoli Prefecture; Tcheinti-Nabine (2000) in the Central Region of Togo; Houngbo (2008) in the Adja Plateau in Southern Benin; Soussou (2012, 2013) in the Litimé; Gnongbo (2013) in the Litimé; Tatoni & Soussou (2013) and Djeri (2019) in the Assoli Prefecture.

12.6 CONCLUSION

This study evaluated spatiotemporal dynamics of land use patterns and vegetation cover in the Assoli Prefecture. It was carried out from the interpretation of three Landsat images from 1987, 2001 and 2015. Four land use/vegetation cover classes have been used. These are open and dry forest, forest-savannah mosaic, fields and fallows, bare soils, and urban areas. The land use maps for three annual time slices over almost 30 years have been processed and resulted in absolute and relative statistics of land cover. Open dry forest and forest-savannah mosaic evidenced a regression rate of 29% between 1987 and 2015 while fields and fallows and bare soils and urban areas experienced a 30% increase rate. The regional analysis of the land use maps highlighted regional focus areas of land use and interconnected landscape degradation in Assoli Prefecture. Therefore, traditional farming is a main factor of the spatiotemporal dynamics of land use in northern Togo.

REFERENCES

Alfa, A.T. 2007. *Koumondè, un terroir montagneux soumis à une crise de l'espace agricole (préfecture d'Assoli)*. Mémoire de Maîtrise en Géographie. Université de Lomé, Lomé.

Badjana, H.M., Selsam, P., Wala, K., Flügel, W.-A., Fink, M., Urban, M., Helmschrot, J., Afouda, A. & Akpagana, K. 2014. Assessment of land-cover changes in a sub-catchment of the Oti basin (West Africa): A case study of the Kara River basin. *Zbl. f. Geol. u. Paläont., Teil I*(1): 151–170.

Bouraima, A. 1996. *Densité de population et évolution du couvert végétal dans la préfecture d'Assoli*. Mémoire de Maîtrise en Géographie. Université du Bénin, Lomé.

Djeri, I. 2014. *La crise agricole en pays Tem: cas du canton de Bafilo*. Mémoire de Maîtrise en Géo--graphie. Université de Kara, Kara.

Djeri, I. 2019. *Pratiques agricoles et dégradation du couvert végétal dans le canton de Daoudè (pré-fecture d'Assoli) au Nord-Togo*. Mémoire de Master en Géographie. Université de Kara, Togo.

DRDAT/Kara (Direction Régionale du Développement et de l'Aménagement du Territoire/Région de la Kara). 2009. Monographie de la Préfecture d'Assoli.

Gnongbo, T.Y. 2013. Mise en valeur agricole et évolution du milieu naturel dans la zone forestière du Litimé (Togo). *Cahiers d'outre-mer, paysannerie africaines et développement* 224: 289–301.

Houngbo, E. 2008. *Dynamique de pauvreté et pratiques agricoles de conservation de l'environne-ment en milieu rural africain: le cas du plateau Adja au Sud-Bénin*, Thèse de Doctorat en Gestion de l'Environnement. Université d'Abomey-Calavi, Abomey-Calavi.

Koumoï, Z., Abdourazakou, A., Djangbedja, M., Boukpessi, T. & Kouya, A. 2013. Dynamique spa-tio-temporelle de l'occupation du sol dans le centre-Togo. *Revue Ahoho de l'Université de Lomé* 10 (7): 163–172.

Koumoï, Z., Boukpessi, T. & Kpedenou, K.D. 2017. Principaux facteurs explicatifs de la dynamique de l'occupation du sol dans le centre-Togo: apport des SIG et des statistiques spatiales. *Revue Ivoirienne de Géographie des Savanes* 3: 252–273.

Kpedenou, K.D., Ahe, P. & Boukpessi, T. 2016a. Dynamique spatio-temporelle de l'occupation/ utilisation des terres dans le Sud-Est Togo: essai de cartographie à l'aide de la télédé-tection spatiale. *Revue Ahoho de l'Université de Lomé* 16 (10): 131–143.

Kpedenou, K.D., Boukpessi, T. & Tchamiè, T.T.K. 2016b. Quantification des changements de l'occupation du sol dans la préfecture de yoto (sud-est Togo) a l'aide de l'imagerie satellitaire Landsat. *Rev. Sc. Env. Univ.* 13: 137–156.

Nourou, T.I., Ousseni, A. & Ismaila, T.I. 2016. Cartographie de la dynamique spatio-temporelle des parcours naturels des troupeaux transushumants dans les communes de Banikoara et de Karima-ma au Bénin (Afrique de l'ouest). *European Scientific Journal* 12 (32): 251–268.

RGPH4-Togo (4e Récensement Général de la Population et de l'Habitat). 2010. Résultats définitifs détaillés. DGSCN, Lomé.

Sidi, I.A., Djangbedja, M., Kpedenou, D.K. & Tchamiè, T.T.K. 2018. Dynamique spatio-temporelle de l'occupation du sol dans les sites d'exploitation de calcaires au sud-est du Togo. *Revue Ivoirienne de Géographie des Savanes* 4: 181–200.

Soulama, S., Kadeba, A., Nacoulma, B., Traore, S., Bachmann, Y. & Thiombiano, A. 2014. Impact des activités anthropiques sur la dynamique de la végétation de la réserve partielle de faune de Pama et de ses périphéries (sud-est du Burkina Faso) dans un contexte de variabilité climatique. *Journal of Applied Biosciences* 87: 8047–8064.

Soussou, T. 2009. *Dynamique forestière dans la plaine du Litimé sous pression anthropique au Togo*. Thèse de Doctorat. Lomé, Marseille: Université de Lomé & Université Paul Cézanne Aix-Marseille 3.

Soussou, T. 2012. Pratiques agricoles et dynamique des agroforêts dans le Litimé au Togo. *Revue Ahoho de l'Université de Lomé* 8: 110–119.

Soussou, T. 2013. Etude des changements morphostructuraux des modes d'utilisation du sol du Liti-mé dans le sud-ouest du Togo (Afrique de l'ouest). *Revue de Géographie de l'Université de Oua-gadougou* 2: 39–53.

Takou, W.P., Boukpessi, T., Djangbedja, M. & Mama, A. 2012. Apports de la télédétection et des sys-tèmes d'information géographiques dans l'étude de la dynamique des paysages végétaux de l'ouest de la Région des Plateaux au Togo. *Rev. Sc. Env. Univ.* 9: 29–48.

Tatoni, T. & Soussou, T. 2013. Déforestation et impacts globaux. Agathe, E. (ed.), Le développement durable à découvert. 110–111. Paris: CNRS Editions.

Tchamiè, T.T.K. & Bouraima, M. 1997. Les formations végétales du plateau Soudou-Dako dans la chaîne de l'Atakora et leur évolution récente (Nord-Togo). *J. Bot. Soc. bot. Fr.* 3: 83–94.

Tchamiè, T.T.K. 1998. La dégradation de l'environnement dans l'extrême nord-Togo: Conditions naturelles et facteurs anthropiques. *Revue togolaise des Sciences*: 28–54.

Tchenti-Nabine, T. 2000. *Pratiques paysannes et dynamiques des systèmes agraires dans la Région Centrale du Togo*. Thèse de Doctorat. Université de Lomé, Lomé.

World Bank Togo 2010. *Analyse environnementale du Togo avec emphase sur les secteurs Agriculture, Energie et Mines*: 5. Washington, D.C.: World Bank.

Yonkeu, S., Da Dapola, E. & Soule, M. 1995. Activités agropastorales et dégradation du couvert vé-gétal dans le centre-nord du Burkina Faso: cas du bassin versant de Bourzanga. *Revue de Géographie*, Université de Ouagadougou 1–15.

CHAPTER 13

Vulnerability of agricultural systems towards rainfall variability between 1965 and 2014 in northern Togo

T. Lamboni & S. Hombre
University of Lomé, Lomé, Togo

ABSTRACT: Tropical agriculture in West Africa is confronted to several constraints of ongoing climate change. Variations in rainfall with disturbed onset and length of rainy reasons are major challenges in the Savannah Region of northern Togo. The agricultural sector, traditional and modern, is especially vulnerable to such climate irregularities. The spatiotemporal analysis of rainfall variability between 1965 and 2014 and the study of physiological needs of selected plants (millet, sorghum, maize and rice) was examined. It was possible to highlight the inter-annual and spatial variability of precipitation, and to detect plant growth stage during which water deficit occurred according to the growth cycle of the crops. The results show that seasonal rains do not always correspond to water requirements of crops during the vegetative cycle. The ecological requirements of crops are not met and their biological cycles are often disturbed. Crops' vulnerability is noticed through the delay in growth and the rate of appearance of tillers, the wilting of plants, the early leafy stage of the aged leaves. Correlation between rainfall and agricultural yields showed a positive relationship for all the crops studied. In conclusion, food production and supply of the consumers are undermined by changes in regional rainfall patterns and disturbed rainfall distribution.

13.1 INTRODUCTION

Sub-Saharan Africa is particularly vulnerable to climate change. The synoptic situation of this geographical area is characterised by disturbance of rainfall deficits amounted between 20% to 30% from the mean values (Diomande et al. 2008). Changes in rainfall 'weaken' ecosystems by making them more vulnerable to disturbance and increase the rate of degradation of biological resources. Agriculture is severely undermined by the random nature of seasonal rains and the decrease in vegetative season duration (Bognini 2011). Therefore, farmers' activities are faced with climate change challenges. These include the destruction of crops by drought and/or by erosion on the fields and the proliferation of weeds. In northern Benin Dimon (2008) reveals that climate threats on agricultural systems lead to more than 50% of crop losses, reduced yields and farm incomes.

In Togo, fluctuations in rainfall (variability) also constitute a risk factor for agricultural production. Traditional and modern agriculture is facing problems characterised by irregularity and decrease in rainfall (Lemou 2014). The Savannah Region in the extreme north of Togo (10°-11° N, Figure 1), is characterised by a unimodal Sudanese tropical climate. Variability in rainfall associated with rising

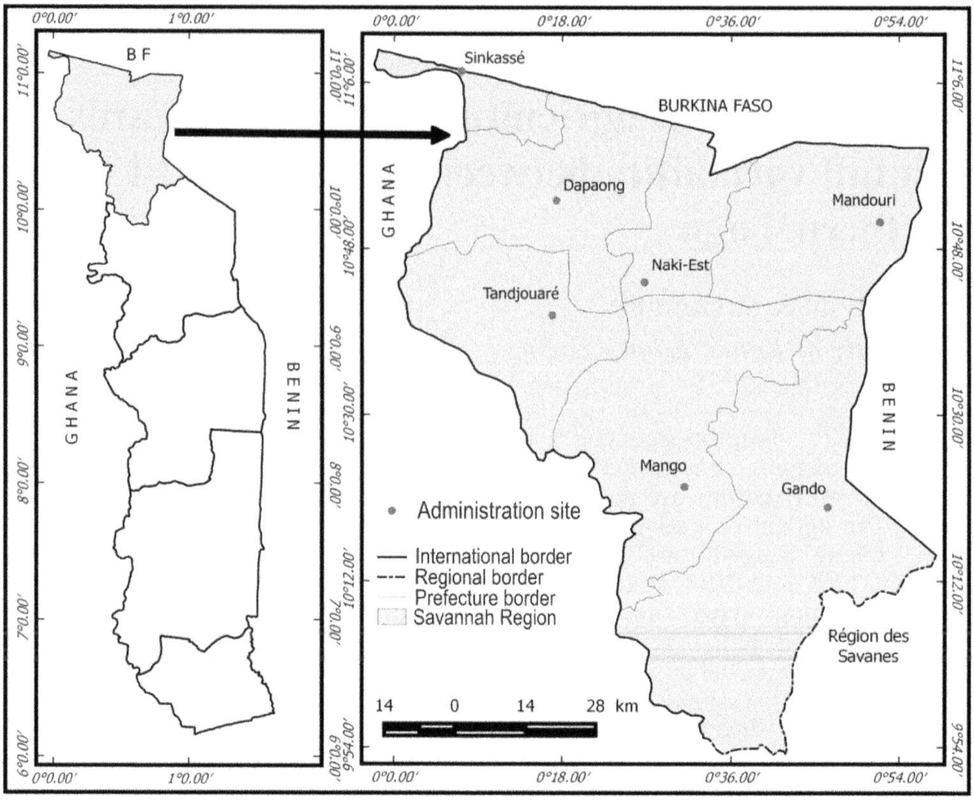

Figure 1. Location map of the study area in northern Togo.

temperatures causes disturbances in crop life cycles. The shortening of the vegetative period is one of the major agroclimatic constraints of rural production. Rain lags, spatial and temporal shifts in rainfall intensities, which are particularly sensitive at the beginning and end of the rainy season, impact the growing cycle of crops. The tendency to dry up and the recurrence of extreme weather make agriculture vulnerable. Water deficits cause, for example, the melting of seeds, poor germination, early drying of foliage. Excess in water causes, among other things, bad plant yeast, crop rot, cracking of flowers and poor grain filling. Whether it is the stress of deficit or excess of water, crops are affected in both cases. This deterioration of agroclimatic conditions affects agricultural production, and consequently food security of the peasant communities.

13.2 METHODS

First of all, scientific literature on the importance of climate for agriculture in the tropics was consulted. Subsequently, data on rainfall in Togo were collected and processed. Data were obtained from the Direction Générale de la Météorologie Nationale (DGMN) in Lomé. Rainfall data originated from the synoptic meteorological stations in Mango and Dapaong. Agricultural data on crop yields were collected from the Togolese Direction des Statistiques Agricoles, de l'Information et de la Documentation (DSID). The parameters used for the analysis of inter-annual variability of rainfall were:

$$\bar{X} = \frac{1}{N} \sum_{i=1}^{n} Xi \tag{1}$$

where $\overline{X}$= Mean; N = number of years; Σ= sum; and X_i = annual rain value, and the formula Anomaly Indices:

$$AI = \frac{(Xi - \overline{X})}{\sigma} \tag{2}$$

where X_i = annual rain value; $\overline{X}$ = series mean; and σ = series standard deviation.

The spatiotemporal rainfall variability (isohyets' shifts during time slices of 10 years from 1965-2014) was illustrated by help of Surfer 12 software (Figure 3).

The Bravais-Pearson correlation coefficient was used to determine the binding intensity or degree of correlation between rainfall variations on agricultural productivity. It is expressed as follows:

$$r = \frac{\frac{1}{N}\sum (x_i - \bar{X})(y_i - \bar{Y})}{\sigma(x) \cdot \sigma(y)} \tag{3}$$

where r = correlation coefficient; x_i and y_i represent the values of the x and y series for year i; $\bar{X}$ and $\bar{Y}$ are the means of X and Y variables; σ(x) and σ(y) are the standard deviations of X and Y series values; N is the total number of observations.

By convention, the relationship between X and Y is perfect if r = 1; very strong if r > 0.8; strong if r is between 0.5 and 0.8; of average intensity if r is between 0.2 and 0.5; low if r is between 0 and 0.2 and nil if r = 0. Graphical representations of the correlation between rainfall and crop yields were developed using EXCEL software. Precipitation maps were designed by using Surfer 12 and QGIS 2.12.0 software (Figure 3).

Social surveys were also carried out to better understand farmers' perceptions of the impact of rainfall variations on crops. A questionnaire was administered to a sample of 115 randomly selected farmers. Twelve persons were interviewed among the Togolese Agricultural Research Institute (ITRA) and the Institut de Conseil d'Appui Technique (ICAT): agricultural technicians, rural development officers, and customary chiefs. This was complemented by empirical field observations. The social survey's findings are not especially highlighted in this paper.

13.3 RAINFALL AND CROP GROWTH

Extreme weather conditions have negative effects on crop growth. Excess water or rainfall deficits lead to poor plant lifting, cracking of flowers, early drying of foliage, and poor seed filling. These physiological disturbances associated with phytopathology highlight crops' vulnerability under difficult humidity conditions when the plant cannot meet its minimum ecological needs. The phenophases are described as follows: G (Germination); L (Lifting); S/F (Spiking/Flowering); F (Flowering); LR/C (Lactic Ripeness) and M (Ripeness) (Figures 4-7). The vulnerability of these crops to water deficit stress is analysed with years of increased rainfall in Dapaong and Mango. This choice shows that even in a very rainy year, it is not uncommon to observe prolonged dry sequences that disrupt crop growth cycle. Data for 2004 were considered for the synoptic weather stations of Dapaong and for 2008 of Mango (Figures 2, 4-7).

13.3.1 Inter-annual changes in rainfall

The temporal variation in rainfall reflects the irregular succession of annual precipitation events. The results of the analysis of the standardised indices (i) in Figure 2 indicate the seasonal rainfall characteristics. They indicate a trend marked by a succession of years to the accumulation of deficit rains when $i < -1$, for the years of excess rains $i > 1$ and years of normal accumulation rains when $-1 < i < 1$. This figure describes non-cyclical alternation of dry and wet episodes. As a result, the studied stations do not have years of deficits and excess rainfall at the same time. The trends and the curve of the moving averages allow us to observe this dissimilar trend which illustrates the phenomenon of variation in cumulated rains between Dapaong and Mango (Figure 2).

Trends and averages define two main rainfall phases at the station of Dapaong: the first phase is marked by a rainfall deficit from 1967-1995 although with slight excesses which occasionally occur from 1965-66, 1974-76, and 1981-82. The second is characterised by a gradual increase in precipitation as reflected in the inverted position of the moving average and linear trend. This period ranges from 1997 to 2014 but sometimes interspersed with years of normal rainfall or slight deficits. From 1965 to 2014, the slightly increasing trend line at Dapaong does not necessarily imply better rainfall conditions for crops. For this reason, it was important to study the vegetative cycles of some crops according to their water requirements by phenophase. At the station of Mango, there were no long dry or wet episodes that were quite distinct but rather a close alternation of rainfall deficits and surpluses with values closer to the average. The trend of Mango describes a very slight decrease in annual rainfall (Figure 2).

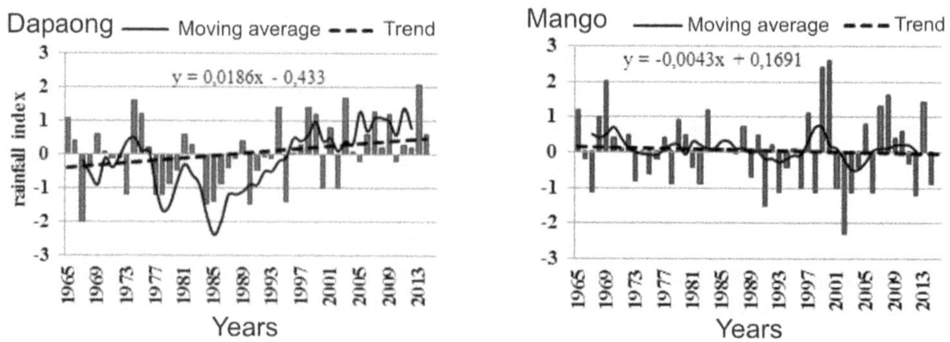

Figure 2. Reduced centred indices and changes in annual precipitations from 1965-2013 for Dapaong and Mango in northern Togo (Source: DGMN Lomé).

13.3.2 Spatial variability of precipitation

The study of isohyets over a 10-year time period since 1965 (Figure 3) led to the decoding of rainfall variations over almost 50 years. The change in decade's averages defines the level of fluctuations in the spatial distribution of precipitations. This variation is significant since the respective averages of the two stations have never been homogeneous for all the studied decades. Time slices in Figure 3 illustrates the change in rainfall from isohyet migration in northern Togo.

Figure 3 shows a 10-year spatial distribution of mean precipitation accumulations. The isohyets of the decades 1965-1974, 1975-1984 and 1985-1994 indicate a North-South decline in the values of isohyets. They range respectively from 1034; 980 and 925 mm at Dapaong while Mango records 1100; 1060 and 1065 mm. However, over the past two decades (1995-2004 and 2005-2014), the trend has reversed. Thus, when the station of Dapaong showed higher isohyets (1090 and1120 mm); the station of Mango was marked by slightly lower isohyets (1040 and 1050 mm). Generally, the spatial distribution shows a more sensitive local variation of rainfall with a decrease in the isohyet values ranging from 1120 mm to 1050 mm at Dapaong and Mango respectively. In fact, the spatial change in rainfall was indicated by the spatial mobility of isohyets which leads to modifications in agricultural calendar.

13.4 IMPACT OF RAINFALL CHANGES ON CROPS

Rainfall disruptions or excess water (flooding) are detrimental to the efficiency of agriculture production. They cause the rotting of seeds, the poor lifting of the plants, decrease in growth, shedding of flowers, the premature drying of foliage, the inhibition of flowering, poor filling of grains, the proliferation of insect pests and bacteria responsible for wilting and plant abortion, yellowing of foliage, and death of plants.

13.4.1 Water needs and vulnerability of millet cultivation

Millet is a well adopted crop for the Sudanese-Sahelian tropical environment. It has a high tolerance to drought and low pH, requires high temperatures during germination and growth, but it is very sensitive to excess water. Although its preferred growing area lies around the 430 mm to 500 mm isohyets, its large genetic diversity allows it to grow between a rainfall range of 200 mm per growing season to 800 mm per growing season (Siband 1983). Table 1 shows water requirements of millet according to the different phases of plant growth.

Millet cultivation shows various forms of vulnerability to moisture abundance or deficiency depending on the stage of growth which is exposed to moisture variability. For example, water deficit stress at the flowering stage of the panicles causes crop abortion and even drying of the panicles that remain wrapped in the terminal leaf of the pampa. There are also repeated flooding of crops surface in the farms of some zones of water accumulation or wetlands where crops are either uprooted or rotted.

The growth of plants is subsequently disrupted and the development of the normally expected size is flawed. The results shown in Figure 4 determine the phenophases during which water deficit stress has disrupted the development of millet cultivation in 2004 in Dapaong.

Water deficit observed in the ripeness phase was quite high because the needed rainfall amount (110 mm) accounted for 50.3 mm only, a deficit of ~50%. The rainfall deficiency between the 1st and 3rd decades in October coincided well with the

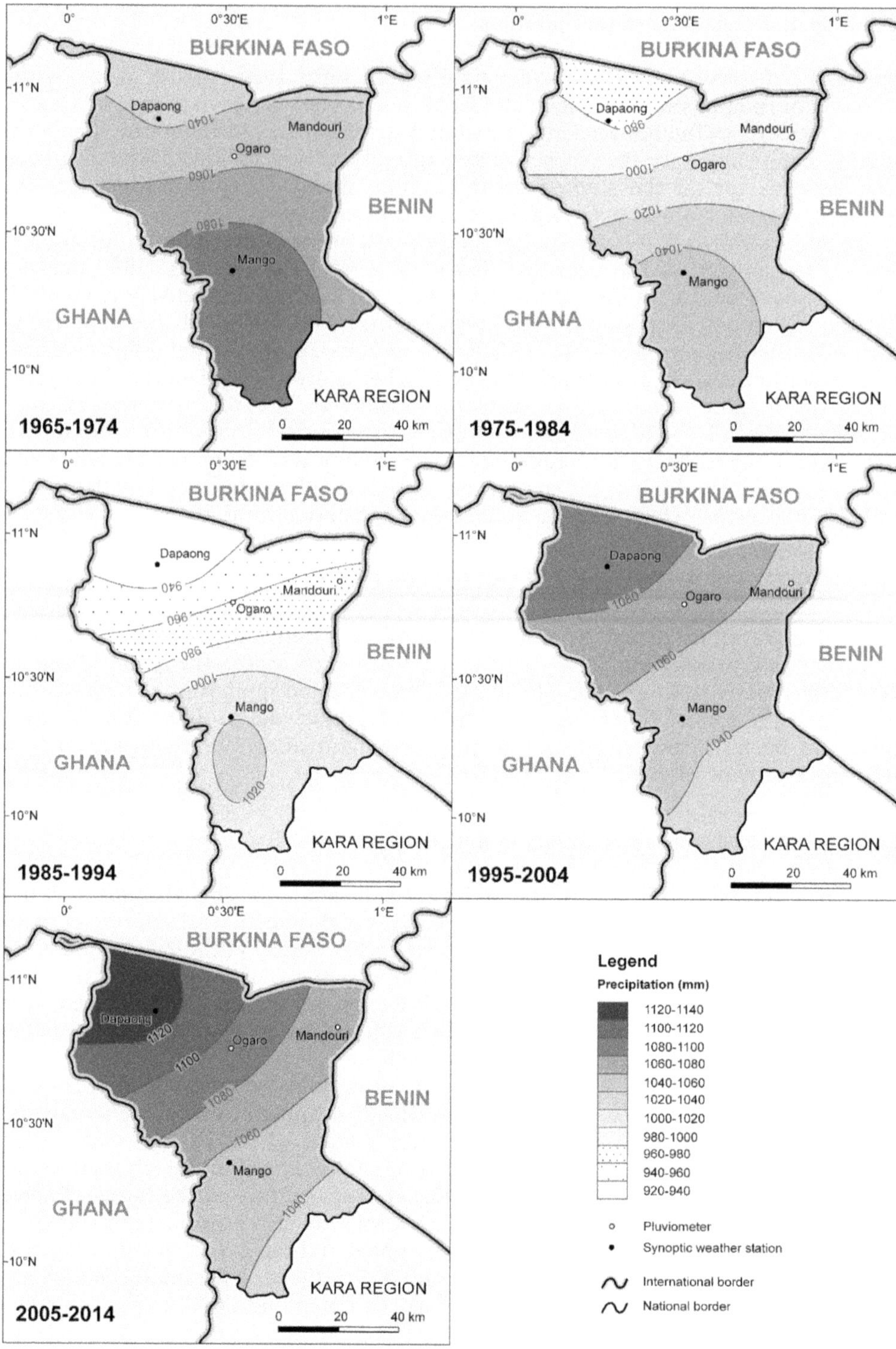

Figure 3. Spatiotemporal variations of precipitation in ten-year time slices since 1965 in northern Togo (Sketch: J. Runge; Cartography: J. Eisenberg, Source: DGMN Lomé).

Table 1. Vegetative cycle, phenophases and water requirement of millet.

Sowing data	Phase 1 (sowing)		Phase 2 (lifting)				Phase 3 (spiking/flowering)						Phase 4 (ripeness)		
3rd decade May – 1st decade June	1	2	3	4	5	6	7	8	9	10	11	12	13	14	15
Decades	2		4					6						3	
Need for water (mm)	6,5	17,2	19,4	23,2	25,3	28,7	31,5	33,1	36,5	37,9	34,6	26,3	18,4	17,2	14,7
Water/ phase	35 mm		130 mm				280 mm						110 mm		

(Source: ITRA-Savannah Region 2011)

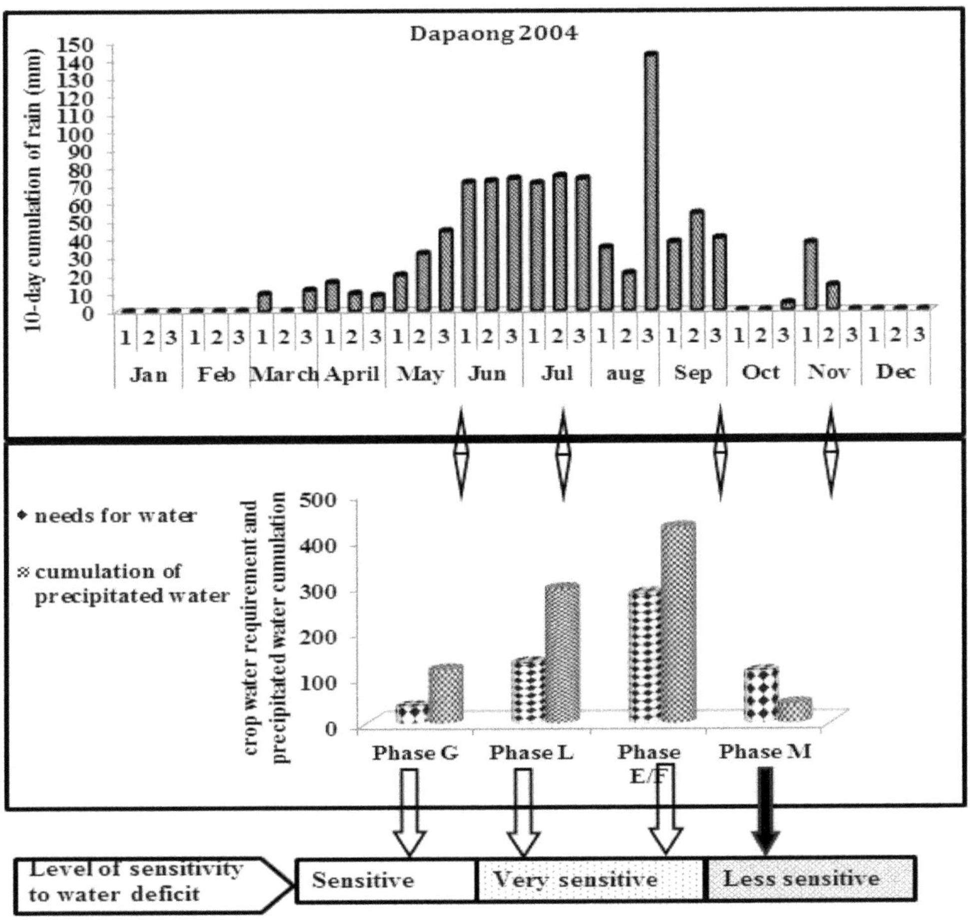

Figure 4. Water stress of millet during ripeness phase in Dapaong 2004 (Source: DGMN, ITRA).

ripeness period. However, this water stress could not have a great impact on the profitability of millet since it occurred during a phase that was not sensitive to the lack of water. In addition, the crop suffered from the stress of excess water during phases G, L and S/F where the quantities of water precipitated are much higher than the water requirements of the crop. In all the cases, be it water deficits or excess, strong rainfall variability significantly impacted the crops.

13.4.2 Vulnerability of Sorghum under moisture variability

Sorghum is mainly grown in areas where annual rainfall ranges between 600 mm to 900 mm per rainy season (Diatta 2007). Sorghum is more resistant to drought due to its ecological requirements, but very vulnerable during the planting/flowering phase. Sorghum's needs for water for a maximum yield range from 380 mm to 630 mm depending on the plant variety (Doorenbos & Kassan 1980). Table 2 shows the actual need for water of sorghum per decade and per phenophase.

Two critical episodes were observed including the flowering and ripeness phases (Figure 5). Water deficit during flowering slows the elongation of the stem and accelerates the challenge of the crop with weeds. As a result, water deficit at the ripening stage can disrupt the filling of sorghum grains. Water deficits in the cycle have had an impact on yield since sorghum is very sensitive to the lack of water during flowering. According to DSID (2012), the profitability of sorghum decreased in 2004 to 0.911 t/ha compared to 0.97 t/ha in 2003 and 0.927 t/ha in 2005. This decline in yield shows the vulnerability of sorghum cultivation in relation to water stress. In addition, drought sequences that occur during sowing and crop emergence cause fungal attack responsible for seeds rotting and poor germination. Decreasing water potential in the soil during the grain fortification phase can also reduce the weight of seeds. Sorghum is a plant that is vulnerable to water saturation in the soil. According to technicians of ICAT (Institut de Conseil d'Appui Technique) in Savannah Region, vulnerability is manifested in this case by the appearance of diseases on sorghum plants.

13.4.3 Maize cultivation: Water requirements and vulnerability factors

Maize is a plant adapted to warm and humid regions but it is rather sensitive to rainfall variations. Maize requires a well distributed rainfall of 400 mm to 800 mm per rainy season to complete its cycle (Doorenbos & Kassan 1980). The ability of maize crop to tolerate water stress under soil water deficit conditions depends on its ability to prevent perspiration while maintaining photosynthesis. However, when the dry sequence lasts longer, the plant eventually loses its physiological growth capacity and productive potential. It withers and eventually dries up.

The stress impacts of water deficit differ according to the stages of development of the plant. The early stress that occurs before the beginning of the cob causes a poor growth, a winding of the leaves, a reduction in leaf surface and thus a loss in number of feet/m^2. Stress at flowering strongly affects the number of seeds per cob and therefore decreases the final yield. It can also promote disease attacks. The stress of water deficit when filling seeds accelerates the senescence of the leaves and the development of diseases (Table 3). It also causes mites to proliferate on the leaf of the cob and cause scalding of up to 2/3 of the seeds. Table 4 summarises the vulnerability symptoms of maize cultivation under excess water stress.

Table 2. Vegetative cycle, phenophases and water requirement of sorghum.

Sowing data	Phase 1 (sowing)		Phase 2 (Flowering)						Phase 3 (Lactic ripeness/waxy)					Phase 4 (Ripeness)		
3rd decade May - 1st decade June	1	2	3	4	5	6	7	8	9	10	11	12	13	14	15	16
Decades	2		6						5					3		
Need for water (mm)	17,3	16,9	19,4	22,5	24,7	26,6	33,9	34,6	34,9	37	37,6	39	39,7	30,8	26	19,2
Water/phase	55 mm		330 mm						330 mm					160 mm		

(Source: ITRA-Savannah Region 2011)

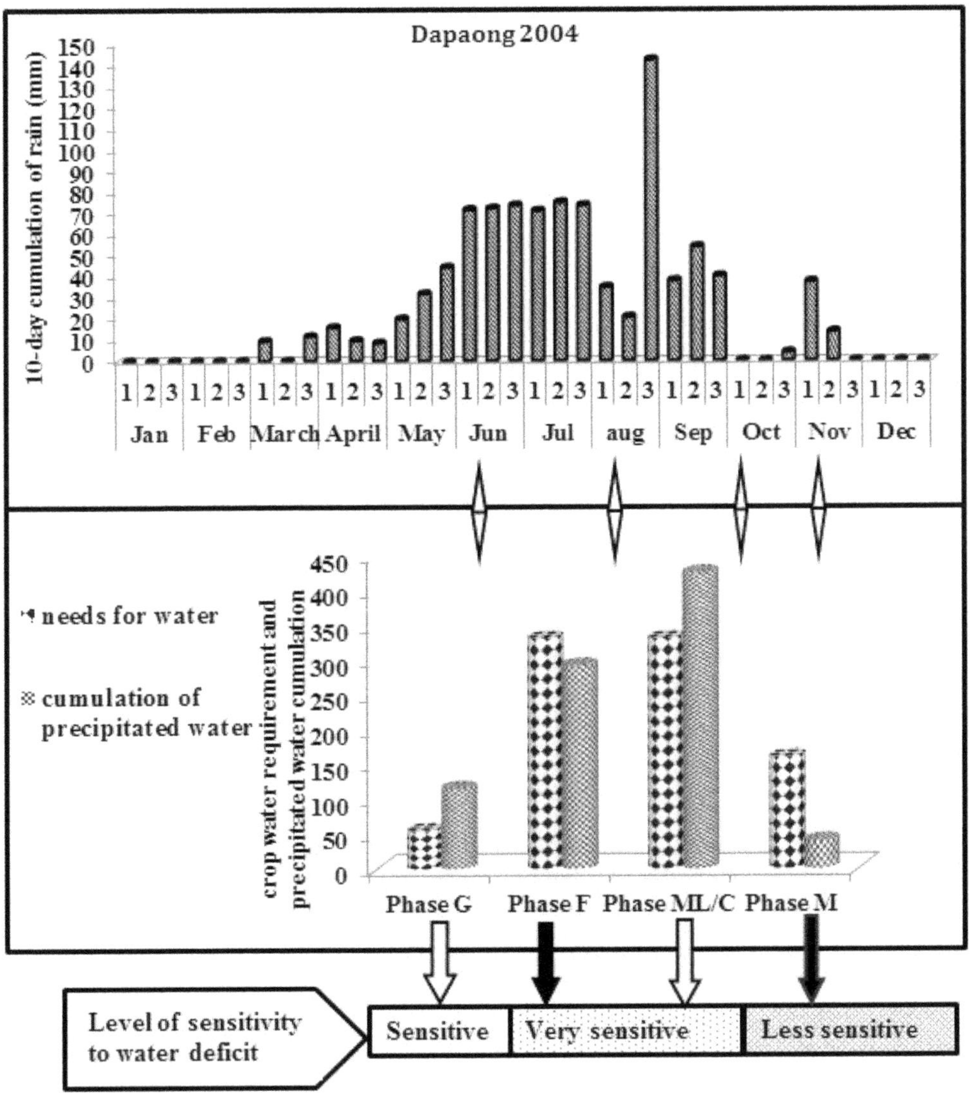

Figure 5. Sorghum stress at flowering and ripeness phases in Dapaong 2004 (Source: DGMN and ITRA-Savannah Region).

Figure 6 detects stages of water stress during the growing cycle of maize.

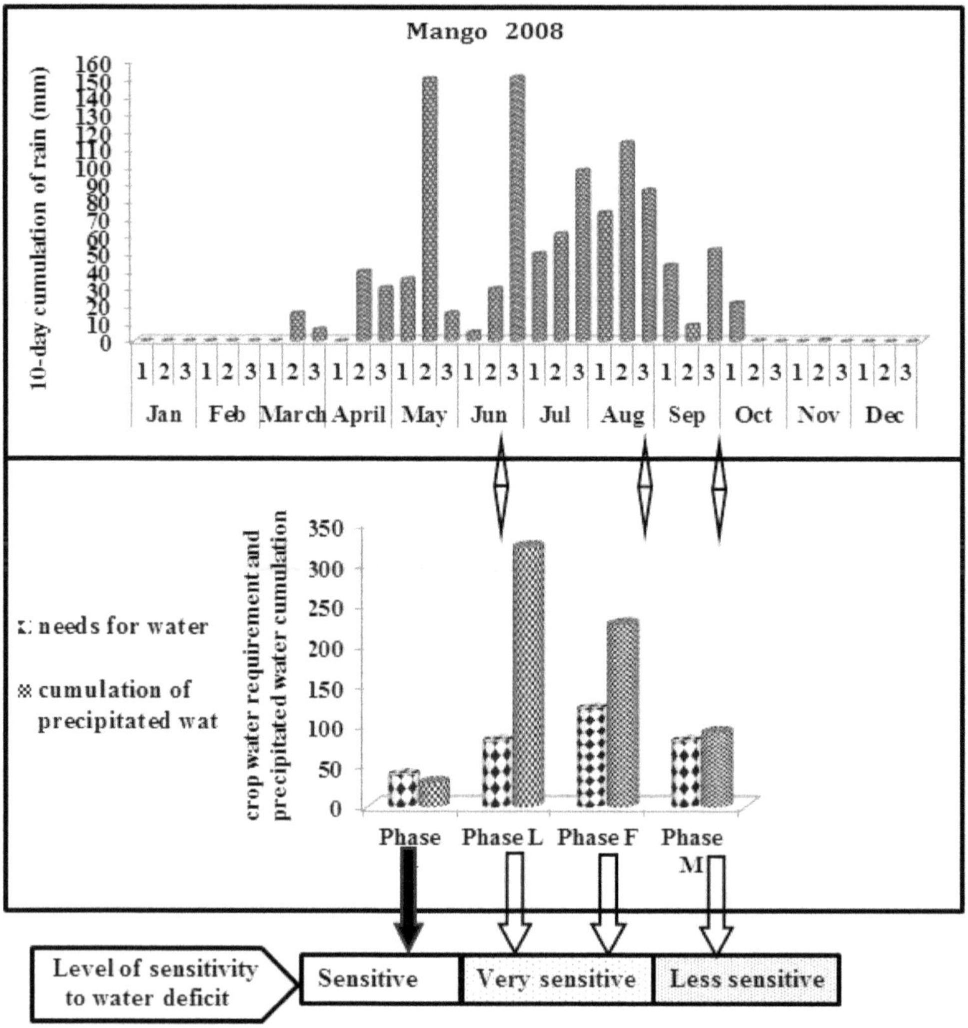

Figure 6. Maize water stress at germination phase in Mango 2008 (Source: DGMN and ITRA 2012).

The period covering the first and second decade of June corresponding to the germination phase unfortunately experienced a low rainfall where the recorded rainfall is slightly inferior to the maize need for water. In 2008, maize crop in the locality of Mango experienced a slight water deficit stress during germination. Furthermore, the situation of waterlogging in the soil is also a major risk for maize cultivation. Excess in water causes oxygen deficiency in the roots and this has negative impacts on plant growth such as poor cell functioning, poor anchorage, roots rotting and leaves yellowing. For example, the case of total loss of maize cultivation after two days leads to crop losses.

13.4.4 Rice cultivation: Water requirements and vulnerability factors

Rice is one of the cereals that provide the basic household food in the region. Unfortunately, yields fluctuate because of the uneven seasonal distribution of rainfall and the shortening of the agricultural season. Water requirements of rice cultivation are high (1000-1800 mm) and a dry sequence of short duration leads to wilting and abortion of the plants. Table 5 determines the duration of the vegetative cycle and water requirements by phenophase of rice cultivation.

Data allowed to identify two phases during which rice cultivation experienced the stress of insufficient water in 2008 in Mango (Figure 5). Significant water deficits were observed during flowering and ripeness phases (Figure 7).

Flowering phase is the second critical period of the growing cycle where rice cultivation is very sensitive to water deficit. This phase has a length of 4 decades ranging from the 3rd decade of August to the 3rd decade of September. Water requirement for the crop was 380 mm at flowering and 170 mm at maturity with deficits of 188.9 mm and 147.6 mm respectively. Water stress observed in these two phases is so significant that its impact on rice yield is obvious. Then, the yield of paddy rice valued at 5688 t/ha in Mango in 2007 decreased to 5488 t/ha in 2008 before increasing again to 5999 t/ha in 2009 (DSID 2012). This decline in yield in 2008 shows the vulnerability of rice cultivation to water deficit. Table 6 summarizes the symptoms of vulnerability of rice cultivation to stresses of water excess and deficit during the phases of full vegetative cycle.

Table 3. Vegetative cycle, phenophases and need for water of AMEN maize.

Sowing data	Phase 1 (Germination)	Phase 2 (Lifting)		Phase 3 (Spiking/Flowering)				Phase 4 (Ripeness)	
1st to 2nd decade of June	1	2	3	4	5	6	7	8	9
Number of decades	1	2		4				2	
Need for water (mm)	18,7	16,3	28,7	37,6	36,8	35,2	35,7	34,3	20,7
Water/phase (mm)	38	75		220				80	

(Source: ITRA-Savannah Region 2011)

Table 4. Water excess stress symptoms for maize cultivation.

Vegetative cycle	Phenological aspects of maize crop in a flooding state
Germination	• Seed swollen from 4 days of flooding, high losses. • If flooding exceeds than 10 days, the germinating grains die.
Lifting	• Poor heterogeneous lifting and yellowing of lifting plants.
Flowering	• Yellowing of older leaves. • Slow growth: slowing down of tiller formation. • In the case of prolonged excess water: the plant withers and dies. • Decrease in tillers and often in the number of cob/m2.
Ripeness	• Early foliar senescence of older leaves. • Poor growth, decrease of tillers and low biomass. • The number of grains/cobs is affected and the cobs are less fertile.

Source: fieldwork 2018

Water deficits that occur during germination lead to weathering in the root system. The main roots become brownish and their ends have a typical, sinuous face devoid of rootlets.

13.5 CORRELATION BETWEEN RAINFALL AND AGRICULTURAL YIELDS

The analysis of yield data shows several uneven trends in agricultural production for different crops and under different precipitation regimes. Figure 8 illustrates correlations between the trend in annual rainfall and the agricultural yields for millet and maize in the environs of Dapaong and in Mango. It makes it possible to assess the influence of rainfall variations on the level of agricultural profitability.

Interestingly, correlations between rainfall, rainfall trends and agricultural yields between 1983 and 2010 do not fit perfectly (Figure 8). The correlation coefficient calculations show a variation in indices across crops. The correlation of millet and sorghum yields describe a relationship of average intensity both for Dapaong (r = 0.56 and r = 0.50) and for Mango (r = 0.40 and r = 0.42). Despite the increase in rainfall at Dapaong, a poor correlation is established for maize yields (r = 0.22) while it is significant at Mango with r = 0.78. The correlation for rice is medium with r = 0.51 and r = 0.40 at Dapaong and Mango. These results underline a general positive relationship. This implies that rainfall variations influence agricultural yields but at varying intensities depending on the eco-physiologic bandwidth of the crop.

Table 5. Rice phenophases and rice needs for water of IR 841, TGR 1, TGR 203 varieties.

Sowing date	Phase 1 (Sowing)			Phase 2 (Lifting)			Phase 3 (Flowering)				Phase 4 (Ripeness)	
3rd decade June, 2nd decade July	1	2	3	4	5	6	7	8	9	10	11	12
Number of decades	3			3			4				2	
Need for water (mm)	42,4	42,7	47,2	49,6	52,3	55,1	58,7	60,3	61,9	58,2	53,1	46,5
Water/ phase	160			260			380				170	

(Source: ITRA-Savannah Region 2011)

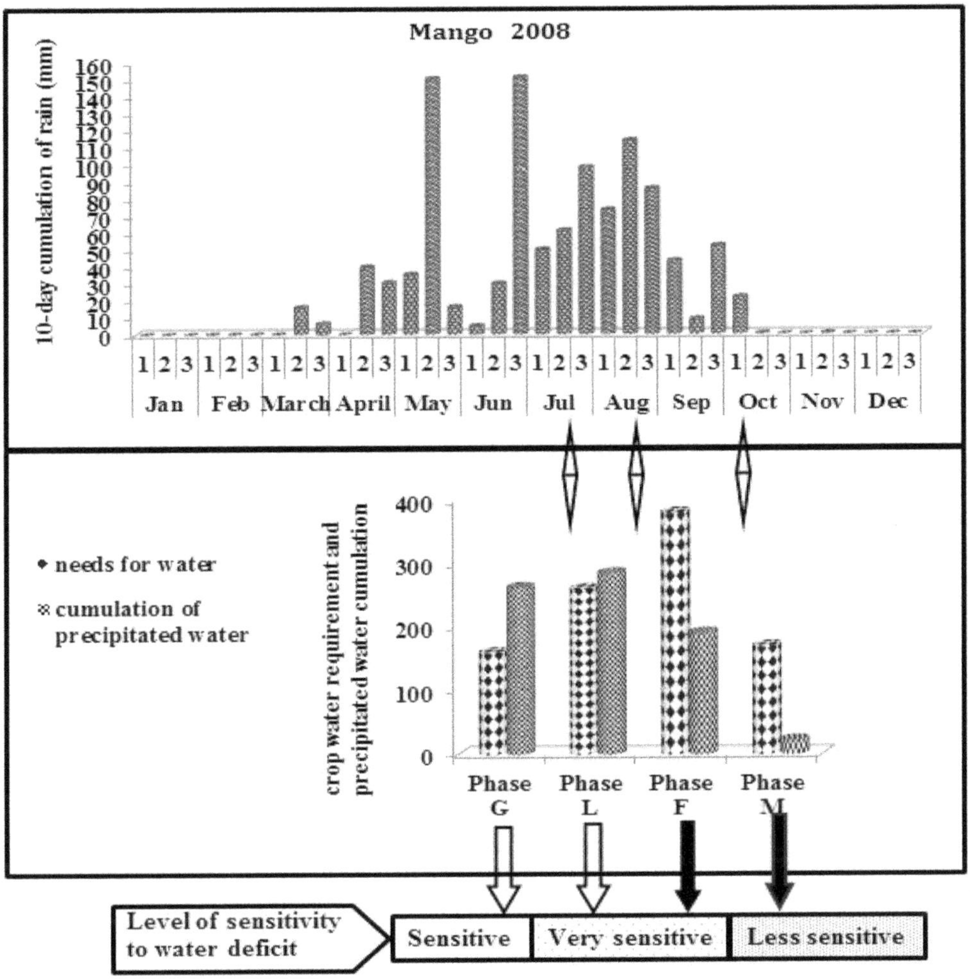

Figure 7. Water stress of rice at flowering and ripeness phases in Mango 2008 (Source: DGMN and ITRA 2011).

13.6 DISCUSSION

The major agricultural risk for farmers and agricultural yields is the spatiotemporal variability of rainfall characterised by the uneven alternation of rainfall deficits and excess, the delay or early beginning and the shortening of the rainy season. Water deficits cause severe water stress at critical plant growth stages such as germination, flowering and grain filling. These deficits sometimes bring about crops drying-up, while excess and flooding lead to crop decay, erosion in the fields, multiplication of pests and the excessive development of weeds. Changing rainfall patterns, disturbed rainfall distribution, recurrence of drought are the main vulnerabilities of crops in the Savannah Region of Togo. Similar results had already been presented by De Haan (1993) illustrating that the irregularity of seasonal rainfall and the first rains often followed by dry sequences causing challenges to the agricultural sector in northern Togo. These results are very similar to those found in Benin by Ogouwalé (2006) and Issa

Table 6. Symptoms of vulnerability of rice to stresses of water excess and deficit.

Phenophases	Stress effects of water deficit	Symptoms of excess water stress
Germination stage	• Damping-off • Germination defects on dry crusted parts	• Silt sedimentation on seeds and seedlings, seedling lifting and blackening of the root.
Lifting stage	• Slow or no lifting • Attacks of pests causing poor lifting or death of seedlings after germination • Necrotic plants with reduced size and leaf roll • Dried plants and appearance of silky white filaments	• An excess of water hinders the lifting: the seed is floating and anchored in the mud by only a few small radishes; the lodging at maturity, thin leaves that become yellow before emerging. • Unallocated seed reserves making young plants vulnerable to pest attacks.
Stage of spiking/ flowering	• Insect attacks causing intense pouring • High proportion of unproductive flowers and empty panicles	• Submerged and asphyxiated seedlings. • Anaerobic fermentations and cessation of adventitious roots formation.
Stage of pasty ripeness	• Total escapement and significant displacement at the end of the run • Parasitism and crop loss • Poor grain filling	• Unfitness for tilling • Development of parasite culms: early attacks resulting in crop losses following sterile tillers and the displacement of the formed flowers.

Source: field surveys, 2018

(2012). Their results reveal that rainfall variations, seasonal rainfall deficits and changes in the duration of the crop season affect the eco-physiological processes of crops. It disrupts crop cycles and 'weakens' agricultural systems. More recently, the study by Adéwi (2012), indicates that the vulnerability of the agricultural sector in Togo is mainly linked to spatiotemporal variability of rain. There are weather extremes such as floods and droughts. Dugué et al. (2012) identified in Mali the shift in the agricultural calendar as a consequence of the shortening of the rainy season and the decline in annual rains that result in the displacement of isohyets by about 200 km towards the south of the country between 1951 and 2000. According to Hounkpè (2013) changes in the hydro-meteorological rhythm and the dependence of agriculture in the Oti Plain have increased the vulnerability of crops in recent decades.

Analysis of the survey results and rainfall - yields graph correlation shows that the level of crop profitability has very often been weakened by water deficits or excess and early cessation of rainfall. Flooding of fields that last several days damage crops and decrease harvests. These results are in line with Lemou (2014) who revealed that rainfall disturbances cause agricultural yields to fall by 17% to 36% with significant socioeconomic repercussions for the population in the Kara Region. The work of Amadou (2004) confirms the same for Niger. His results show that the lack of soil moisture at the end of the crop cycle in the Aguié area led to a decrease in yields by 14.7% for millet and 19.6% for sorghum. However, the same study notes that the effect of water shortage or excess on the growth and yield of the plant depends on the species and variety cultivated. For Dimon (2008) climate variability in northern Benin is often a source of the multiplication and expansion of crop pests (insects, weeds) that increase the risk of post-harvest losses.

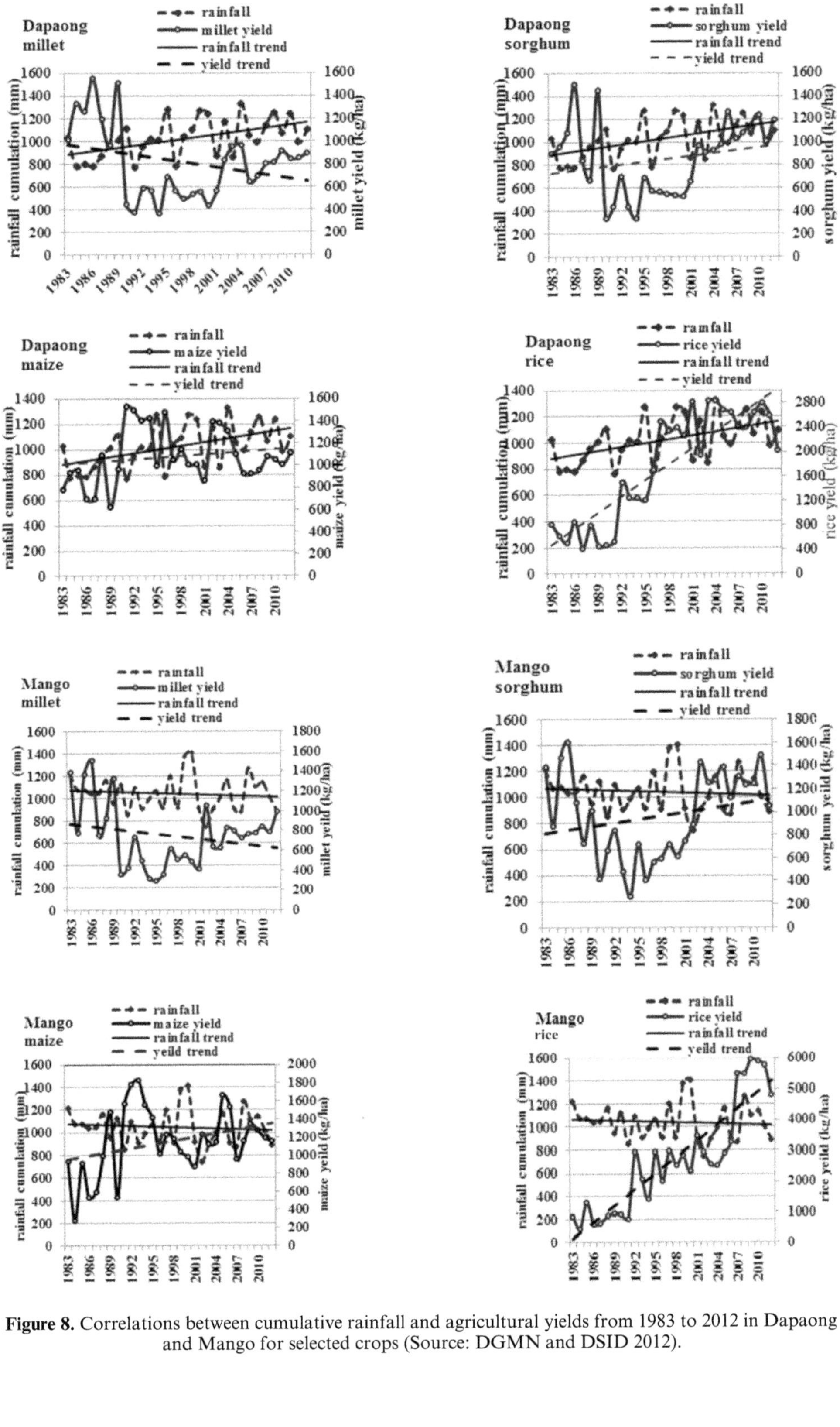

Figure 8. Correlations between cumulative rainfall and agricultural yields from 1983 to 2012 in Dapaong and Mango for selected crops (Source: DGMN and DSID 2012).

13.7 CONCLUSION

Inter-annual and intra-season rainfall fluctuations which are difficult to predict, are usually characterised by poorly distributed rainfall over time and space. The tendency to dry up and the recurrence of extreme weather events make agricultural production vulnerable in the Savannah Region of Togo. The sudden cessation of rainfall before ripeness of some crops is also a major agroclimatic constraint for agricultural productivity. Dry sequences that occur during sowing periods cause mould that lead to the decay and poor seed germination. Excess water or rain breaks lead to poor lifting of plants, short berries of flowers, early drying of foliage, and poor filling of grains. The study of the vulnerability of some crops in relation to the stress of water deficits has shown that even in the year of high rainfall, it is not uncommon to observe prolonged dry sequences that disrupt crop growth at certain phenophases of the vegetative cycle. The graph correlations between rainfall and agricultural yields are all positive for the crops studied. This implies that rainfall irregularities effectively influence agricultural yields. The various climatic obstacles of yield namely excess or water deficit stress, diseases and pests can cause a significant decline in agricultural yields from year to year when they occur at critical crop growth stages such as germination, flowering and grain filling. These constraints have a strong impact on farmers' harvests and incomes.

REFERENCES

Adéwi, E. 2012. *Les stratégies agricoles de gestion de la péjoration pluviométriques au Togo*. Doctoral thesis: 319. Togo.

Amadou, M.L. 2004. *Impact de la variabilité climatique sur les systèmes de production au Niger: cas de zones de Gaya et d'Aguie*. Mémoire de DESS: 78. Niamey: Université Abdou Moumouni de Niamey.

Bognini, S. 2011. Impacts des changements climatiques sur les cultures maraîchères au nord du Bur-kina Faso: cas d'Ouahigouya. *Réseau National des Agro-sylvo-pasteurs du Faso (RENAF)*: 38.

De Haan, L. 1993. *La Région des Savanes au Togo, l'Etat, les paysans et l'intégration régionale (1885-1985)*. Paris: Editions Karthala.

Diatta, I. 2007. *Impact des fluctuations pluviométriques sur la production agricole dans la région de Thionck-Essyl en Basse Casamance*. Dakar: Université Cheikh Anta Diop.

Dimon, R. 2008. Adaptation aux changements climatiques: perceptions, savoirs locaux et stratégies d'adaptation développées par les producteurs des communes de Kandi et de Banikoara, au Nord du Bénin. Doctoral thesis: 209. Abomey-Calavi: Université d'Abomey-Calavi.

Diomande, M., Dongo, K., Koné, B., Cissé, G., Biémi, J. & Bonfoh, B. 2008. Vulnérabilité de l'agriculture pluviale au changement de régime pluviométrique et adaptation des communautés rurales du « V-Baoulé » en Côte d'Ivoire. *14e colloque international en évaluation environnementale Niamey-Niger 26 au 29 mai 2009*: 11.

Doorenbos, J. & Kassan, A.H. 1980. Réponse des rendements à l'eau. *FAO Bulletin d'irrigation et de Drainage*: 33. Rome: FAO.

DSID, Direction des Statistiques Agricoles, de l'Information et de la Documentation 2012. *Résultats des enquêtes statistiques des campagnes agricoles*. Lomé.

Dugué, M.-J., Delille, H. & Malgrange, S. 2012. Caractérisation des stratégies d'adaptation au changement climatique en agriculture paysanne. *Etude de capitalisation réalisée sur les terrains de coopération d'AVSF*: 50.

Hounkpè, K. 2013. *Impacts des fluctuations climatiques et des activités anthropologiques sur les écosystèmes de la plaine de l'Oti (Togo)*. Doctoral thesis: 142. Lomé: Université de Lomé.

Issa, M.-S. 2012. *Changements climatiques et agrosystèmes dans le Moyen-Bénin: impacts et stratégies d'adaptation.* Doctoral thesis: 278. Abomey Calavi: Université d'Abomey-Calavi.

ITRA, *Togolese Agricultural Research Institute* 2011. *Fenêtre sur les activités de recherche: recherche sur les cultures et gestion intégrée de la fertilité des sols.*

Lemou, F. 2014. *Dynamique climatique et production agricole dans la Région de la Kara (Nord-Togo).* Doctoral thesis: 1–302. Lomé: Université de Lomé.

Ogouwalé, E. 2006. *Changements climatiques dans le Bénin méridional et central: indicateurs, scénarios et prospective de la sécurité alimentaire.* Doctoral thesis: 286. Abomey-Calavi: Université d'Abomey-Calavi.

Siband, P. 1983. Nutrition minérale des plantules de mil, maïs et sorgho au cours des premiers jours de végétation. *Agronomie tropicale* 34(3): 242–249.

CHAPTER 14

Vegetation cover, land use dynamics (1988-2015) and future perspectives of sustainable management strategies in the Oti catchment in Benin

K.T. Souberou, I.F. Yabi & E. Ogouwale
University of Abomey-Calavi, Cotonou, Benin

ABSTRACT: This research analyses current and future dynamics under pessimistic and optimistic hypotheses of land use in the Oti catchment in Benin. The IDRISI SELVA and Arc GIS mapping software enabled the processing of satellite images with field reference data to study changes in vegetation cover over ~25 years. Field work allowed to obtain data relating to land use units, and particularly the impacts of sustainable management strategies for the near future (2050). The results indicate a strong change in land use and vegetation cover by anthropogenic pressure between 1988 and 2015. The average annual change rates of land cover ('landscape/vegetation dynamics') were estimated at –0.25% decline for the first period (1988-2000), and at –0.8% per year during the second period (2000-2015). In 2050 the fragmentation of the landscape will have negative consequences on biodiversity, the hydrological cycle and soil quality.

14.1 INTRODUCTION

West Africa is one of the most exposed and vulnerable regions in the world to the effects of climate change (rising temperatures, less frequent and irregular rain, flooding, drought) on rain-type agriculture (IPCC 2007, 2013). Variations in climate parameters in combination with anthropogenic pressure are the reason for poor performance and degradation of natural resources (Grouziz 1986, Kabré 2008, IPCC 2019). Environmental change has had repercussions on hydrology and socio-economic involvements that constitute a constraint for integrated management of water resources (Vissin 2007). These changes are behind the tremendous shift in the interactions between the landscape surface and the atmosphere (Groot & McLaughlin 2000). According to Issa (2012), these seasonal irregularities are a limiting factor for the expansion of agriculture. Soil fertility has become inadequate for agriculture over the years, due to soil erosion and the practice of degrading environmental techniques such as slash-and-burn shifting cultivation (Ouorou Barré 2014).

Over the past five decades, with strong population growth and technological change, humanity has evolved as a major actor in environmental change (Oloukoi 2012). This increase has led to a gradual reduction in the length of fallows, the use of environmentally degrading equipment and techniques, and the need of increasing agricultural yields by producers. These factors determine aspects of land use and the state of the natural vegetation cover. They form the basis of spatiotemporal changes observed in Benin and have a direct impact on other environmental problems (Bamba et al. 2008).

The dynamics of land use and soil cover remain a reality today in the southern part of the Oti catchment as a result of the extension of agricultural areas, particularly towards low-lying areas (Souberou et al. 2017). However, this change of state of the landscape units is not the same throughout the basin. Altitude, slope and valley bottom areas do not evolve in the same way over time. This paper analyses the current and future dynamics by considering pessimistic and optimistic scenarios of land use and vegetation cover in the Oti watershed in Benin by application of remote sensing and Geographic Information Systems (GIS).

14.2 STUDY AREA

The study area is located in northwestern Benin in the prefectures of Atacora and Donga. It occupies 47.20% of the Oti River basin in Benin. It lies between 09°19′6” and 10°54′08" northern latitude and 0°45′34" and 1°41′48" eastern longitude. Administratively, it belongs to the municipalities of Boukoumbé, Cobli, Ouaké and parts of Bassila, Copargo, Djougou, Kouandé, Matéri, Natitingou, Tanguiéta and Toucountouna. It is subdivided into fifty districts (Figure 1).

Covering an area of 8034 km^2, its relief is hilly with altitudes ranging from 118 to 667 m asl (Digital Altitude Model, ASTER DEM 2007). Three major geomorphological units share the territory. Those are the Atacora mountain range (high altitude), the levelled peneplain (medium altitude) and the Gourma alluvial plain (low altitude) (Figure 2).

The climate is Sudanese-type, characterised by two main seasons: a rainy season (May to October) and a dry season (November to April). Annual rainfall ranges from 967 to 1255 mm (Ouorou Barré 2014). Mean annual temperature is about 27.3 °C with an annual thermal amplitude in the order of 18 °C (Idiéti 2012). The hydrographic network is dense: a total of nine sub-watersheds are identified in the southern part of the Oti/Pendjari River basin in Benin. These are the water resources management units (Birnao, Kara, Kéran, Koumangou, Mango, Lower Oti, Middle Oti, Upper Oti and Pendjari) drained by the tributaries of the Pendjari River (which became the Oti in Togo) and then some other tributaries of the Oti (Kéran and Kara) which have their sources at the foot slopes of the Atacora chain.

From a pedological point of view, several types of soil are identified and grouped into five main groups, with respective proportions of 8%, 4%, 17%, 46%, 25%. These are raw mineral soils and poorly developed hydromorphic soils found in the Oti depressions, the poorly developed lithic soils found in the Atacorian massif, the sesquioxide soils of iron and manganese (tropical ferruginous soils) constituting the large part of the south of the Oti catchment and the ferralitic soils (ORSTOM 1976, Ouorou Barré 2014). The vegetation of the landscape is marked by a varied range of plant communities, including dense dry forests, riparian forests, bare surfaces, shrub savannahs and mosaics of crops and fallows.

The local population of 723,762 inhabitants (INSAE 2014), has a large sociocultural diversity made up of 32 ethnic groups. This heterogenous groups, mainly peasants are in search of fertile land for agriculture. They all contribute to the development of the lowlands, today facing challenges of climatic variability. The number of agricultural households increased from 38,041 in 1992, to 63,781 in 2002, and to 93,272 in 2013 (census RGPH2-4, INSAE 2014).

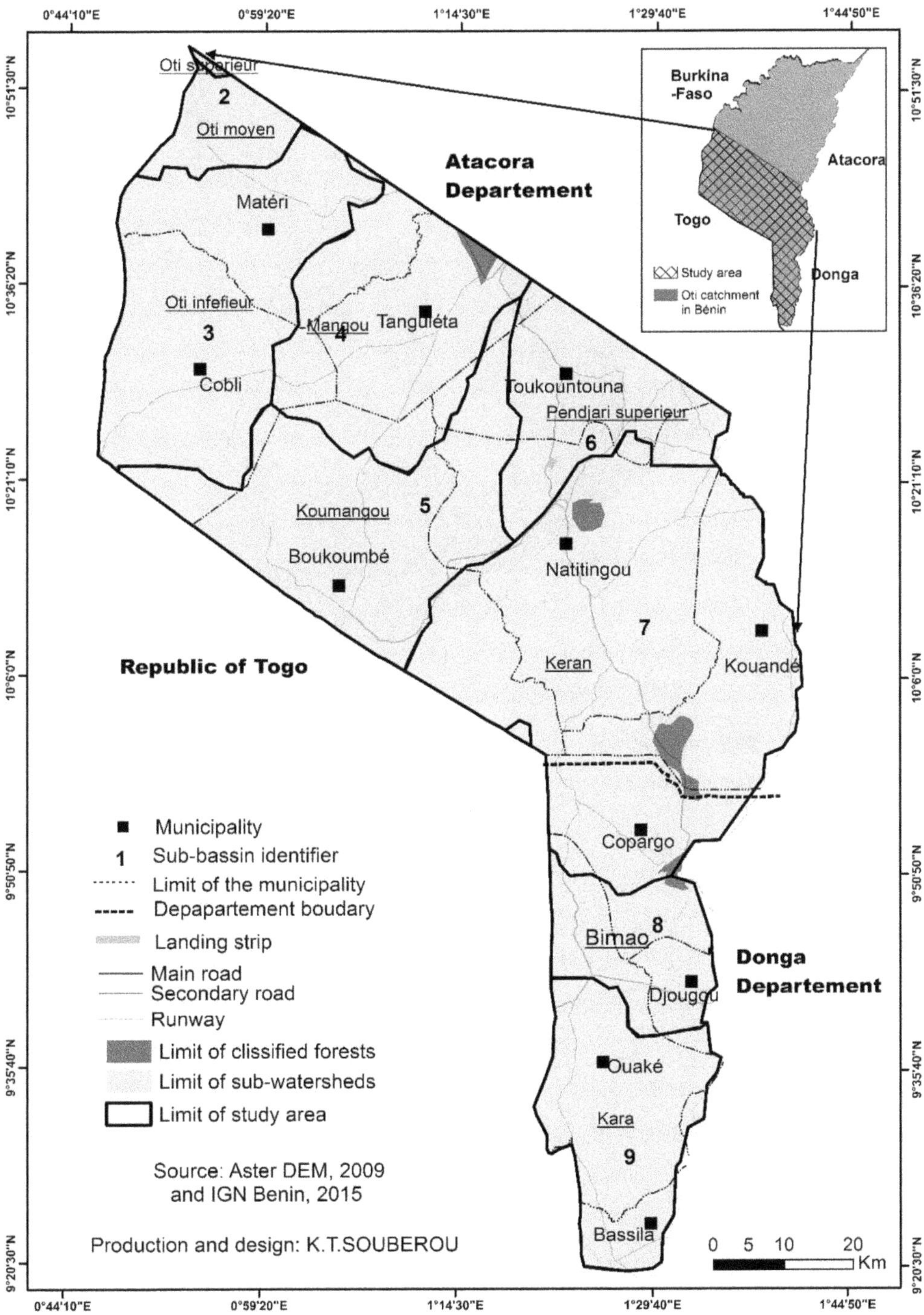

Figure 1. Geographical and administrative location of the study area in northern Benin.

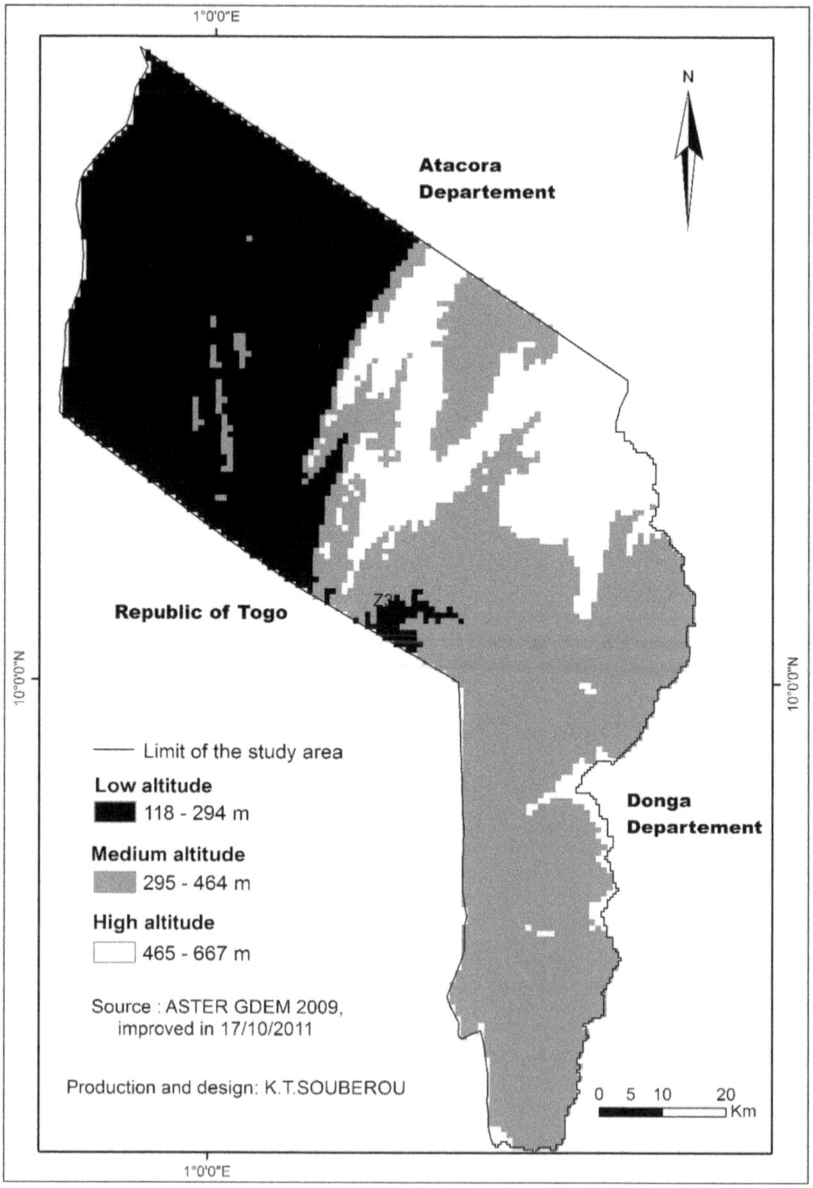

Figure 2. Major topographic units (metres above sea level, m asl) of the Oti/Pendjari catchment in northern Benin.

14.3 METHODS

14.3.1 Data types and data collection

Satellite images and basic cartographic documents for geographical coordinates of the lowlands, and socio-anthropological data have been used in this study. These include, in particular:

- Landsat TM images from 1988, ETM+ from 2000 and OLI (Operational Land Imager) from 2015 acquired for the spatiotemporal analysis of the vegetation cover and land use in the Oti watershed were obtained from the University of Maryland website (Global Land Cover Facility (GLCF, http://glcfapp.glcf.umd.edu:8080/esdi/).
- ASTER GDEM 2009 satellite image (improved in 17/10/2011) was acquired from the United States Geological Survey (USGS, http://earthexplorer.usgs.gov/) for topographic analysis (relief)
- Topographic map sheets (Djougou, Natitingou, Sansanné-Mango) at a scale of 1/200,000 of former French West Africa (AOF), covering northwestern Benin; these maps were reproduced by the Institut Géographique National de France (IGN), Paris, and published between 1960 to 1990.
- Land use maps produced as part of the Woodfuel Project, Phase II, for the National Forest Inventory (IFN) in 2007 (Orékan 2007), including the map sheets of Djougou, Natitingou and Sansame-Mango at a scale of 1/200,000.
- Ground Control Point of the alluvial plain and in different land use units were taken by GPS (Global Positioning System) in the field;
- Information on the impacts of land use dynamics and management measures was collected through ground truthing and field investigations.

Thus, field visits and exchanges with the heads of development (e.g. NGOs involved in environmental management and decentralizing of state services) at the municipal level allowed to get more socio-ecological data.

14.3.2 Software, data processing and hypotheses

The implementation of land use and land cover mapping in the southern part of the Oti catchment followed the processing different satellite images. Remotely sensed data (Landsat TM, ETM+, OLI) followed first a pre-treatment (enhancement of image contrast, interpretation of image histogram, colour composition and visual interpretation), then the supervised classification in the IDRISI SELVA software environment to extract land use information. Results were improved by field work and finally the validation of the classification by conventional methods of evaluating the thematic accuracy of the classification results (confusion matrix, Cartographic Validation Index (CVI), Class Purity Index (CPI) and Classification Accuracy Index (CAI) were applied). Classified images have been undergoing spatial generalization by filtering the classification (isolated pixels are eliminated) following the vectorization operation of these images (transition from raster to vector format). A total of seven land use units were identified (dense forest, gallery forest, open forest and wooded, tree and shrub savannah, mosaic of crops and fallows, rocky surface and bare soil and urban areas).

The analysis of land use and occupation of dynamics was based on the change detection approach to compare the different land use classes from 1988 to 2000 and from 2000 to 2015. Change detection consists in identifying differences in the states of an object or phenomenon by observing it at different time periods (Hountondji 2008, Oloukoi 2012, Tohozin 2016). It uses the GIS approach through the determination of transition matrices and visual interpretation to understand the different types of conversion between these occupation units. Land use changes were assessed on the basis of the annual rate of change of each occupation unit expressed as a proportion. This is the average annual rate of spatial expansion between 1988 and 2000 and between 2000 and 2015. It is calculated using formula (Puyravaud 2003, Schulz et al. 2010, Oloukoi et al. 2014, Tohozin 2016):

$$T = \frac{1}{t2 - t1} \cdot \ln \frac{S2}{S1} \tag{1}$$

where T = the annual rate of change; $t1$ = the year of departure; and $t2$ = the year of arrival; $S1$ = the area on date $t1$; and $S2$ = the area on date $t2$; ln = the natural logarithm.

The R indicator has made it possible to characterize the evolution of land use units across the three topographic elevation levels (high, medium and low) and reflects the temporal rate of observed changes. Expressed in percentage terms, a negative result indicates a phenomenon of savannization or anthropization, while a positive result reflects afforestation (Vigneau 2013). Its formula is:

$$R = \left(\left(\frac{S2}{S1} \cdot \frac{1}{t2 - t1} \right) - 1 \right) \cdot 100 \tag{2}$$

Spatial modelling of land use and occupation by 2050 was carried out with pessimistic and optimistic views, based on a set of assumptions relevant to the explanation of the future. (Godet 1997). The Land Change Modeler (LCM) incorporated into the IDRSI SELVA software was scouted about to implement the socio-economic scenario based on the Markov chain and external data. This software offers large flexibility thanks to the programming tools (predefined operations) available for the execution of a multitude of commands (tasks) in its graphical environment. It has been validated by several studies with interesting results (Pontius et al. 2008, Eastman 2009, Mas et al. 2011).

The results of land use and land cover in the Oti catchment area from 1988 to 2000 and from 2000-2015 were taken into account for a simulation of spatial dynamics in 2030 and 2050. This spatial projection to 2050 was made because FAO (1997) conducted a survey of world agriculture over the 2080 temporal horizon and forecasts a stabilization of fertility in developing countries in response to climate change at that time. In addition, the time required for climate adaptation requires that the time horizon 2050 not be too close to the base year of the projections. This scenario is based on hypotheses that focused on previous work (Orékan 2007, Arouna 2012, Oloukoi 2012, Boko 2012, Ouorou Barré 2014). Changes in lowland occupation are a function of biophysical and socio-economic factors. The following modelling hypotheses were considered:

- the increasing population trend has led to a significant degradation of natural resources and a disruption of environmental balances (Sounon et al. 2007);
- populations use environmentally friendly techniques (harness farming, slash-and-burn shifting cultivation, soil conservation and soil fertility restoration techniques, uncontrolled deforestation, reduction of fallow time);
- the rural population's enthusiasm for low-wetland areas is strong and their development is a strategy for adapting to climate variability (Souberou 2013).

These basic considerations of the socio-economic scenario have made it possible to identify future trends in the various occupation and land use units in the basin in order to propose measures for sustainability, especially in the context of climate change.

Also, the development of an optimistic scenario based on the hypotheses inspired by prospective studies of land use dynamics in Benin (Alafia 2025 and

IMPETUS scenarios in relation to those of the IPCC) will provide sufficient data for its sustainable management. The basic assumptions are that the economic situation of the population has improved and the economic growth rate at the national level is 12.4%. as forecasted in the Alafia 2025 scenario (République du Bénin 2000). Then, agriculture is mechanized and agricultural intensification techniques are adopted by producers and the rate of forest loss is reduced from 100,000 ha/year to 50,000 ha/year and also the rate of land tenure dynamics is reduced by half compared to the rate obtained between 1986 and 2006. Finally, the share of wood in meeting people's domestic energy needs is reduced by 50% in favour of other energy sources such as domestic gas. This ecological analysis of changes in land use in the Oti catchment area has been carried out in a context where plant forming will be restored with a climate favourable to production in the high grounds (interfluves) and population growth, will reduce the strong anthropogenic pressure on resources and will contribute to increase the lifetime of plant formations.

The land use maps and positioning points of the gathered land use units were used to pre-treat and validate the supervised classification of the 1988, 2000 and 2015 images. However, to validate the results of the simulation by the LCM model, a comparison is made between the map simulated by the model in 2015 and the (actual) reference land use map of the same date. Statistical analyses social data.

14.4 RESULTS

The evolution of the vegetation cover in the Oti catchment area is assessed by the analysis the satellite images classification, the transition matrix of the land use units and their prospective modelling.

14.4.1 Accuracy of classification TM, ETM+, OLI TIRS

From the analysis of the matrices of confusion of the three classified images, the confusions between the land use units are slight (riparian forests and dense forests and open forests and wooded and shrubby savannah; mosaic of crops and fallows). The high rates of the CPI (96.72%) and CVI (97.01%) clues at the level of the different classes indicate that a large proportion of the occupation classes of grounds has been identified with an accuracy clue of the mean classification of 97.3% for the 1988, 2000, 2015. Cohen's (1960) mean agreement kappa coefficient is estimated at 0.97 and the mapping accuracy at 96.95%, which allows the results of this classification to be accepted.

14.4.2 Evolution of land use and vegetation cover 1988 to 2015

The evolution of the seven occupation units is analysed successively from 1988 to 2000 and from 2000 to 2015 through their areas and the calculation of the rates of change. Table 1 presents the different transformations within the vegetation cover that are characterised by conversions of the occupation units between them.

Figures 3-4 show that between 1988 and 2000, nearly 26,958 ha of dense forests were converted to open woodlands, trees and shrubs savannahs. Open forests and wooded savannahs have been transformed into wooded and shrub savannahs and mosaics of crops and fallows, about 27,071.39 ha. More than 147,000 ha of wooded and shrub savannahs into mosaics of crops and fallows is an important and destructive conversion of plant forming. In addition, a significant district of stability of the

Table 1. Transition matrix of lowland land use units between 1988-2000 and 2000-2015 (ha).

	Classes	Ag	Fcsb	Fd	Fg	Mcj	Saa	Sr
1988	Ag	**4663.56**	0	0	0	16.249	64.99	0
	Fcsb	259.99	**97,662.31**	15583.13	243.74	714.97	26,356.45	0
	Fd	32.498	3867.35	**516.24**	0	3314.86	23,090.33	48.74
	Fg	32.05	0	1500.08	**13,500.50**	4062.33	0	0
	Mcj	406.23	20,004.28	0	0	**52,742.45**	12,214.71	0
	Saa	324.98	9424.62	276.23	424.62	147,604.04	**355,700.01**	64.84
	Sr	68.76	0	**0**	0	1283.88	204.39	**7136.37**
	2000							
	Classes	Ag	Fcsb	Fd	Fg	Mcj	Saa	Sr
2000	Ag	**5788.08**	0	0	0	0	0	0
	Fcsb	0	**70,384.01**	6500	937.28	9404.90	43,521.13	211.24
	Fd	0	4351.29	**4363.01**	0	536.22	8625.17	0
	Fg	0	4234	0	**9176.46**	0	758.40225	0
	Mcj	286.88	0	0	0	**187,824.59**	20000	1627.33
	Saa	211.24	822.61	0	0	48,077.04	**368,385.44**	134.56
	Sr	206.39	0	0	0	17.45	0	**7026.12**
	2015							

Ag: Agglomeration; Fcsb: Open forest and wooded savannah; Fd: Dense forest; Fg: Riparian forest; Mcj: Cultivation mosaic and fallow; Saa: Tree and shrub savannah Sr: Rock surface and bare ground.

seven classes between 1988 and 2000 (areas in bold) as well as a conversion of more than 9424 ha of wooded and shrub savannahs into wooded savannahs, is taken into account, then the evolution of the open forests and wooded savannahs in dense formations is about 15,583 ha (0.03%).

Between 2000 and 2015, changes in occupation units continued (Table 1). From the analysis of this matrix, the agglomeration class did not undergo any conversion compared to the previous period (5788.08 ha stable). While 13,512.68 ha of dense forest areas have been reduced to open and wooded savannah, wooded and shrub savannah and fallow; more than 52,926 ha of open and wooded savannah to wooded and shrubby savannah and mosaic fields and fallows; and dense forests have been largely transformed between 1988 and 2000. During the period 2000 to 2015, nearly 246,399,805 ha of plant forming (forests and savannahs) transitioned to these anthropogenic forming (crop mosaics and fallows, rocky surfaces and bare soils). On the other hand, the opposite was very slight, 20,000 ha of mosaics of crops and fallows in wooded and shrub savannahs.

The results of the matrices (1988-2000 and 2000-2015) reflect the changes in the state of the occupation units in the study domain, which led to two conversion processes: savannization and anthropization. The conversion relationships between the observed land use units are as follows: dense forest' open forest and wooded savannah' tree and shrub savannah' crop mosaics and fallow' rocky surface and bare soil' agglomerations, in terms of the biomass present. Despite these transformations noted in the units, an average stability of nearly 58% of plant forming (forests and savannahs) is observed.

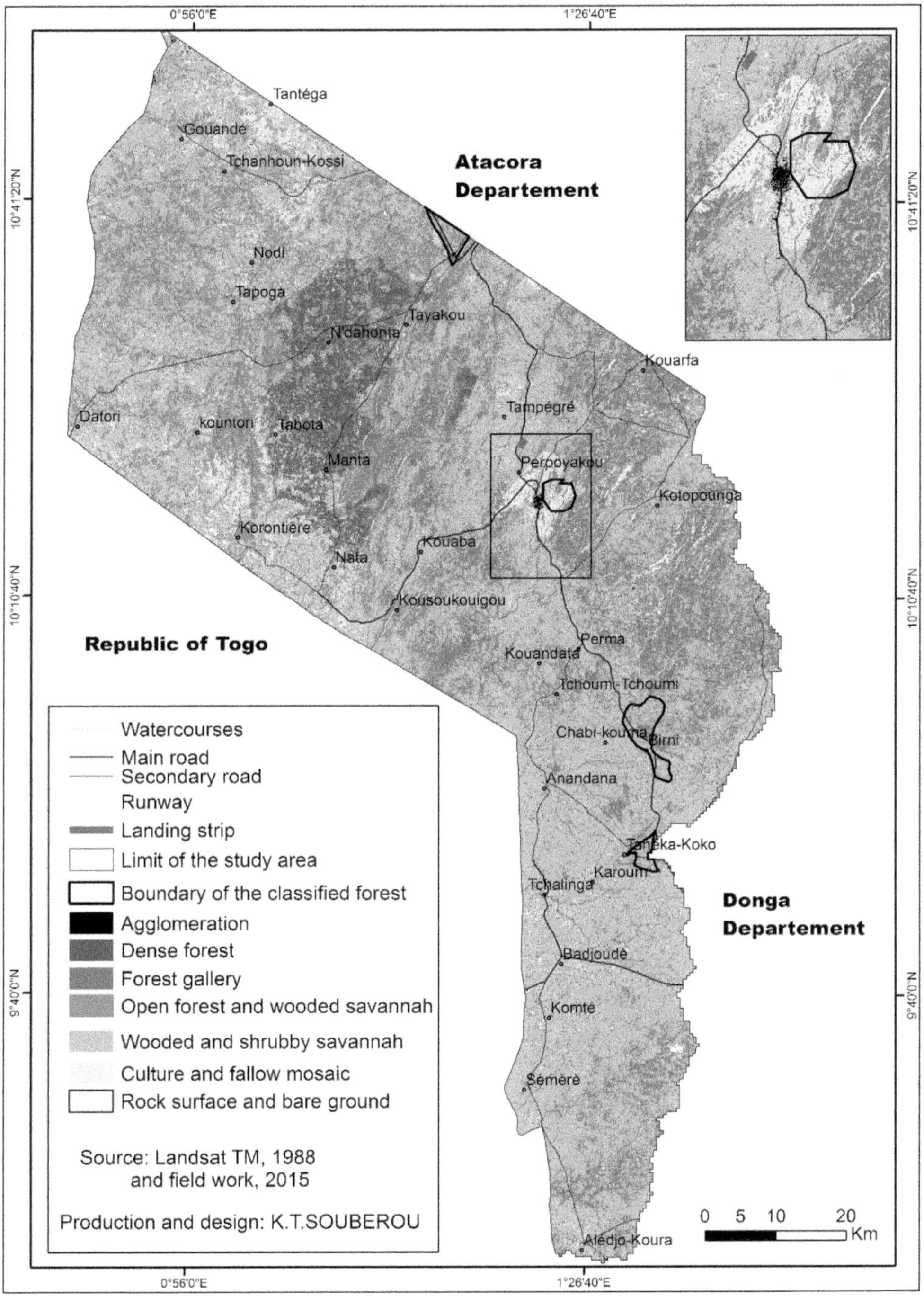

Figure 3. Land use and land cover in the southern Oti catchment area in 1988.

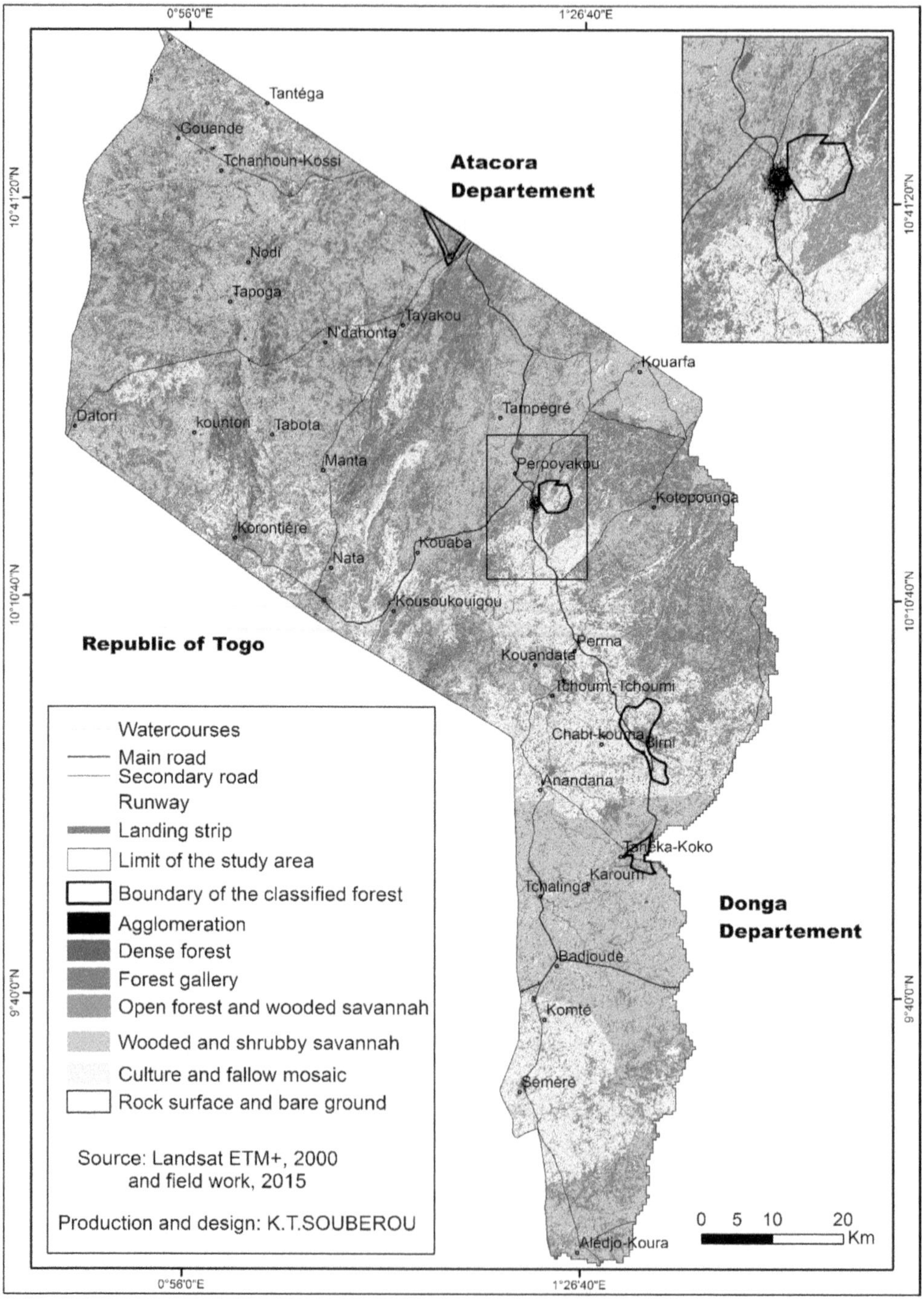

Figure 4. Land use and land cover in the southern Oti catchment in 2000.

Table 2. Evolution of land use in the Southern part of the Oti catchment from 1988 and 2015.

	Areas						Annual rate of change (%)	
	1988		2000		2015			
Classes	ha	%	ha	%	ha	%	1988-2000	2002-2015
Ag	4744.81	0.6	5788.08	0.72	6492.59	0.81	1.67	0.88
Fcsb	140,820.6	17.52	130,958.57	16.3	79791.92	9.93	-0.61	-3.81
Fd	30,870.04	3.84	17,875.7	2.23	10,863.01	1.35	-4.55	-3.83
Fg	19,094.97	2.38	14,168.87	1.76	10,113.75	1.26	-2.47	-2.59
Mcj	85,367.68	10.63	209,738.8	26.11	245,860.21	30.6	7.49	1.22
Saa	513,819.36	63.95	417,630.89	51.98	441,290.14	54.93	-1.73	0.42
Sr	8693.41	1.08	7249.96	0.9	8999.25	1.12	-1.51	1.66
Total	803,410.87		803,410.87		803,410.87		-1.71	-6.05
Average							-0.25	-0.86

Ag: Agglomeration; Fcsb: Open forest and wooded savannah; Fd: Dense forest; Fg: Riparian forest; Mcj: Cultivation mosaic and fallow; Saa: Tree and shrub savannah Sr: Rock surface and bare ground.

14.4.3 Dynamics of land use and vegetation cover in the Oti catchment 1988 to 2015

Evolution of the areas of different land use units South of the Oti catchment is presented in Table 2.

Table 2 shows that during the two research periods (1988-2000; 2000-2015), urban district (1.67%; 0.88%) field and fallow mosaics (7.49%; 1.22%). The area of the other units is reduced, especially that of dense formations, with a rate of change of (-4.55%; -3.83%). The average annual rate of change is estimated at -0.25% for the first period and -0.86% for the second. Land use dynamics have therefore declined by 0.56% annually from 1988 to 2015. Figure 3 illustrates the initial state of analysis of the vegetation cover of the Oti catchment area in 1988.

From the analysis in Figure 3, the southern part of the Oti catchment area is covered with more than 60% of its surface area in wooded and shrubby savannah with less than 11% of the area under cultivation and fallow. Figure 4 shows the land use and occupation from 2000 to 2015 in the study area.

The examination of Figure 4 shows an extension to the centre and southwest of the study area, agglomeration units and especially crop and fallow mosaics from 1988 to 2000, which more than doubled in area (26.11%). To the northwest, an increase in the surface area of wooded and shrubby savannahs in 2000 is noted due to the dependence on dense formations, open forests and degraded wooded savannahs during this period. These processes of savannization and anthropization continue until 2015 (Figure 5).

The analysis in Figure 5 shows the same trends as in the previous period, this time with a strong extension of crop and fallow areas (30.6%) and bare soils (1.12%) with about 55% of savannah with trees and shrubs.

14.4.4 Evolution of land use units in accordance with topography

The analysis of the areas of land use units in the southern part of the Oti catchment area on a short scale (Table 3) made it possible to assess trends in the spatial dynamics of land use in the three high-altitude sectors (high, medium and low altitude).

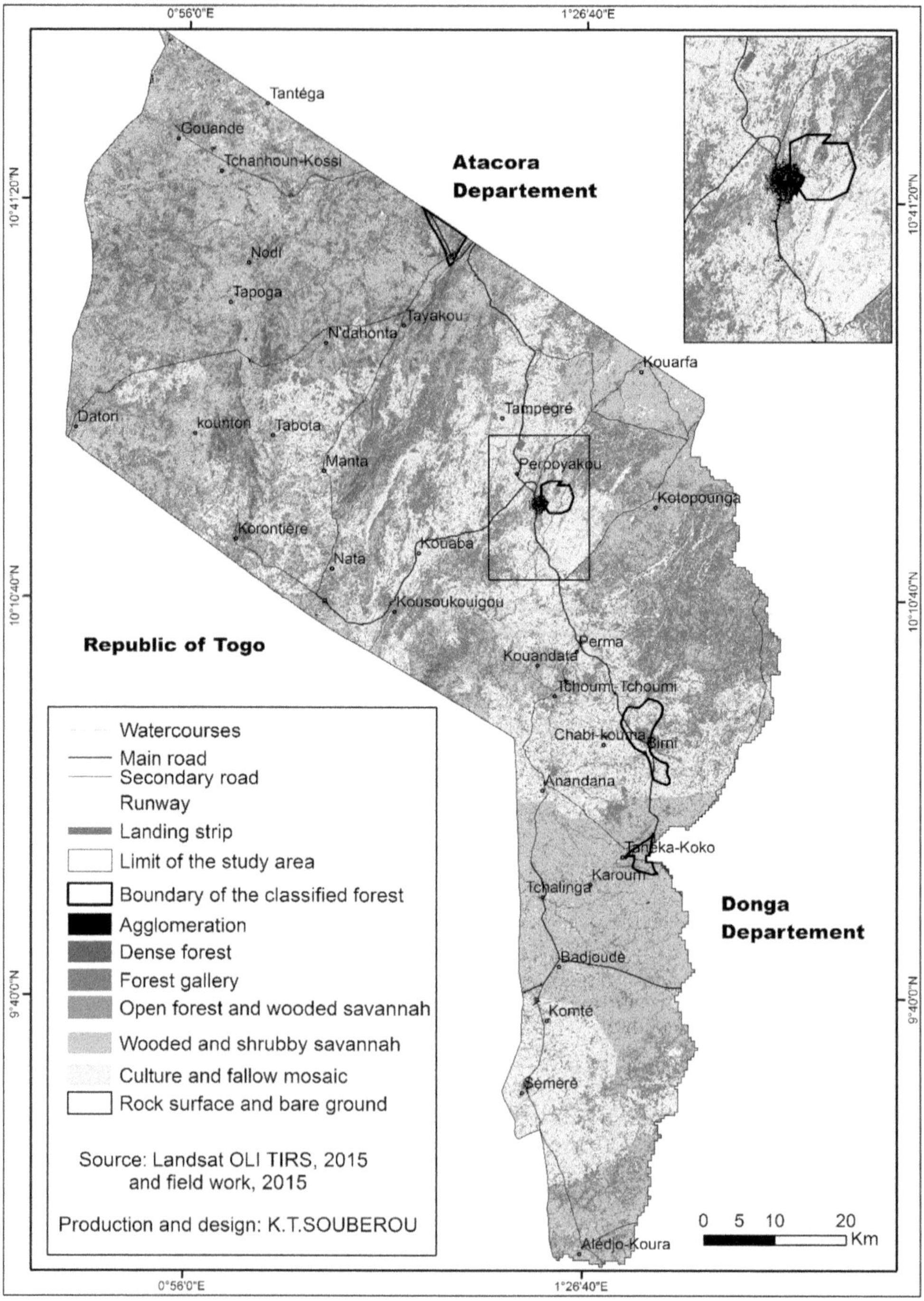

Figure 5. Land use and land cover in the southern part of the Oti catchment in 2015.

Table 3 indicates an increase of urban areas and mosaic of crops and fallows during the two periods and a reduction in forest forming. Confirming the processes of savannization and anthropization that are displayed through the conversion of land use and land cover classes. The obtained R indicator proves an anthropization over the entire study area. Negative indicators reflect an opening of environments (degradation of forest formations) under humans. There has been an acceleration since 1988 in the three altitude sectors: high (-91.09% in 1988-2000 to -95.79% between 2000-2015), medium (-92.03% in 1988-2000 to -96.32 between 2000-2015) and low (-89.66% in 1988-2000 to -95.7 between 2000-2015). Between 1988 and 2015, 19.77% of the high ground, 39.58% of the interfluves and 25.75% of the low height areas are occupied by crops. Concerning the low altitude sector, between 1988-2015, there was a more rapid anthropization than that of the high and medium altitude sectors. The low-lying sector is increasingly experiencing a mosaic profit in crops and fallows during the two periods whilst the two others stabilize between 2000-2015. Similarly, the high altitude sector began to recolonize in open forest and wooded savannah during the same period. Consequently, anthropization would migrate to the low altitude sector with hydromorphic soils, after having taken place mainly in the medium altitude sector. The scalable trend of agricultural spaces emanates the importance assigned to agriculture in the municipalities' economy constituting the research sector, the cultivable land depletion at high and medium heights and the putting in fallow of cultivated land over a long period (0 to 5 years). It is important to prospect the possible futures of land use and occupation in the Southern part of the Oti catchment in accordance with current trends.

Table 3. Land use statistics in the three altitude sectors (high, medium, low) and R as indicator of vegetation cover dynamics.

		Areas (ha)			Indicator R (%)	
Altitudes	Classes	1988	2000	2015	1988-2000	2000-2015
High (465-667 m asl)	Ag	502.04	635.61	682.14	-91.09	-95.78
	Fcsb	24,857.53	17,234.49	18,673.77		
	Fd	5596.28	3206.40	2606.40		
	Fg	2962.04	2408.36	2220.27		
	Mcj	10,976.80	20,088.64	23,886.26		
	Saa	72,687.28	69,646.46	70,310.64		
	Sr	3226.75	3351.15	3626.87		
Average (294-464 m asl)	Ag	3140.50	3658.45	4228.90	-92.03	-96.32
	Fcsb	45,650.48	29,987.71	8799.34		
	Fd	15,486.25	6778.02	2347.77		
	Fg	5542.17	3758.45	1991.41		
	Mcj	54,820.68	126,975.81	143,000.17		
	Saa	232,118.12	187,802.81	197,493.11		
	Sr	3845.62	2546.96	3779.64		
Low (118-293 m asl)	Ag	1102.27	1494.02	1581.56	-89.66	-95.77
	Fcsb	70,312.59	83,736.37	52,318.81		
	Fd	9787.52	7891.27	5908.83		
	Fg	10,590.76	8002.07	5902.07		
	Mcj	19,570.20	58,876.72	82,771.40		
	Saa	209,013.96	160,181.62	170,486.40		
	Sr	1621.03	1351.85	1651.85		

Ag: Agglomeration; Fcsb: Open forest and wooded savannah; Fd: Dense forest; Fg: Riparian forest; Mcj: Cultivation mosaic and fallow; Saa: Tree and shrub savannah Sr: Rock surface and bare ground.

14.4.5 Modelling of land use and land cover dynamics by 2050

14.4.5.1 *Validation of the simulation results*

The simulation of land use and occupation in 2015 based on real situations (1988 and 2000) made it possible to make a comparative analysis with the real situation in 2015 pre-classified.

For this model, at least nine out of ten pixels was correctly predicted and about 96.83% of the area observed between 2000 and 2015. This prediction highlighted the consistency of the transition rules for land use units. Following the analysis of the probability matrix (simulated), notwithstanding the observed confusions, the mapping accuracy is estimated at 94.56% and the proportions of the resulting areas are compared to the actual ones (Figure 6).

The analysis in Figure 6 shows that the proportion of areas (crop mosaic and fallow land; open forests and wooded savannahs, rocky surface and bare soil) calculated after the information from the projection is slightly higher than in 2015. This indicates that the level of reliability of the simulation is acceptable.

Table 4 presents the areas of land use units in 2050 under both ecological and socio-economic scenarios. It shows a progression of the mosaic of crops and fallows for the first scenario of 5.73% and a regression of 5.78% for the tree and shrub savannah class. As for the others (dense forest, open forest and wooded savannah), there is a conservative dynamism of forest ant savannah.

As far as the second scenario is concerned, it reveals that plant forming is in decline, taking advantage of crop and fallow mosaics and urban areas in 2050. The savannah will go through a reduction in its surface area (about 7.84%) while the cultivation and fallow areas will expand further. Thus, crop and fallow mosaics, which occupied 30.6% in 2015, will increase by 9.01% in 2050. At the same time, the agglomerations will increase from 0.81 to 0.87 for the same period. Equally, forests (dense, open and wooded savannahs) and wooded and shrub savannahs will be gradually reduced.

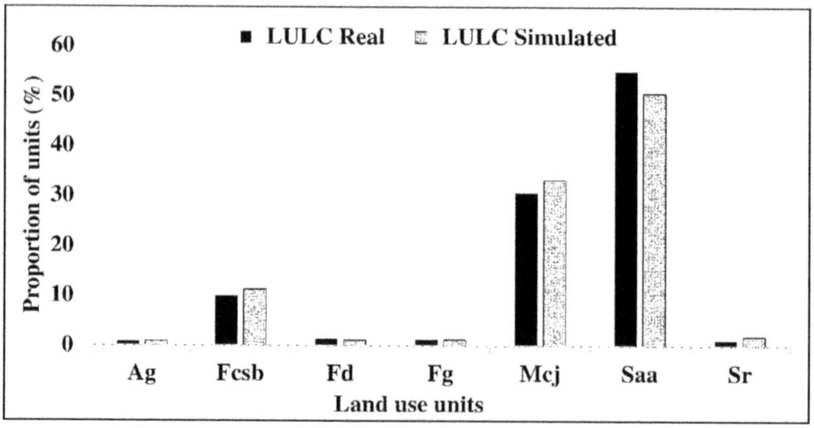

Figure 6. Distribution of the dominant land use, real and simulated in 2015 in the southern part of the Oti catchment area. Ag: Agglomeration; Fcsb: Open forest and wooded savannah; Fd: Dense forest; Fg: Riparian forest; Mcj: Culture mosaic and fallow; Saa: Wooded and shrubby savannah Sr: Rock surface and bare ground.

Table 4. Land use classes by 2050.

	Area					
	2015		2050 (1. optimistic)		2050 (2. pessimistic)	
Classes	ha	%	ha	%	ha	%
Ag	6492.59	0.81	6858	0.85	7033.26	0.87
Fcsb	79,791.92	9.93	80,029	9.96	75194.13	9.36
Fd	10,863	1.35	10,940	1.33	10,660.2	1.32
Fg	10,113.74	1.26	10,037	1.25	9966.17	1.24
Mcj	245,860.21	30.6	291,617	36.33	313,735.1	39.05
Saa	441,290.14	54.93	394,919	49.15	379,972.4	47.04
Sr	8999.25	1.12	8849.66	1.10	9010.87	1.13

Ag: Agglomeration; Fcsb: Open forest and wooded savannah; Fd: Dense forest; Fg: Riparian forest; Mcj: Culture mosaic and fallow; Saa: Wooded and shrubby savannah Sr: Rock surface and bare ground.

14.4.5.2 *Optimistic scenario in land use unit changes*

Projection of land use and occupation in relation to the environmental stability scenario to lead to a significant increase in plant forming (Figure 7).

From the observation of Figure 7, an extension of dense, open and wooded savannah forests in the low and high elevation areas emerges. Gallery forests, wooded and shrubby savannahs and mosaics of crops and fallows will experience a slight decline as well as bare soils.

14.4.5.3 *Pessimistic scenario in land use unit changes*

Figure 8 Illustrates land use and occupation in the southern Oti catchment area according to the socio-economic scenario by 2050.

The examination of Figure 8 shows a remarkable extension of crop and fallow mosaics throughout the study area in 2050 with a decline in tree and shrub savannahs and open and wooded forests. This prediction of the future of the area has shown that future conversions will allow an increase in the area of fields and fallows, especially in the northwestern part of the study area, specifically in the low-lying area. The unchanged land use units in the centre and south (mid- and high-altitude sector) are areas that have not yet been exploited or areas where the terrain is very uneven. This phenomenon of anthropization of the environment at this rate requires decision-makers to take measures to manage and conserve natural resources in order to anticipate irreversible degradation and enable it to play its role as a regulator in the hydrological cycle of the Oti River basin.

14.4.6 Measures to manage the vegetation cover of the Oti catchment

Agriculture has led to the progressive transformation (clearing, itinerant cultivation on burn) of land use outside the variation of climatic parameters and the decline in soil fertility. In the same way, the intensification of wildfires for hunting and agricultural fires leads to a change in the state of the various land use units.

According to field investigations, the dynamics of its land use units contribute to the degradation of biodiversity (depletion of certain forest and plant species,

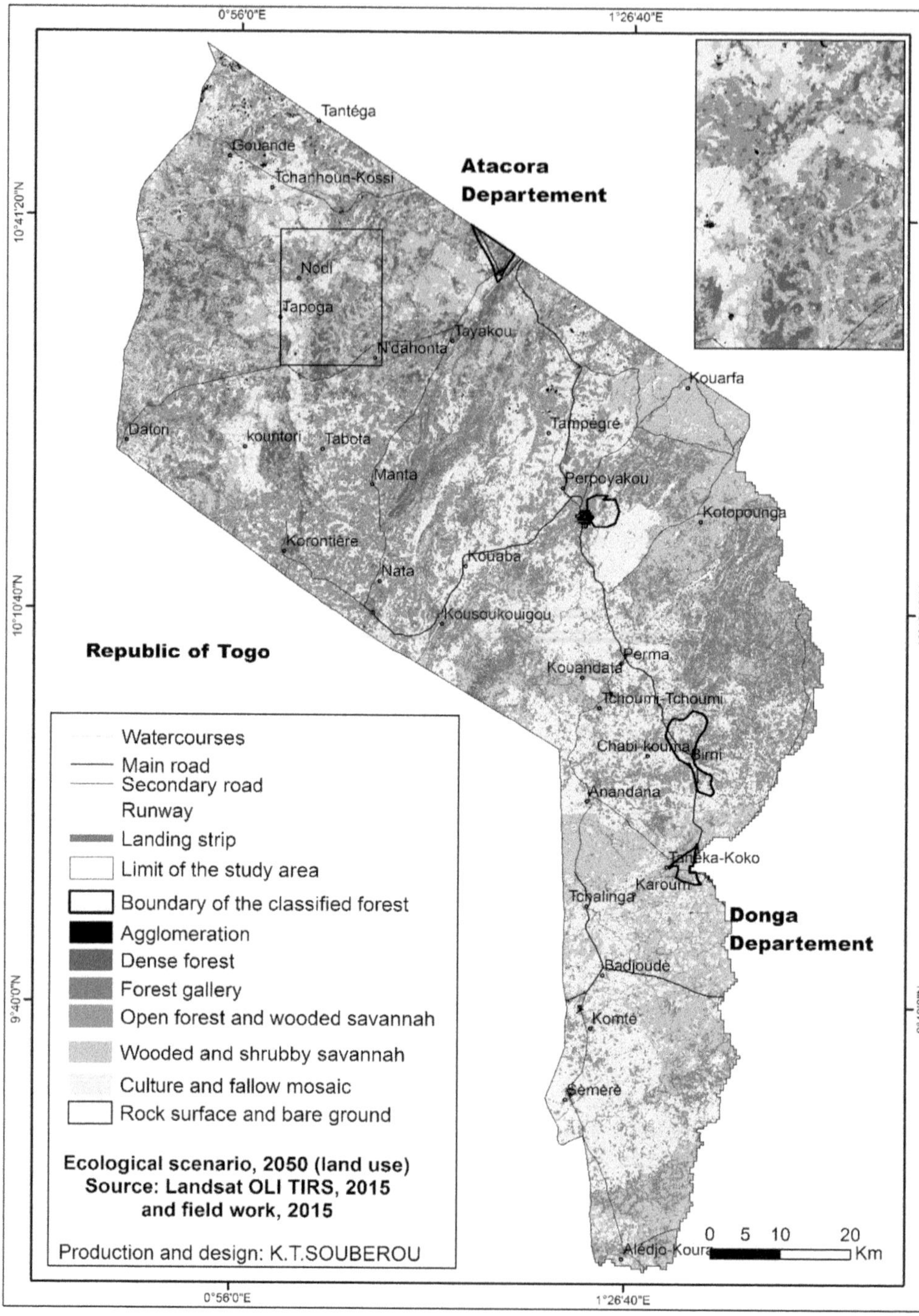

Figure 7. Projected land use and land cover in the southern Oti catchment according to the ecological scenario in 2050.

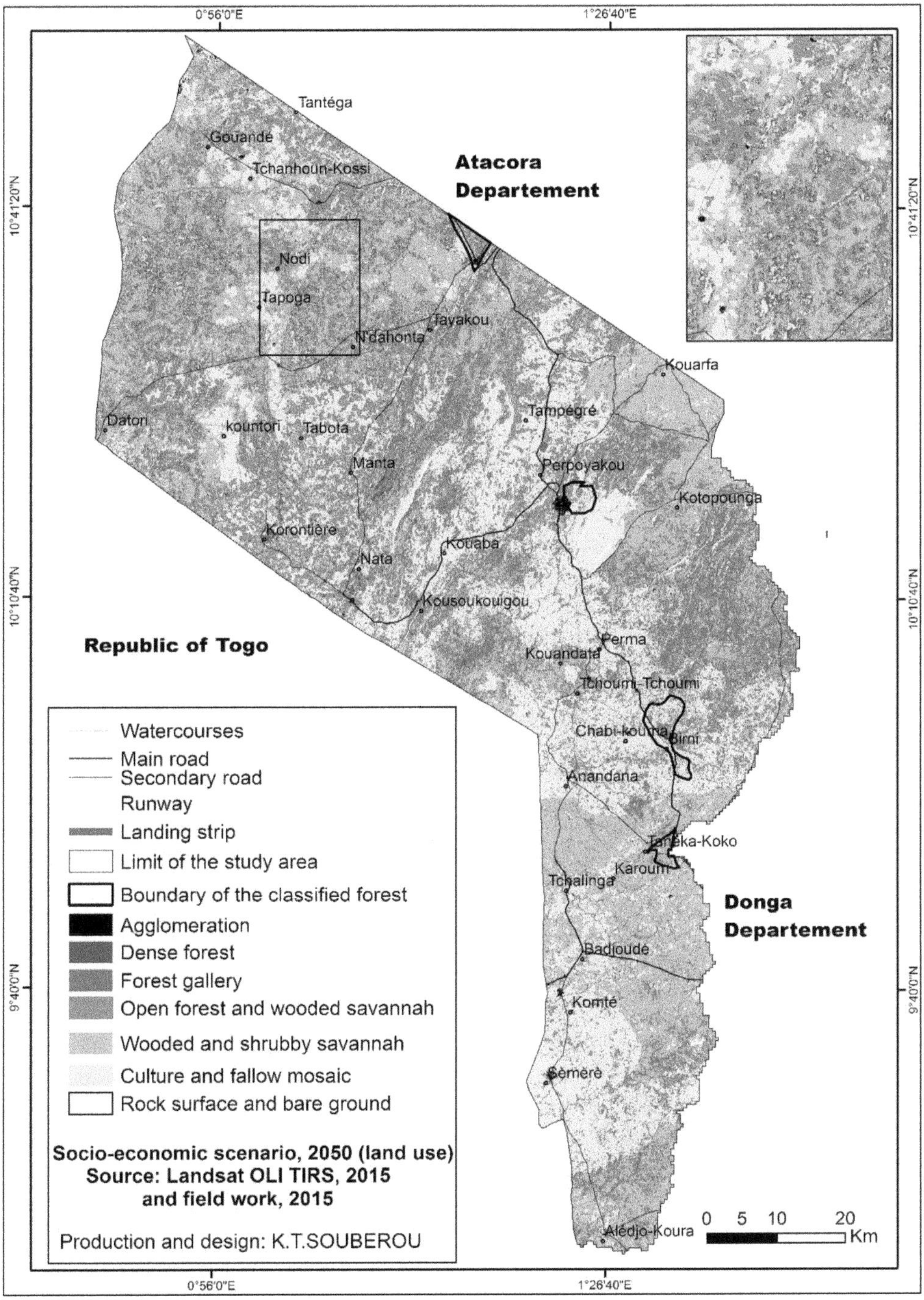

Figure 8. Projected land use and land cover in the southern Oti catchment according to the socio-economic scenario in 2050.

disappearance of animal species, etc.), the accentuation of erosion phenomena (creation of gullies, especially in the areas of the Atacora mountains and the peneplain and silting of areas alluvial zones) and the indirect allocation of climate parameters. Many actions are carried out by development agencies and NGOs when implementing projects in the context of natural resource management and conservation, soil erosion control and reforestation.

The results of the current analysis of the dynamics of land use and occupation in the Oti catchment area from 1988 to 2015 and prospective according to the socio-economic scenario (2050) condition a number of measures, including strict compliance with forest management laws, awareness-raising among the stakeholders, the establishment of rules for the extension of agricultural areas and deforestation and the implementation of reforestation campaigns to mitigate degradation. But, it will be necessary for the managers of the decentralized services of the state as well as NGOs, based on the results of this scientific research to do more planning in the integrated management of its plant formations so that they can experience a real progress more than envisaged by the ecological scenario.

14.5 DISCUSSION

In the southern part of the Oti catchment, land use and occupation units such as tree and shrub savannahs, crop and fallow mosaics and urban areas increased in size from 1988 to 2015 at the expense of dense forests, open forests and wooded savannahs and gallery formation. This analysis confirms that of Agbanou et al. (2018) on the dynamics of land use in the Natitingou-Boukoumbé agricultural extension sector with a decrease in forested areas and an increase in the area of anthropized units. Similarly, this observation is made by several scientific actors in different research sectors (Arouna 2012, Oloukoi 2012, Abdoulaye 2015, Azandégbé & Toko Imorou 2018, Raogo 2018) through the modelling of temporal spatial changes in land use. This degradation of plant formations recorded during the two study periods occurred with an average annual decline of -0.56%. This rate is lower than that obtained for all forest formations in Benin over the period 1990 to 2005 (Hountondji 2008). The expansion of agricultural areas and the exploitation of wood resources are leading to a loss of biodiversity, as Zakari (2015) states. According to Ouédraogo et al. (2015), agriculture under the expulsion of population growth is responsible for the direct impacts on land use and landscape configuration of the landscape. In this context, a prospective analysis of the evolution of the occupancy units is an essential prerequisite for the main orientations for the sustainable development and management of the Oti catchment as well as for agricultural policies. This analysis is similar to the one of Vigneau (2013) who believes that the application of ecological management hypotheses leads to the maintenance of open-ended forest formations until the reconstitution and stability of anthropic spaces or a clear reduction in their surface area.

14.6 CONCLUSION

The methodological approach to map land use and vegetation cover in the southern part of the Oti catchment has led to the establishment of a spatial database and the results confirm trivial and prospective modelling with the LCM.

The dynamics of vegetation cover in the study area are essentially marked by a decline in plant forming and an extension of crop and fallow mosaics from 1988 to 2015. This trend will continue on the basis of the socio-economic scenario until 2050.

This situation calls on the actors in charge of environmental management issues to take appropriate measures and preserve plant resources. But with the ecological scenario, it will have a low progression of plant forming (reconstitution) and a stability of anthropogenic formations.

REFERENCES

Abdoulaye, D. 2015. *Dynamique de l'occupation des terres et ses incidences sur l'écoulement dans le bassin versant de l'Ouémé à l'exutoire de Bétérou (nord-Bénin)*. Doctoral Thesis: 270. Cotonou: University of Abomey-Calavi, EDP/FLASH.

Agbanou, B.T, Orékan, A.O.V., Abdoulaye, D., Paegelow, M. & Tente, B. 2018. Dynamique spatio-temporelle de l'occupation du sol en zone d'agriculture extensive: cas du secteur Natitingou-Boukoumbe au nord-ouest du Benin. *Mélanges en hommage aux professeurs Thomas Omer, Houssou Sègè Christophe et Houndaga Cossi Jean. La géographie au service du développement durable*, Sep 2018, *Abomey-Calavi, Bénin*: 22–34.

Arouna, O. 2012. *Cartographie et modélisation prédictive des changements spatio-temporels de la végétation dans la Commune de Djidja au Bénin: implication pour l'aménagement du territoire.* Doctoral thesis: 252. Cotonou: University of Abomey-Calavi, GEN/FLASH/EDP/UAC.

Azandégbé, E.J. & Toko Imorou, I. 2018. Suivi diachronique de la consommation des terres agricoles par imagerie spatiale dans la commune de Djakotomey, Benin. *Mélanges en hommage aux professeurs Thomas Omer, Houssou Sègè Christophe et Houndaga Cossi Jean. La géographie au service du développement durable*, Sep 2018, *Abomey-Calavi, Bénin*: 107–126.

Bamba, I.A., Mama, A., Neuba, D.F.R., Koffi, K.J., Traoré, D., Visser, M., Sinsin, B., Lejoly, J. & Bogaert, J. 2008. Influence des actions anthropiques sur la dynamique spatio-temporelle de l'occupation du sol dans la province du Bas-Congo (R.D. Congo). *Sciences & Nature* 5(1): 49–60.

Boko, G.J. 2012. *Trajectoires des changements dans l'occupation du sol: déterminants et simulation. Cas du bassin versant de l'Alibori (Bénin, Afrique de l'Ouest)*. Doctoral thesis: 291. Cotonou: University of Abomey-Calavi.

Cohen, J. 1960. "A coefficient of agreement for nominal scales". *Educational and Psychological Measurement*. 20(1): 37–46.

Eastman, R. 2009. *Idrisi Taiga, Guide to GIS and Image Processing, manual version 16.02*. Worcester: Clark University.

FAO 1997. *Stratégie de la FAO relative au changement climatique*. Rapport d'étude, Rome, Italie.

Godet, M. 1997. *Manuel de prospective stratégique: une discipline intellectuelle*. Paris: Dunod.

Groot, R. & McLaughlin, J. 2000. *Geospatial data infrastructures: Concepts, cases and good practice*. Oxford: Oxford University Press.

Grouziz, M. 1986. *Péjoration climatique au Burkina-Faso: effet sur les ressources en eau et les productions végétales*. Paris: Colloque Nord est-sahel.

Hountondji, Y.C.H. 2008. *Dynamique environnementale en zones sahélienne et soudanienne de l'Afrique de l'Ouest: Analyse des modifications et évaluation de la dégradation du couvert végétal*. Doctoral thesis: 131. Liège: University of Liège.

Idiéti, E.M. 2012. *Les hydro-écorégions du Bassin de la Pendjari au Bénin: Analyse des déterminants socio-économiques et environnementaux de la dynamique des écosystèmes naturels*. Doctoral thesis: 229. Cotonou: University of Abomey-Calavi.

INSAE (Institut National de la Statistique et de l'Analyse Economique) 2014. *Résultats provisoires du recensement général de la population et de l'habitation. (RGPH2-4)*. Cotonou: INSAE.

IPCC (Intergovernmental Panel on Climate Change) 2007. *Climate change: Impacts, Adaptation and Vulnerability, Summary for policymakers*. Cambridge.

IPCC (Intergovernmental Panel on Climate Change) 2013. *Climate Change 2013: The Physical Science Basis*. Contribution of report (Stocker, T.F, D. Qin, G.K. Plattner, M. Tignor, S.K. Allen, J. Boschung, A. Nauels, Y. Xia, V. Bex and P.M. Midgley (eds.). Cambridge University Press, Cambridge, United Kingdom and New York, NY, USA.

IPCC (Intergovernmental Panel on Climate Change) 2019. *Oceans and Cyospheres: Dooming prospects for the planet, special report on climate change*. 52nd IPCC plenary assembly, 20 to 23 September, Monaco.

Issa, M.S. 2012. *Changements climatiques dans le moyen Bénin: impacts, analyse prospective des agro-systèmes et mesures d'adaptation/alternatives*. Doctoral thesis: 250. Cotonou: University of Abomey-Calavi, EDP/FLASH.

Kabré, M. 2008. *Les stratégies d'adaptations des populations aux changements climatiques dans le Sahel burkinabé (cas de Belgou dans la province du Seno)*. Master thesis: !. Ouagadougou: University of Ouagadougou.

Mas, J.F., Kolb, M., Houet, T., Paegelow, M. & Olmedo, M. 2011. Éclairer le choix des outils de simulation des changements des modes d'occupation et d'usages des sols. Une approche comparative. *Revue Internationale de Géomatique* 21: 405–430.

Oloukoi, J. 2012. *Utilité de la télédétection et des systèmes d'information géographique dans l'étude de la dynamique spatiale de l'occupation des terres au centre du bénin*. Doctoral thesis: 305. Cotonou: University of Abomey-Calavi.

Oloukoi, J., Oyinloye, R.O. & Yadjemi, H. 2014. Geospatial analysis of urban sprawl in Ile-Ife city (Nigeria). *South African Journal of Geomatics* 3(2): 128–144.

Orékan, V. 2007. Traitement et analyse des données LANDSAT 7. Elaboration de la carte forestière, Exécution d'un Inventaire Forestier National (IFN). *Projet Bois de Feu, phase II, Cotonou, Bénin.*

ORSTOM 1976. *Soil map for Benin (Carte pédologique de reconnaissance de la République du Bénin à 1: 200.000), feuille de Djougou, Natitingou et Sansame-Mango*. Paris: ORSTOM.

Ouédraogo, B., Ouédraogo, I. & Kaboré, O. 2015. Fragmentation de l'espace et conflits d'usage au sahel: cas du bassin versant de Yakouta (Burkina-Faso). *International Journal of Biological and Chemical Sciences* 9(6): 2727–2739.

Ouorou Barré, F.I. 2014. *Contraintes climatiques, pédologiques et production agricole dans l'Atacora (Nord-Ouest du Benin)*. Doctoral thesis: 241. Cotonou: University of Abomey-Calavi, EDP/FLASH.

Pontius, R.G., Boersma, W., Castella, J-C., Clarke, K., de Nijs, T., Dietzel, C., Duan, Z., Fotsing, E., Goldstein, N., Kok, K., Koomen, E., Lippitt, C.D., McConnell, W., Mohd Sood, A., Pijanowski, B., Pithadia, S., Sweeney, S., Trung, T.N., Veldkamp, A.T. & Verburg, P.H. 2008. Comparing input, output, and validation maps for several models of land change. *Annals of Regional Science* 42(1): 11–27.

Puyravaud, J.P. 2003. Standardizing the calculation of the annual rate of deforestation. *Forest Ecology and Management* 177(1–3): 593–596.

Raogo, N.G. 2018. Dynamique du couvert végétal et implications socio-environnementales à la périphérie du parc W/Burkina Faso. *VertigO - la revue électronique en sciences de l'environnement* 18(1).

République du Bénin 2000. *Bénin 2025: Alafia. Stratégies de développement du Bénin à long terme*. Cotonou: République du Bénin.

Schulz, J., Cayuela, L., Echeverria, C., Salas, J.& Benayas, R. 2010. Monitoring land cover change of the dryland forest landscape of Central Chile (1975–2008). *Applied Geography* (30): 36–447.

Souberou, K.T. 2013. *Contribution de la télédétection et du SIG à la caractérisation des bas-fonds de la Commune de Matéri (Nord Ouest, Bénin)*. Master thesis: 135. Ile-Ife: Universtiy of Obafèmi Awolowo, Centre Régional de Formation aux Techniques des Levés Aérospatiaux (RECTAS).

Souberou, K.T., Toko Mouhamadou, I., Ogouwale, E. & Agbossou, K.E. 2017. Dynamics of the Inland Valley Occupation in the Oti Watershed in Benin: Diagnosis and Prospective Analysis. *International Journal of Sciences* 12: 8–21, DOI: 10.18483/ijSci.1484.

Sounon, B., Sinsin, B. & Goura Soulé, B. 2007. Effets de la dynamique d'occupation du sol sur la structure et la diversité floristique des forêts claires et savanes au Bénin. *Tropicultura* 25(4): 221–227.

Tohozin, C.A.B. 2016. *Etalement urbain et restructuration de la ville de Porto-Novo, Bénin*. Doctoral thesis: 317. Cotonou: University of Abomey-Calavi, EDP/FLASH.

Vigneau, C. 2013. *Cartographie et modélisation des changements d'occupation des sols dans le Haut-Vicdessos 1942-2008*. Master thesis: 75. Toulouse: University of Toulouse.

Vissin, E.W. 2007. *Impact de la variabilité climatique et de la dynamique des états de surface sur les écoulements du bassin béninois du fleuve Niger*. Doctoral thesis: 267. Burgundy: University of Burgundy.

Zakari, S. 2015. *Vulnérabilité des parcours naturels aux changements climatiques dans le bassin de la Sota à l'exutoire de Coubéri (Bénin)*. Doctoral thesis: 203. Cotonou: University of Abomey-Calavi.

CHAPTER 15

Governance of the Pendjari Biosphere Reserve in view of climate variability

H.B.M. Gbénahou
University of Abomey-Calavi, Cotonou, Benin

K. Sokemawu
University of Lomé, Lomé, Togo

ABSTRACT: The economic importance for tourism of the Pendjari Biosphere Reserve (PBR) in northern Benin raises the question of its governance in view of climate variability and ongoing climate change. This paper examines stakeholders' perceptions, forms, effects and consequences of climate variability in and around the PBR. It identifies strategies used by the management people and surrounding local populations in adapting to climate change. It also analyses current and future governance in nature conservation in view of environmental change. Recent indicators of climate change are documented by delayed rainfall, shorter and irregular rainy seasons, changes in rainfall over time, excessive heat, and increased wind speed. The main strategies adopted are varietal adaptation, extension of planting, adoption of open seedlings and diversification of activities. Consequently, the sustainability of such governance requires the adoption of new paradigms.

15.1 INTRODUCTION

Climate change has been a reality since the 1970s. Because of its visible effects it has become a major issue for political decision-makers, development partners and scientists (Niang 2009). This engagement is based on the fact that climate variability presents obstacles for human development (Boko 1988, Brown & Crawford 2008). In spite of the necessary contextualization of their manifestations, the scientific community generally retains major foci to characterize climate change by rainfall variability and rise in temperature (Sircoulon 1990; Bergonzini 2004). Climate variability and climate change give rise to fears of important disturbances in the linkage of community and environment. These disturbances represent a potential source of tension and conflict because they affect both, socio-economic activities and natural resources. The imbalances they carry, if not urgently controlled, cause profound upheavals in natural ecosystems and produce social disarray, all of which compromise "the capacity of ecosystems to provide, maintain and regulate their essential functions and services, including resistance to climate variability and natural disasters" (Distefano 2012: 5). It is therefore a truism to assert that climate variability influences the spatiotemporal dynamics of natural resources and poses serious threats to their sustainability. These also affect the Pendjari Biosphere Reserve (PBR) in northern Benin.

This paper aims to: i) study, by category of stakeholders, perceptions and manifestations of climate variability; ii) list the effects and consequences they have on ecosystems of the PBR and surrounding local populations; iii) identify strategies used by the PBR management people and the local populations in adapting to the effects and constraints of climate variability, and finally, it analyses the ecology as local, global and future governance of the PBR to see whether or not it will integrate the new challenges imposed by climate variability and change. Two fundamental questions will be tackled: (i) How are the manifestations and effects of climate variability and change perceived and understood in and around the park, and how do the present and future modes of governance integrate them; and (ii) Which roles do territorial communities play?

15.2 CLIMATE VARIABILITY AND GOVERNANCE: FROM CONCEPTUALISATION TO EXPERIMENTED MEASUREMENTS

15.2.1 Conceptualisation of Climate Variability (CV)

Olivier de Sardan (1995) suggests a multidimensional empirical study of climatic variability by social groups that are exposed to effects of interactions which are created, or which they were obliged to create, practices and representations which are formalized, or which they were forced to formalize, and this in a diachronic and prospective perception. "Climate change is a change that is attributed directly or indirectly to human activity altering the composition of the global atmosphere and adding to the natural variability of the climate observed over comparable periods of time" (United Nations 1992: 4). In general, climate variability refers to natural variations within the atmospheric system (Al Hamndou & Requier-Desjardins 2008). These are variations that can be detected, and that can be significant within the same ecological zone (intra) or in between different zones (inter). With regards to the stability of the notion "variation" in the manifestations of climate change (rainfall, temperature, wind) and the articles on the phenomenon as well as the difficulties encountered when trying to dissociate variability and climate change, it is the concept of "Climate Variability (CV)" that is used in this article. Understood as the significant inter-annual variability of different climate parameters compared to long-term average values, whether natural or anthropogenic (Niasse et al. 2004). Therefore, as changes are affecting climate parameters, the CV concept considers all aspects related to significant disruption of relations between humans and the environment. To address issues and impacts of CV, it is important to distinguish between current CV and medium to long-term CV. The latter are generally much larger than the specific questions related to current CV. The need for anticipation and preparation time are the best option (C2A 2017).

Two types of CV manifestations are distinguished: CV measured with appropriate scientific tools (instruments) and CV individually felt (perception). The latter can, according to C2A (2017), be smaller than the measured changes, either because the ecosystems are able to reduce these changes, or because CV does not really create new constraints, especially for agricultural populations. When CV, even limited ones, have significant effects because of high vulnerability of the ecosystem, or the income of a family (e.g. a life in poverty), the perceived CV can be more significant than the meteorologically measured CV.

Individuals, communities and the biophysical environment response to CV induced changes by adaptation. Adaptation consists of an adjustment to a climatic stimulus or its current or intended effects (Burton et al. 2002). Adaptation of individuals or a community consists of a process that first takes root in socialisation, social and political learning, and is expressed through mechanisms. This socialisation subsequently influences the choices of groups and individuals to deal with climate stress. Adaptation can also be based on learning resulting from periods of upheaval (Adger 1999; Adger & Kelly 1999; Brooksa and al. 2005). Adaptation is a multiple and variable concept whose content and perception differ from one population to another and depends on several factors. Adaptation discusses the strategies, initiatives and measures imagined and taken individually or collectively to reduce CV induced vulnerability to people and their environment. These strategies can be anticipatory or reactive and be influenced by "the evolution of average climatic conditions (increase in the random nature of climatic parameters, increase in the frequency and intensity of extreme events or accidents (droughts, heavy and/or severe precipitation, cyclones, or extreme heat/cold, and/or longer ones)" (C2A 2017: 10-11).

15.2.2 Characterisation of Governance System

Governance, although now a key term in public development aid policies (Olivier de Sardan 2009), is a strongly impaired notion that needs to be clarified. For Benko and Lipietz (1995), governance is a mode of regulation that is not based either on market mechanisms or on state intervention, but rather on civil society. The increasingly frequent use of this concept implies, according to Contandriopoulos (2008: 193-195), "changes at the three levels in which regulation is exercised: macroscopic (that of state structures); mesoscopic (that of the organisations) and microscopic (the relations between the individual actors)".

At the macroscopic level, it leads to the ideas of greater direct participation of citizens in decisions, increased decentralisation to better utilise and empower skills and the adaptation of the governance system to the local particularities (Contandriopoulos 2008). It also refers to greater openness to new forms of financing and the increased use of inter-sector dynamics and accountability to assess the consequences of public policies and actions.

At the mesoscopic level, it brings up to the idea that, in order to eliminate the barriers and rigidities of the current system, it is necessary to operate in networks, introduce internal market mechanisms, decentralise decisions and generalise systems of evaluation and continuous improvement of quality.

At the microscopic level, it raises the need to organise the spaces of expression and tension that exist, so that the decisions made are understood and perceived as legitimate by all stakeholders. To operationalise, governance relies on three systems that do not work on their own: a management, an information and a financing system. The management system is a set of rules that define the distribution of power and responsibilities. The information system consists of all the data and their operating system needed to make the organized system of action intelligible and transparent at all times for professionals, managers, planners, patients and the public. The financing system includes all the incentives conveyed by the financing modalities of the system, the mechanisms for allocating budgets and the payment mechanisms of the actors. To activate these systems, it is not enough to direct, manage and evaluate them (Saltman & Ferroussier-Davis 2000).

As a fundamental point of governance, the need for management implies the existence of a multi-stakeholder melting pot, formalised or not, with a view to integrated, dynamic, and coherent perfomances (capacity to achieve its mission or goal achievement). It also requires (i) adaptation to the environment (acquiring and controlling resources, meeting the expectations of the population), (ii) producing quality services with productivity, and (iii) maintaining or developing common values (organisational culture). Governance thus connotes the action of piloting a system, coordinating a collective action (Paquet 1999: 23-24) in which the organisational order and power are negotiated between the involved stakeholders. This configuration decentralizes and multiplies the decision spaces. Ultimately, it is a question of rehabilitating of certain categories of actors who were formerly marginalized and excluded by the governors. It is therefore in itself a process of "permanent construction" (Nguinguiri 2003: 21) which involves a plurality of actors and, by extension, a plurality of power authorities and which obliges the hybridization of the coordination of individual actions, imposed or impulsed by the State, and forms of collective action of public interest or community interest (Leyronas et al. 2016). Governance, "is neither a system of rules, nor an activity, but a process. It is not based on domination, but on accommodation. It is not formalized, but focuses on continuous interactions and involves both private and public actors" (Nguinguiri 2003: 22). It is clear that the empirical documentation and analysis of governance makes it necessary to focus on the modalities of governance and thus to focus on the concrete forms of action commonly referred to as mode of governance. According to Olivier de Sardan (2009: 8-9), local governance "includes all the ways in which an institution (formal or informal local, public or not) delivers such public or collective goods or services and manages the symbolic and material resources that it controls for this purpose in the name of certain conception of its interests and the public or collective interest. Each mode of local governance has its specific forms of authority and legitimacy, whether or not accepted or contested, whether or not effective with regards to the delivery of goods and services".

The socio-anthropological study of governance in the Pendjari Biosphere Reserve (PBR) in times of greater risk of climate variability requires the examination of onsite and offsite conditions, namely the differences between official administration and practical standards on the ground (Olivier de Sardan 2008).

15.2.3 Study area, context and methods

15.2.3.1 *Pendjari Biosphere Reserve*

The Pendjari Biosphere Reserve (PBR) is located on the northwestern edge of the Republic of Benin between 10°30 to 11°30 northern latitude and 0°50 to 2° eastern longitude (Figure 1). It is part of the W-Arly-Pendjari ecosystem (WAP), one of the largest (~35,000 km^2) protected area complexes in West Africa (Pendjari Biosphere Reserve, Transboundary Biosphere Reserve "W" [Benin, Burkian Faso, Niger], and protected areas of Burkina Faso [Reserves Arly and Singou]). The PBR is located in the Sudanian climate-ecological zone characterised by alternation of a dry season (October to April) and a rainy season (May to September). The average annual rainfall varies between 800 and 1000 mm (north-south gradient). Air temperature annually oscillates between 18 °C and 37 °C. The vegetation of PBR is savannah, shrub savannah (dominant), tree savannah ('woodland'), open forests and gallery forests. Its faunal diversity makes the PBR one of the most important sites in West Africa and a viable international destination for ecotourism. In Benin, it covers an area of 4711.4 km^2 including 2750 km^2 for the PBR, 1800 km^2 for the hunting zone of the Pendjari, and 251 km^2 for the Atacora hunting area (Konkombri) (UNESCO 2002). It extends over the communes of Tanguiéta, Materi and Kérou (Figure 1). Between 1996 and 2017, PBR was governed by the National Center of Management of Wildlife Reserves (CENAGREF 2015), together with the Association of Wildlife

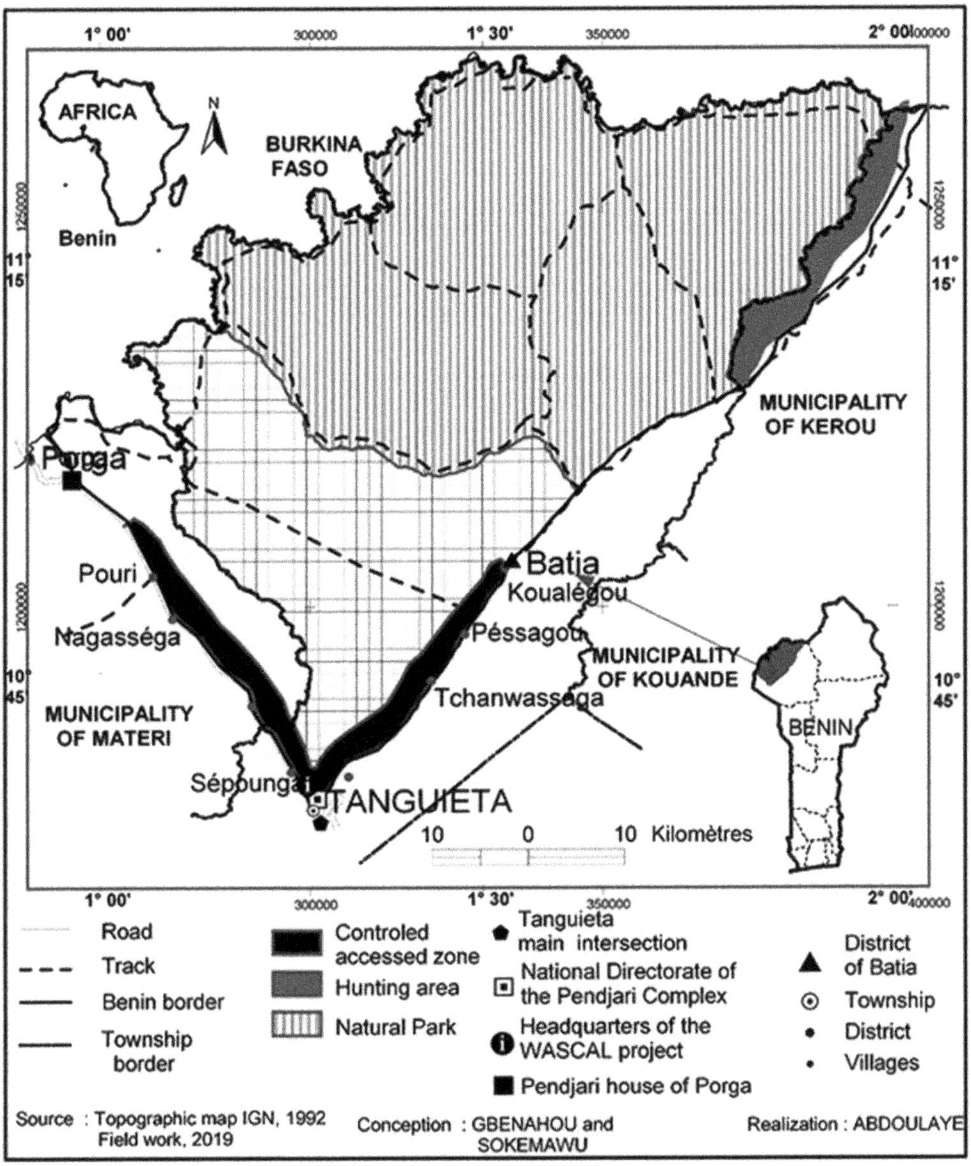

Figure 1. Location of the Pendjari Biosphere Reserve (PBR) in northern Benin.

Management (AVIGREF), a community organisation through which populations learn responsibility for management, ecological monitoring and development.

In July-August 2017, governance of the Pendjari Biosphere Reserve (PBR) was transferred to the African Parks Network (APN), an international non-governmental organisation (NGO) specialized in sustainable management of natural resources (rehabilitation of animal habitats, reproduction of animals, and others). The Benin government has granted APM the management right for a period of 10 years. APN will have to invest 26 million US dollars in order to improve governance of quite complex faunal and floristic resources. However, since the arrival of APN, involvement of AVIGREF and other stakeholders has not been taken into account (shortcomings in participation and governance).

15.2.3.2 *Data collection*

Socio-historical and micro-analytical documents were first consulted. Then, empirical data collection combined open interviews (individual, group), 'free' observations and mapping of stakeholders followed. In order to highlight the perceptions on climate variability, data were collected from institutional, organisational, community and household sources.

At the institutional level, the actors involved were: National Agency for Heritage and Tourism Development (ANPT), National Center for the Management of Wildlife Reserves (CENAGREF 2015), Pendjari Biosphere Reserve National Office (PBR-NO), African Park Network (APN), German Cooperation (GIZ), Departmental Office of Craft and Tourism, Communal Services of Water and Forests, and Local Communities Federation (LCF). Especially LCFs of Tanguiéta and Materi were selected because the study considers stakeholders with an official entry permit to the PBR, depending heavily on its natural resources. At the level of Local Communities four types of actors have been met: Natural Resources (NR) focal point, the President of the Communal Commission for Environmental Affairs, and the Chief of Financial and Economic Affairs Services (C/FEAS). At the organisational level, the stakeholders met were the Association of Wildlife Management Union (U-AVIGREF), the fishermen's association, the tourist guide associations, the various service providers, etc. With institutional and organisational actors, twenty individual interviews were conducted.

Local communities' federation of Tanguiéta and Materi were selected because the study especially considers locals with an official entry permit to the PBR. At the household level, the collected data related to the perceptions of CV and to individual adaptation strategies. In assigning scores to these manifestations and effects of CV, the experiences of the populations were taken into account. These experiences were systematized over time through direct observations and practices. Such experiences and feelings generally change discourses, practices (park governance and agricultural practices) and lifestyles. Perceptions and understandings legitimize by actors constitute local knowledge likely to help apprehend and explain the recent evolution of climate parameters. With populations, five group interviews were implemented into two phases.

First, the phase of identifying CV events went through two sub-phases. During the first sub-phase, all events were recorded without analysis but during the second sub-phase, refocusing and categorisation work were implemented. This work made it possible to analyse the plausibility of the manifestations of the CV evoked, to cross-check those that intersected and to eliminate those, which are not real or less important. This made it possible to systematize with the actors nine indicators of change in the climatic parameters showed: delay of the rains, variation of the rain over time, variation of the rain in the space, shortening of the duration of the rainy season, increased high winds, excessive heat, increased hot days, lower water levels and flooding. These indicators are used to address the results and the revealed frequencies shown in Figure 2 were obtained after triangulation of collected data.

Then, there was the evaluation phase itself, during which actors in the interview groups had to assign scores of "1 to 3" which correspond respectively to weak/low, strong/high and very strong/very high to each indicator of climate change. This is

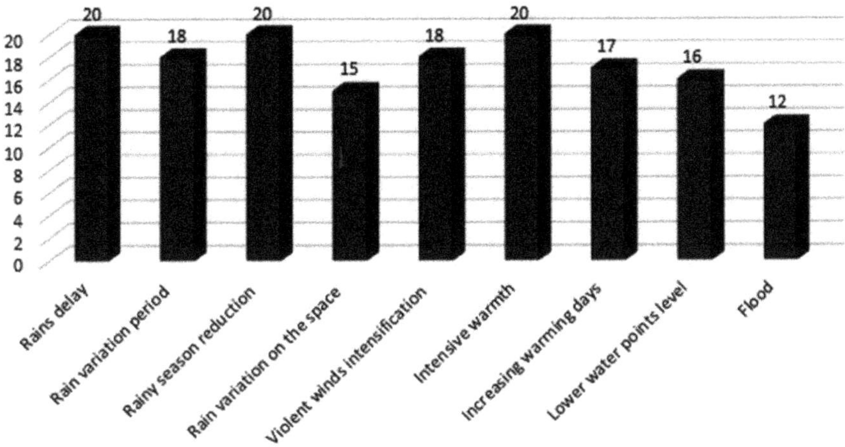

Figure 2. Revealing indicators frequency to climatic parameters change.

a bivariate evaluation. Indeed, two evaluation criteria were defined: the intensity and the persistence of the indicator. The first criterion "the intensity" makes it possible to evaluate the severity of the indicator of change to climatic parameters. The question that underpinned the discussion of the evaluation criterion is whether the mentioned phenomenon via the indicator affects to a great or a small extent the governance of the park, the activities that generate income and livelihoods. The second criterion "persistence" makes it possible to evaluate the seasonality or the regularity of the indicator of change of climate parameters considered. It offers the opportunity to make an analysis of the dynamics over time of each of the indicators. The question that underlay the evaluation of this criterion is whether if it is a phenomenon that occurs every year or sporadically?

15.3 LOCAL PERCEPTIONS, UNDERSTANDING AND ADAPTATION TO CLIMATE VARIABILITY

15.3.1 Manifestations of climate variability (CV)

15.3.1.1 *Local perceptions*

Most of the surveyed actors emphasise changes in time and space, in climate parameters that they perceive as upheavals. According to them, CV is felt through modifications of climatic parameters such as rainfall, temperature and wind (Figure 2). Based on their observations and experiences, they emphasise the disturbances that occurred with respect to these parameters:

> *"When we talk about climate variability, what comes to mind from what we see is that we have a kind of disruption in the number of climate-related factors. To be more precise, there are for example, the questions related to rainfall: there is the rainfall regime which changes the distribution of rainfall in time and space that varies seriously, the intensity of the rain also changes (for a very short number of days). There is a large amount of rainfalls, then there are porches of drought more and more extensive observed during the rainy season). Apart from the rainfall factor, there is temperature that is changing. There is an increase in*

temperature, that is, an exacerbation of excess heat, an increase in the number of hot days and an extension of drought periods. Only those phenomena allow me to say that we are facing a change in the climate regime" (Respondent no. 11, Tanguiéta, September 2019).

The majority of respondents agreed that Tanguiéta and Materi are experiencing higher temperatures. The dominant perception highlights condensation in the atmosphere of two types of heat: the direct heat coming from the sun, and the heat that the earth emits. According to them, there is during the day a recrudescence of heat due to CV. Generally, nights are known to be colder than days. However, for more or less five years, the nights are also hot because the earth emits the heat that was stored during the day.

Apart from the climatic parameters they observed a lengthening of the dry season and shortening of the rainy season have had a severe impact on agricultural activities and reforestation activities in the PBR. They also refer to the PBR as a natural barrier that seriously mitigates the manifestations of CV currently felt although they increase extreme intensity.

"The peasants have to be very intelligent in knowing when to start growing crops. When we try to reforest, we are facing a big problem because the rainy season is not long enough while the dry season is too long. Sometimes, you have the plantation that emerges well, but the dry season being so long and the dry days in the rainy season growing more and more, it ends up burning. In view of the changes in climatic factors observed in recent years, we believe that the PBR has a great 'chance' because it has an advancement of the desert and reduction of the severity/intensity of the manifestation barriers. In fact, despite the feelings people have, the PBR helps to create micro climates. If it had not been there, the phenomena we feel in Tanguiéta would be more acutely and extend well beyond the commune" (Respondent no. 1, Tanguiéta, September 2019).

Among the indicators of changes in climate parameters, delay of rains, shortening of the duration of the rainy season, excessive heat, variation of the rains in time and intensification of violent winds were most frequently mentioned (Figure 2). These results are consistent with those obtained by recent studies (Bambara et al. 2013, Sanou et al. 2018).

Crosschecks made by taking into consideration the assigned scores for persistence and intensity allow the establishment of four categories (Figure 3, Table 1). The first category includes three indicators of change in climate parameters, the second category only one indicator, the third category four indicators and the fourth category also one indicator. The results of the evaluation according to actors involved in the management of the Pendjari Biosphere Reserve are illustrated in Figure 4 (members of African Parks Network and Association of Wildlife Management).

According to the actors involved in the governance of the park, the delays of the rains, the variation of the rains in time, the accentuation of the violent winds, the excessive heat and an increase in the number of hot days are the most significant indicators of change in climate parameters. The actors involved in the direct management of the Biosphere Reserve pointed out five indicators of climatic parameters change under category "Very high intensity and very strong persistence". Under category "High intensity and very strong persistence", they put only one indicator and under the category "High intensity and strong persistence" they post three indicators (Table 2).

When triangulating the two evaluations, two categories: "Very high intensity and very strong persistence" and "High intensity and very strong persistence" overlap. Considering the first category "very high intensity and very strong persistence", it is obvious that three indicators were accommodated on both sides,

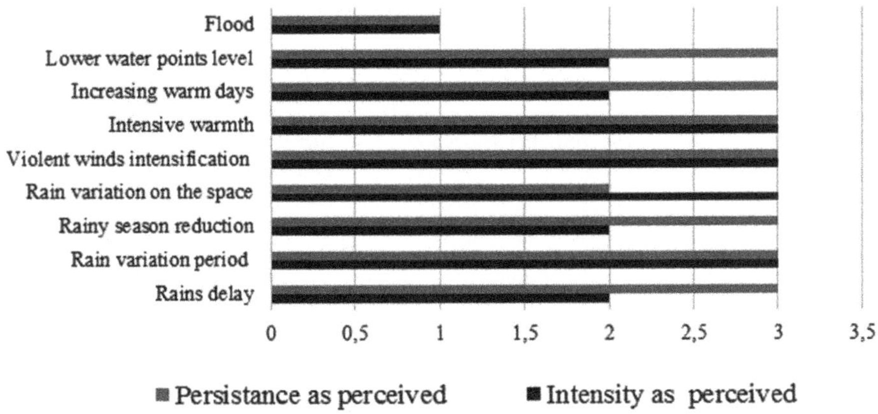

Figure 3. Climate change assessment indicators by non-involved stakeholders in Pendjari Biosphere Reserve (PBR) direct management.

Table 1. Categorisation of assessment carried out with non-involved stakeholders in the Pendjari Biosphere Reserve management.

Category	Indicators of change in climatic parameters
Very high intensity and very strong persistence	Excessive heat; Intensification of violent winds; Variation of rainfall over time.
Very high intensity and strong persistence	Variation of rainfall in space.
High intensity and very strong persistence	Rainy Delay; Shortening of the duration of the rains; Increase in the number of hot days; Lower streams level.
Low intensity and weak persistence	Flood

Source: From the results of the evaluation according to actors, October 2019.

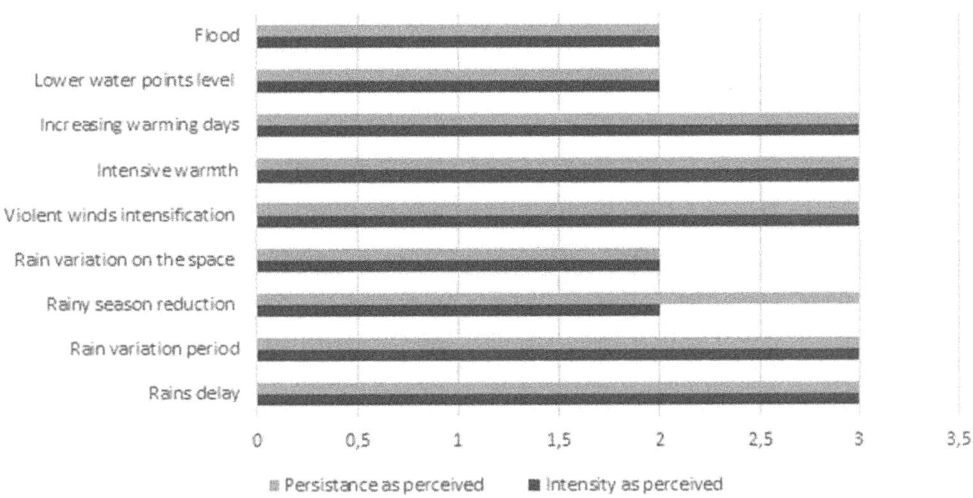

Figure 4. Evaluation of climate variability (CV) event indicators by stakeholders directly involved in the Pendjari Biosphere Reserve park management.

Table 2. Categorisation of assessment carried out with stakeholders directly involved in the Pendjari Biosphere Reserve management.

Category	Climate Change Indicators
Very high intensity and very strong persistence	Increase in the number of hot days; Excessive heat; Accentuation of strong winds; Variation of rainfall over time; Rain delay.
High intensity and very strong persistence	Shortening duration of the rains.
High intensity and strong persistence	Variation of rainfall in space; Lower water levels; Flood.

Source: From the results of the evaluation according to actors, October 2019.

namely: excessive heat, increased high winds and changes in rainfall over time. However, when considering the second category "Strong intensity and very strong persistence" only the indicator "shortening of the rain duration" is mentioned on both sides. This overlap gives evidence of the importance of these indicators. Moreover, the indicators that are quoted in the two overlapping categories are similar to the most evoked indicators (Figure 2) by the surveyed stakeholders. It was evidenced by these evaluations that stakeholders clearly perceive impacts of CV.

15.3.1.2 *Measured impacts of climate variability (CV)*

In April 2018 the NGO *Action-Conseils pour le Développement Durable* (ACDD) conducted a survey of WAP parks and measured rainfall and temperature to indicate changes overtime due to CV.

According to ACDD (2018) analysis of rainfall relating to CV in the PBR region over the period 1950-2015 shows two striking sub-periods (Figure 5). The first from1950-1980 is generally characterised by an excess tendency of rainfall heights.

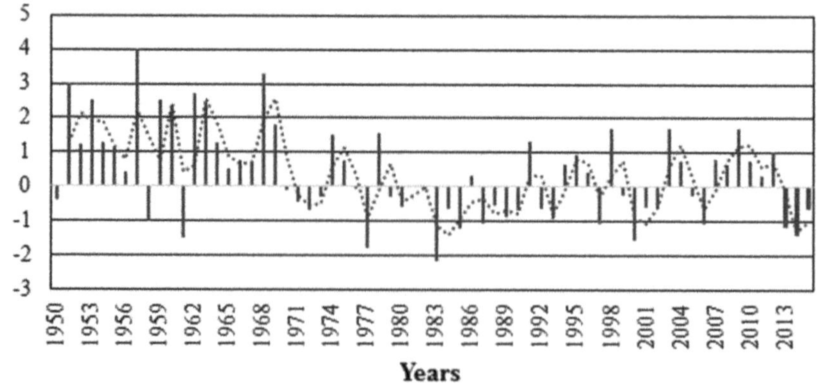

Figure 5. Rainfall anomalies 1950-2015 for Natitingou and Tanguiéta in the area of the PBR (Source: ACDD 2018: 25).

This sub-period has nine years of deficit compared with twenty years of surplus. The second sub-period from 1981-2015 is characterised by increased inter-annual rainfall variability with twenty deficit years and fifteen surplus years. This underlines that in the Natitingou/Tanguiéta area a general decrease in rainfall and multiple years of rainfall deficiency is recognisable.

The minimum temperature anomalies (Figure 6) presented also combine two sub-periods. In the 1960-1998 period, the trend is clearly characterised by low minimum temperatures (minimum annual temperature below the average). Over this sub-period, 84.6% of the years were 'cold'. However, over the sub-period 1999-2015 there is a tendency of increasing annual average temperature (70.5% of 'hot' years were registered).

'Cold' and 'warm' years are shown in Figure 7 compared to the maximum air temperature in Natitingou/Tanguiéta. It generally reveals that the period 1960-2015 includes two sub periods: 1960-2004 with a tendency to lower maximum temperatures ('cooling trend', 60.7%) and 2005-2015 with a clear warming trend. Thirty-four 'colder' years were identified against twenty-two 'warmer/hotter' years.

Figure 6. Minimum temperature anomalies 1960-2015 for Natitingou/Tanguiéta in the area of the PBR (Source: ACDD 2018: 25).

Figure 7. Maximum temperature anomalies 1960-2015 for Natitingou/Tanguiéta in the area of the PBR (Source: ACDD 2018: 25).

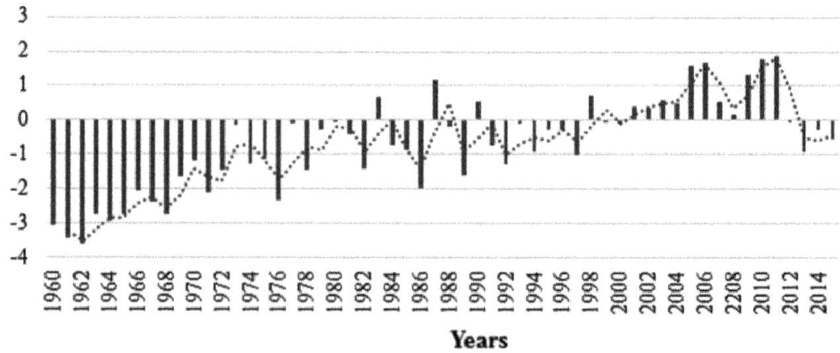

Figure 8. Mean temperature anomalies 1960-2015 for Natitingou/Tanguiéta in the area of the PBR (Source: ACDD 2018: 25).

The comparison of cold and warm years' anomalies 1960-2015 are illustrated in Figure 8: the average temperature of Natitingou/Tanguiéta shows two sub-periods: one for 1960-2003 characterised by a cooling trend, and another 2004-2015 period showing a short warming trend. Over the entire period, forty-one 'cold' years were identified as opposed to fifteen 'hot' years. The cold years (88.1%) are mainly belonging to the first sub-period, while the hot years (66.7%) are in the second sub-period (2004-2015). The study of ACDD (2018) confirms the results obtained in previous studies that attest a decrease in rainfall from the early 1970s in Benin, and variations of other climatic parameters in northwestern Benin (Afouda 1990, Boko 1988, Houndénou 1999, Vodounou & Onibon Doubogan 2016) (see chapters 9 and 13 in this volume).

15.3.2 Effects of climate variability (CV) on residents and natural resources.

For more than 20 years rural residents have been living in the Controlled Access Zone of the PBR (Figure 1). They carry out rain-fed agriculture and cultivate cereal crops such as *Sesamum indicum* and cash crops (including cotton). Production units are small and feed one family in average (shifting cultivation). Land surveys conducted in 2013 in the Controlled Access Zone (Figure 1) identified 2955 farmers cultivating 13,061 hectares of land. On average a single farmer occupies about 3.5 ha of land, but some farms also cover an area of 10-20 ha (CENAGREF 2015). Farmers who are setting up their fields in this zone of PBR are currently facing an upheaval in the agricultural calendar, loss of production, reduced yields, difficulties in applying phyto-sanitary treatments because of the lengthening of the dry days, intermittent rainy season nature and/or early rain stoppages that create an absence of moisture. Applying treatments despite this lack of moisture burns the seeds. As a result, producers face difficulties in feeding their families and getting through the lean periods.

> *"A Farmer who wants to do agriculture needs rain. Formerly, rain started in April to aid agricultural activities but, in recent years, there are many gaps that make the first agricultural activities conducted towards the end of May, or even June. Previously, there were sowing periods for each speculation where the upheavals of the rainfall system mean that this calendar is non-respected. This supposes that there is a shift (modification) in the agricultural calendar. This change means that some farmers are caught unprepared and at the end of the crop year, they cannot get enough*

agricultural products (crops) to meet the basic needs of their families and reserves to manage the welding period. (...) In the context of seed growth, at a given period, it is necessary to have moisture: for example, the phyto-sanitary treatment to be effective needs to be carried out in a period when the atmosphere is charged with little humidity in the rainy season. The rains as a result allow to have these microclimates. The scarcity of rains and the lengthening of dry days in the rainy season make phyto-sanitary treatments worse. Presently, even if there is no moisture, most farmers treat seeds whenever they notice infestation. In these cases, it is a safe bet that the treatment conditions will contribute to crop damage" (Respondent no. 11, Tanguieta, September 2019).

Farmers have to diversify crops to mitigate shocks, but this requires land availability. For this reason, they expand their agricultural fields and agricultural practices which sometimes leads to overexploitation of land. In addition, for all conventional seeds used in the Controlled Access Zone, farmers use huge quantities of pesticides, herbicides and chemical fertilizers especially on cotton. These chemicals do not only destroy the Occupied and Controlled Zone lands (cumulative effect of climate variability and excessive use of chemical inputs) but have a direct and negative impact on biodiversity (flora and fauna). Furthermore, most importantly they are leached by runoff water to the ponds and streams of the Pendjari Biosphere Reserve and pollute them. This pollution exposes the faunal resources of the park to high risks of contamination. To summarize, rain fed and conventional agriculture practiced in the Controlled Occupied Zone were subjected to major constraints such as degradation of productive potential, in particular the disappearance of biodiversity, impoverishment of cultivated lands, high vulnerability to climatic hazards, low use of improved non-chemical or organic inputs. All these aspects drastically lower yields.

"Currently, the Controlled Access Zone is almost 'unoccupied' because people have cut almost all trees for cultural purposes. (...) This land is also currently very poor, so that farmers are constantly expanding the agricultural areas. They are looking for better land that is more fertile. This quest usually leads them to cross the CAZ boundary to farm in the protected zone. (...) What is paradoxical that obviously the Government that should work to safeguard the flora and fauna species of the PBR, which encourages the production in the CAZ conventional speculations even though it consumes many chemical fertilizers. "Who is protesting or campaigning against the Government"? You may say that you want to manage and protect the natural resources of the park, but you are still encouraging the production of conventional crops around the PBR. Such speculations lead to excessive destruction. By producing conventionally, all the chemical inputs used are drained into the PBR and are likely to contaminate animal habitats and streams. Serious impacts are noticed on them" (Respondent no. 15, Tanguiéta, September 2019).

The effects are not only observed in the agricultural areas of the Controlled Access Zone (CAZ), households and animal breeders are also affected. Some actors questioned in Matéri have highlighted effects such as temporary and permanent migrations, indebtedness, temptation to professional alternatives and the development of 'vices'. According to these stakeholders, people who fit into this logic usually have no choice. However, it seems more plausible to consider these parameters as consequences (one goes into debt or one migrates because he/she has no choice) of the desiccation of livelihoods induced by CV. The adverse effects on populations are: upheaval of agricultural calendar, disruption of farming practices, unlimited extension of agricultural land, increase usage of chemical inputs, destruction of biodiversity, or even the desiccation of lands and the degradation of the pastures, declining yields that cause disability to feed family. These contribute to lean periods with difficulties; food and nutrition insecurity; increased water stress (especially for livestock); destruction of materials; accentuation of transhumance and conflicts between farmers and pastoralists. All these effects create and maintain vulnerabilities conducting people to social transformation from the PBR natural resources. Such situation increases infringements and therefore, conflicts between management people of the park and rural populations. The types of offenses often encountered are illegal hunting, illegal grazing, illegal fishing, illegal settlement and movements into the protected areas. As well as trafficking and marketing of protected species; exerting non-authorized agricultural works within the boundaries of the park; and practicing harmful activities (sand wedging, logging, straw and bamboo harvesting, removal of woody forest products). Faced with such offenses, the African Parks Network (APN) opts for intransigence but has not yielded effective promotion and discussion with the population. The effects on the park's natural resources are degradation of wildlife habitats; chemical contamination by runoff; and significant regression of gallery, dense and shrubby forests (see chapter 14 in this volume). Although they are real, consequences of population encroachments on the natural resources of the PBR are less visible now just because of the measures taken by the APN to limit damage.

15.3.3 Adaptation strategies for actors and the governance model

Three types of adaptation strategies were identified depending on individual experiences, feelings and means. Anticipatory adaptation strategies aim to contain the effects of CV. Reactive coping strategies aim to respond to effects in order to reduce the vulnerability created. Planned coping strategies are based on deliberate political decisions deriving from an awareness of current and future changes. They are part of a prospect approach that is mostly documented and analyzed. The assessment of the adaptive capacity of populations and ecosystems has been made but the current question of documenting the strategies is to know whether it contributes in reducing the effects of CV.

15.3.3.1 *Adaptation strategies for local populations*

Adaptation to changes in the agricultural calendar

Farmers developed strategies that are responses to the temporal variations of the rains, to their early cessation and extension of dry days during the rainy season. Traditionally, they could be categorized in three groups regarding the sowing period (see chapter 13 in this volume).

Firstly, there are the early growers who plant during the first rains (April and May) in the hope that the soil will retain moisture and that seeds will germinate. They also hope that the next rains will not be long. Secondly, there are the cautious farmers, who usually sow in June, a month in which rainfall is supposed to be plentiful. Thirdly, there are 'laggard' farmers who, for a number of reasons cannot quickly plant their fields. They sow desperately during the month of August and accompany their seedlings with a lot of prayer according to the actors met. Faced with CV 'outbreaks', including increased dry days during the rainy season, excessive heat, and intermittent rainfall, these practices are not suitable. Some farmers are adopting other planting techniques such as open-seeded seeding to prevent the temperature of the sand used to cover the seeds from drying out. It is also reported the development of increasingly sophisticated producers of observation and reading of natural phenomena with a view to anticipating a certain number of activities.

Thus, they modify the sowing dates in order to plant the crops during favorable periods (early sowing, installation of the fields even before the start of the rainy season). This technique allows them to avoid the seeds of intense periods of water stress. Some farmers return to crop associations. According to Boulier & Jouve (1990), the beginning of the rains and the duration of the rainy season are two essential parameters for rain fed agriculture, because they determine the date of sowing and thus, the position of the cycles and the length of time during which crops can benefit from rainfall. The adaptation strategies developed to comply with the changes in the agricultural calendar in the Pendjari Biosphere Reserve region confirm this. The data collected shows that some farmers are increasingly opting for crops that suffer less water stress, especially *Sesamum indicum* (sesame).

> *"In recent years, farmers are trying to keep up with the agricultural calendar that they largely missed out because of the many discrepancies observed. Here, early farmers sow between April and May, cautious in June and latecomers in August. This year, for example, all those who made the mistake of sowing maize or sorghum in August realized that they would yield absolutely nothing. Only sesame gives a good yield if it is sown during this month"* (Respondent no. 7, Porga, September 2019).

The adoption of new speculations

Most farmers use short-cycle seeds (45 days or two months at most) instead of long-cycle seeds (3-4 months) that are more demanding in terms of moisture. Therefore, even if they sow late, they are sure to harvest. These short-cycle varieties according to the people interviewed, adapt to the shortening of the duration of the rainy season, but seem to have a low density of nutrients and thus, do not guarantee a good concentration of nutrients. This adaptation strategy is documented and analysed in the West African subregion by Kouressy et al. (2008), Roncoli et al. (2000), and Sanou et al. (2018). Farmers also develop mixed farming, which designs the growing of different seedlings with short cycles on several agricultural plots. This intensification of production areas aims at increasing the area under cultivation to compensate for the fall in yields due to water shortage, temporal variations in rainfall and degradation of land caused by the cumulative effects of climate variability and excessive use of chemical inputs. This strategy is adopted in the Controlled Occupied Zone because farmers rely on the Pendjari Biosphere Reserve lands that they assume sufficiently fertile. However, this practice leads to a decrease in the reserves of arable land, a destruction of fallow land which compromises the regeneration of soil fertility and increases their leaching, a destruction of rare plant species and a destruction of wildlife habitats.

Sensitization for a change of behaviour

It is noticed according to the actors met that numerous villages bordering the Pendjari Biosphere Reserve engage in sensitization sessions with the central thematic CV.

> *"Today at the village level, people are becoming aware. In some villages now, village leaders are conducting sensitization sessions to change people's perceptions and practices, including cutting wood for domestic consumption. People in the villages are now collecting lists of species that should no longer be subject or cut for domestic and agricultural use. On the other hand, there is growing evidence that people in the villages are getting used to re-wooding"* (Respondent no. 13, Matéri, September 2019).

This respondent makes clear that the current manifestations of CV, the effects and consequences they induce are not due solely to climate events. There are, of course, anthropogenic practices of agricultural production and cooking that contribute significantly to it. Faced with these practices, there is an awareness, particularly at the level of local elected officials of some villages. This awareness is demonstrated by the organisation of sensitization sessions in the establishment of lists of prohibited plant and extinct species given their importance in eating, nutrition and therapeutic habits. The adaptation strategies developed by the populations are much more reactive. These strategies emanate from dynamic processes that are a function of a number of factors. There are many possibilities for adaptation, but it is imperative to intensify actions to reduce vulnerabilities to climate variability. There are barriers, limitations and costs that farmers do not always fully understand and need to be aware of (Pachauri & Reisinger 2007). Some authors have evoked collective coping strategies such as praying to divinities. In the north, Hounkponou et al. (2009) reported curative measures like traditional rites and specific practices (example, wearing garments upside down to go to the mosque). They indicate that the curative measures evoked by the populations are also mystical like to go to the river to implore the divinities of the rain, or to go around the village with tam-tam while pronouncing incantations to attract the rain. Similar practices have also been reported by Vissoh et al. (2012), following a study carried out in South East Benin.

15.3.3.2 *Adaptation Strategies of the African Parks Network(APN)*

The promotion of income-generating activities, both agricultural and off-farm is a mechanism chosen by the African Parks Network to reduce the direct and indirect effects of CV on park natural resources. Members of the team are trying to strengthen the sustainable agriculture program hosted by the Association of Wildlife Management Union. With this program, the communities have organised themselves to solicit partners who accompany them so that they can definitely stop the use of chemical inputs in the Controlled Occupied Zone. This organic production currently concerns, according to the people interviewed, three cultures: soya, coconut and sesame. However, the option of organic production has not yet been offered to all resident farmers. Beyond this, the APN team implements projects to diversify income-generating activities. The flagship project being launched, is according to the actors met, the beekeeping project.

> *"We are currently proposing a beekeeping project. We are now in the awareness phase to explain to people that in other parks managed by the African Parks Network, people organise beekeeping around nature reserves and because of this activity, a buffer zone has been defined, in which pesticides use is forbidden. Thus, when we make honey in this zone and obtain uncontaminated honey without chemical elements inputs. The honey that will be produced is highly sought after and we are ready to assist them to abandon the pesticides to indulge in this activity. If we find that there is no presence of pesticides in the honey produced, we will buy this honey at a very good price from them. Someone who does well in this activity will have even more income than someone who makes conventional cotton"* (Respondent no. 1, Tanguiéta, September 2019).

Apart from beekeeping, it plans to support the population in a number of income-generating activities so that many farmers will gradually abandon conventional farming in the Controlled Access Zone. This option was apparently made because the growing population in the area poses real risks to the park's natural resources. This diversification of income-generating activities for the benefit of local populations are included in a document called "Community Development Plan". This document is being prepared following a socio-economic study on health and education carried out in 2018. In the same vein, there have been achievements of hydrological works (Dam in Matéri, multifunction drilling in Daga), especially as water source for animals. There are also prospects for the development of fish or multifunctional dams around the Pendjari Biosphere Reserve to allow fishermen to continue their activities without necessarily depending on the watercourses of the park. These implementations will also encourage the development of market gardening activities. Furthermore, the African Parks Network is taking steps to prevent riparian populations from improperly entering the park and encroaching on its resources. The non-governmental organisation is considering the erection of a wire fence between the Controlled Access Zone and the hunting area. The electric grid fence (not harmful to humans) 96 km long, protects residents from animal attacks in villages and therefore should reduce animal-population conflicts. Gates will be installed for that local populations can enter the park when this is necessary. The wire fence aims to control access to the park's natural resources, and to ensure conformity validation between the authorizations offered. Finally, it aims to prevent the encroachments duly 'orchestrated' by the populations on the park's land resources. The actors met, said that they began to raise awareness about this fence. At present, the fence line is delineated with the help of local people who cleaned up the eight-meter wide strip (four in the hunting zone, and four in the Controlled Access Zone). This indicates, according to the team of the African Parks Network, an appropriation of the project by the local populations.

15.3.3.3 *Adapting Pendjari Biosphere Reserve governance to the requirements of climate variability (CV)*

The governance of the PBR offers a field of confrontation of powers that must be coordinated. To establish sustainable governance, the African Parks Network should assume four main functions: "achieve legitimate goals", "adapt to its environment", "produce quality", and "maintain and create values". Achievement of goals reflect the ability of the organisation to achieve its mission. Adaptation is the ability of the organisation to acquire and control the resources it needs (financial and human resources). It is also about the capacity to innovate, to transform, to adapt to the needs of the population and other actors, and to attract customers and mobilize people. Production corresponds to the production processes of individual services and collective interventions. It refers to the coordination and continuity of services, to collaborate between the various actors and the different powers involved. The maintenance and creation of values and norms within the organisation facilitate or constrain the fulfillment of the three previous functions. Values produce meaning, cohesion, or organisational confusion. Values help guide and justify choices, support development projects, and the various demands (Sicotte et al. 1998). These three functions must therefore be provided by the African Parks Network

Achieving legitimate goals and adapting to the environment

Achievement of the missions assigned to the natural resources and for PBR's sustainable governance. According to the people interviewed, the African Parks Network is a specialist in conservation and ecology (re-habilitation of animal habitats, reproduction of animals).

> *"There is no fear of achieving these goals because the African Parks Network will do the job right. However, at the periphery of the park, residents need to understand the issues and the need for sustainable governance of this resource. If they understand, we will be safe"* (Respondent No. 1, Tanguiéta, September 2019).

Beyond the well-tested ability of African Parks Network elsewhere, the postures and logics of local populations are essential. Since the African Parks Network objectives aim at "change", the actors argue that it is possible to work to know them and incorporate them into governance arrangements. According to them, a certain number of postures must be established and certain types of actions must be taken. The postures must be "the art of adjusting the elements of the official model to the local context" (Nguinguiri 2003: 19), that is to say that they must cautiously contribute to the documentation and incorporation of "practical standards" (Olivier de Sardan 2008). With regard to the proposed actions, the first important thing that comes out of the data is the safeguarding and the respect of the established rules. In fact, CV and the constraints they impose on neighboring populations must not according to the actors interviewed, be a pretext for not applying the conservation rules of the PBR natural resources. At the same time, it is urgent, according to the same people, to think about finding and supporting, in collaboration with the local populations, other types of income-generating activities that do not depend too much on the climate and on the park's natural resources.

In these reflections, it will be necessary to continue to insist on the urgency to get out the seeds consuming chemical inputs of the Controlled Access Zone and valorise only the biological seeds ('bio'). It is therefore a question of pursuing and reinforcing the actions already initiated in this direction. In the same vein, a typology on the actors in the different villages bordering the PBR should be carried out in order to know which ones are directly affected by its natural resources. Setting up a typology of income-generating activities conducted by stakeholders would help to understand how they are dependent on the natural resources of the reserve. It will also be necessary to consult the stakeholders to know what sectors in the riparian zone depends very little on the resources of the PBR (Controlled Access Zone and hunting zone) and to identify the level of each actor that is ready to reconvert in these areas and what their needs are.

The finalisation of the communal development planning and its implementation should contribute to this. It also would be helpful to authorize and survey the harvesting of non-timber forest products especially in the time of food shortages. Possibilities of creating, in the absence of multifunctional dams and reservoirs, alternative water supply for the survival of inhabitants and cattle breeders.

Production and maintenance of values creation

The African Parks Network is certainly a conservation specialist, but the participation of all the stakeholders must be effective during its activities. It is by relying on the real or ideal powers of all these actors; on the "inverted meanings in action" (Dubar 2003: 336), on the interests of each other and the logics that they induce that a convergent system of governance can be put in place. The co-management experienced since 1993 and the decentralization process underway in Benin since 2003 provide a favorable context for better reading, internalizing and incorporating the powers in place. The concession granted to the African Parks Network by the Beninese State should not mean disengagement and renunciation but should retain its regulatory role. This is the most important as the PBR has a national character which transcends the territorial limits of a municipality. The current rules applied are issued exclusively by the State and therefore fall under the "control" regulation (Dubar 2003). However, sustainable governance calls for joint regulation!

To avoid having a fragmented world in which stakeholders are perpetually in conflict particularly around usage, status and role, the African Parks Network will have to opt for this type of regulation because its offer can only emerge on the side of the pilotage and must remain sustainable for the future and adhered by the actors relevant to the co-construction, the co-animation of the system of action and environment (Paradeise 2003). Beyond the management rules enacted by the Beninese State and currently strictly enforced by the African Parks Network, the adoption of joint regulation offers multi-stakeholder platforms to decide together the rules of usage and/or access and conduct horizontal negotiations that avoid, or at least significantly reduce infractions and clashes and ensure the coordination and continuity of services, collaboration between different actors and different "worlds" or powers in presence.

This regulation also removes the fixed and descending nature, even the otherness, of norms and rules, status and roles, as well as postures and logics of actions because it reconciles local knowledge and practices with proven knowledge and practices for appropriate conservation and sustainable governance. This joint regulation is part of the hybrid modes of governance, which combine the coordination of actions imposed or impelled by the State under the cover of the rules it has established and forms of collective actions of public interest. Community interest is aimed at safeguarding and enhancing the livelihoods of local populations. For governance of PBR to be sustainable facing CV, it is necessary to work at the level of management systems that Strauss (1992) calls "negotiated social order" that avoids instrumental governance by conducting continuous or periodic mediation of conservation rules and institutional procedures. As can be seen, whether directly or indirectly (via a concession for example), "the presence of the State is required in the conservation of protected areas" (Giazzi & Tchamiè 2007: 82). But, it is not enough for everything since it is not only a panacea for the conservation and sustainable management of protected areas but the presence of the central state legitimizes the presence of other actors, including local communities who are closer to the natural resources under conservation of resident populations involved in the co-management schemes. Decentralization gives Benin an important power in the management of natural resources and forest areas. Law 97-029 on the organisation of municipalities in the Republic of Benin, in Article 94, states that "the municipality (...) shall ensure the protection of natural resources, including forests, soil, wildlife, hydrological resources contribute to their best. As part of the sustainable governance of the PBR, local communities are failing to take the lead in management because of the classified characters of the national PBR. According to the actors met, the management team need appropriate assistance in internalizing and taking the lead in information and education in this particular environment so that the local population understand the merits of safeguarding. They must also work to support the actors in the management of the infractions and conflict resolutions on the uses to revitalize the "Community of riparian communes" of the PBR to safeguard and valorize the means of existence of the populations. The territorial authorities must fully play these roles and thoroughly report the prerogatives conferred to them and should never abandon the post given them for political gains and adopt posts of "developers". The latter should lead them to distinguish the "development time" during which the development actions are conducted objectively and sustainably during the "political time" in the course of which the political basis force to make allegiance to the populations in the view to obtain electoral votes.

15.4 CONCLUSION

The present study has documented and analysed stakeholder's perceptions and effects of climate variability (CV) in and in the surroundings of the Pendjari

Biosphere Reserve (PBR) and the strategies developed by stakeholders involved, more or less, in governance of it.

The results reveal that climate variability and their effects are well-perceived by the stakeholders involved. The most frequently mentioned indicators of change in rainfall parameters are the delays of the rains, the shortening of the rainy season, and the variation of the rains over time. With respect to changes in temperature and wind, the most reported change indicators were excessive heat and increased strong winds. The climatic data used to analyse the temporal variations of the climatic parameters of the zone, confirm the perceptions of the local actors and indicate an inter annual variation as much of the rainfall as of the temperature, but especially a rise of the minimum and maximum temperature since 2005 as well as a declining precipitation trend. The effect of these events is observed as much on the neighboring populations and their livelihoods as on the park's natural resources. The effects on the park are both direct and indirect. The latter are related to the relationships maintained by the resident to the park because of the constraints they face due to CV. These results prove that climate variability is a reality in the PBR area. Documented adaptation strategies aim to reduce the influence of climatic parameters on the livelihoods of populations and on the park's natural resources. They are reactive at the population level and much more proactive at the level of the park management team.

From the analysis of sustainable governance, it is essential to think of the mechanisms that allow not only the rehabilitation of animal habitats, the regeneration of animal and plant species but all the safeguarding and valorisation means of existence of riparian populations while reducing their dependencies on the park's natural resources as well as the misuse of chemical agricultural inputs in the Controlled Access Zone. Sustainable governance also requires the adoption of a management system that allows the creation of a negotiated order, involving an analysis of the present "world", the powers and interests they hold, the logic they develop, the tensions that are created among them and contributions they can make. In this perspective, the sustainable governance of the PBR park must be done through a continuous process of evaluation, adaptation, adjustment and negotiation operated not by the management team alone, but in synergy with international partners. This negotiated order indicates that sustainable governance cannot be decreed unilaterally but through co-construction and mediation. The analyses that have been made in this article show that the ecology of governance, that is, the contexts and actors, must be known and constantly referenced.

REFERENCES

ACDD, *Actions-Conseils pour le Développement Durable* 2018. *Analyse et évaluation de la vulnérabilité de la population et des écosystèmes face au changement climatique*. Observatoire du Sahara et du Sahel.

Adger, W.N. & Kelly, P.M. 1999. Social Vulnerability to Climate Change and the Architecture of Entitlements. *Mitigation and Adaptation Strategies for Global Change* 4 (3-4): 253–266.

Adger, W.N. 1999. Social Vulnerability to Climate Change and Extremes in Coastal Vietnam. *World Development* 27 (2): 249–269.

Afouda, F. 1990. *L'eau et les cultures dans le Bénin central et septentrional : Etude de la variabilité des bilans de l'eau dans leurs relations avec le milieu et la savane africaine*. Thèse de Doctorat nouveau régime, Université de Paris IV (Sorbonne), Institut de Géographie. 428p.

C2A, Commission Agriculture et Alimentation 2017. *Quelles politiques publiques pour promouvoir l'adaptation des agricultures familiales aux changements climatiques (rapport de la C2A)*. 1–84. https://www.coordinationsud.org/wp-content/uploads/rapport-C2A-2017-french-web-MD.pdf

Al Hamndou, D., Requier-Desjardins, M. 2008. Variabilité climatique, désertification et biodiversité en Afrique : S'adapter, une approche intégrée. *VertigO - La revue électronique en sciences de l'environnement* 8(1): 1–24.

Bambara, D., Bilgo, A., Hien, E., Masse, D., Thiombiano, A., et Hien, V. 2013. Perceptions paysannes des changements climatiques et leurs conséquences socio environnementales à Tougou et Donsin, climats sahélien et sahélosoudanien du Burkina Faso. *Bulletin de la Recherche Agronomique du Bénin* 74: 8–16.

Benko, G. & Lipietz, A. 1995. De la régulation des espaces aux espaces de régulation. In Boyer, R. et Saillard, Y. (dir.), *Théorie de la régulation – État des savoirs*: 293–303. Paris: La Découverte.

Bergonzini, J.C. 2004. Changements climatiques, désertification, diversité biologique, et forêts. Nogent-sur-Marne: SILVA.

Boko, M. 1988. *Climats et communautés rurales du Bénin: Rythmes climatiques et rythmes de développement*. Thèse de doctorat d'Etat ès Lettres et Sciences Humaines. CNRS Dijon: Université de Bourgogne.

Boulier, F. & Jouve, P. 1990. *Evolution des systèmes de production Sahéliens et leur adaptation à la sécheresse*. Montpellier: CIRAD-DSA.

Brooksa, N., Adger, W.N. & Kelly, P.M. 2005. The determinants of vulnerability and adaptive capacity at the national level and the implications for adaptation. *Global Environmental Change* 15: 151–163.

Brown, O., & Crawford, A. 2008. *Évaluation des conséquences des changements climatiques sur la sécurité en Afrique de l'Ouest : Étude de cas national du Ghana et du Burkina Faso*. https://www.iisd.org/pdf/2008/security_implications_west_africa_fr.pdf

Burton, I., Huq, S., Lim, B., Pilifosova, O. & Schipper, E.L. 2002. From Impacts Assessment to Adaptation Priorities: The Shaping of Adaptation Policy. *Climate Policy* 2: 145–159.

CENAGREF, Centre National de Gestion des Réserves de Faune 2015. *Plan d'Aménagement et de Gestion Participatif de la Réserve de Biosphère de la Pendjari (2016–2025)*. Cotonou: PAPE-Composante 2: 1–154.

Contandriopoulos, A.-P. 2008. La gouvernance dans le domaine de la santé : Une régulation orientée par la performance. *Revue Santé Publique 20(*2): 191–199.

Distefano, E. 2012. *Integration of climate change dimensions into project activities*. Rome: FAO.

Dubar, C. 2003. Régulation conjointe et double transaction : Réflexions pour un débat. In de Terssac, G. (ed.), *La théorie de la régulation sociale de Jean-Daniel Reynaud. Débats et prolongements*: 331–344. Paris: La Découverte.

Giazzi F., Tchamiè T.T. K. (2007). La participation des populations locales dans la gestion des ressources naturelles. Historique de la création des aires protégées en Afrique de l'Ouest et évolution récente de la conservation. In Fournier, A., Sinsin, B., Mensah G.A. (ed.). *Quelles aires protégées pour l'Afrique de l'Ouest ? Conservation de la biodiversité et développement*. Paris: IRD, 77–85.

Houndénou, C. 1999. *Variabilité climatique et maïsiculture en milieu tropical humide: l'exemple du Bénin, diagnostic et modélisation*. Thèse de Doctorat de Géographie, UMR 5080, CNRS « climatologie de l'Espace Tropical ». Dijon: Université de Bourgogne.

Hounkponou, K.S., Bokonon-Ganta, E., Nouatin, G., Gnangassi, C & Ahounou, M. 2009. Changement climatique au Bénin: Vulnérabilité et stratégies d'adaptation. *Revue sur l'agriculture durable à faibles apports externes (AGRIDAPE)* 24(4): 8–10.

Kouressy, M., Traoré, S. et Vaksmann, M. 2008. Adaptation des sorghos du Mali à la variabilité climatique. *Cahiers Agricultures* 17: 95–100.

Leyronas, S., Rojat, D., Maurel, F. & Giraud, G. 2016. Un cadre d'analyse pour la gouvernance des ressources naturelles. Le cas des eaux souterraines. *Revue d'économie du développement* 24(3): 129–148.

Nguinguiri, J.-C. 2003. Gouvernance des aires protégées : l'importance des « normes pratiques » de régulation de la gestion locale pour la faisabilité des réformes dans le Bassin du Congo. In *APAD. Gestion des ressources naturelles : Participations et médiations. Bulletin de l'APAD 36*: 17–25.

Niang, I. 2009. Le changement climatique et ses impacts : les prévisions au niveau mondial. In IEPF (eds.). *Adaptation au changement climatique*. Liaison Énergie-Francophonie 85: 13–19.

Niasse, M., Afouda, A. & Amani, A. 2004. *Réduire la vulnérabilité de l'Afrique de l'Ouest aux impacts du climat sur les ressources en eau, les zones humides et la désertification : Eléments de stratégie régionale de préparation et d'adaptation*. Gland (Suisse) et Cambridge, UK: UICN.

Olivier de Sardan, J.-P. 1995. *Anthropologie et développement. Essai de socio anthropologie du changement social*. Paris: APAD-KARTHALA.

Olivier de Sardan, J-P. 2008. A la recherche des normes pratiques de la gouvernance réelle en Afrique. *Afrique Pouvoir et Politique* (APP*)-Overseas Development Institute, Working Paper* 5: 1–23.

Olivier de Sardan, J.-P. 2009. Les huit modes de gouvernance locale en Afrique de l'Ouest. *Afrique: Pouvoir et Politique (APP) - Overseas Development Institute, Working Paper* 4: 1–56.

Pachauri, R.K. & Reisinger, A. 2007. *Bilan 2007 des changements climatiques. Contribution des Groupes de travail I, II et III au quatrième Rapport d'évaluation du Groupe d'experts intergouvernemental sur l'évolution du climat*. Genève: GIEC/IPCC.

Paquet, G. 1999. *Governance Through Social Learning*. Ottawa: Ottawa University Press.

Paradeise, C. 2003. La théorie de la régulation sociale à l'épreuve de la pratique. In de Terssac, G. (ed.), *La théorie de la régulation sociale de Jean-Daniel Reynaud. Débats et prolongements*: 41–49. Paris: La Découverte.

Roncoli, C., Ingram, K. & Kirshen, P. 2000. Can farmers of Burkina Faso? *Practicing Anthropology* 22 (4): 24–28.

Saltman, R.B. & Ferroussier-Davis, O. 2000. The concept of stewardship in health policy. *Bulletin of the World Health Organization* 78: 732–739.

Sanou, K., Amadou, S., Adjegan, K. & Tsatsu, K.D. 2018. Perceptions et stratégies d'adaptation des producteurs agricoles aux changements climatiques au nord-ouest de la Région des Savanes du Togo. Institut Togolais de Recherche Agronomique (ITRA) 30 (1): 87–97.

Sicotte, C., Champagne, F., Contandriopoulos, A.P., Béland, F., Denis, J.L., Bilodeau, H., Bremond, M., Le-better-Charles, L., Barnsley, J., Leggat, S. & Barker, R. 1998. Members of the HEALN and Health Care Management Group. A conceptual framework for the analysis of health care performance organizations. *Health Services Management Research* 11: 24–48.

Sircoulon, J. 1990. *Impact possible des changements climatiques à venir sur les ressources en eau des régions arides et semi-arides*. Paris: ORSTOM 1–87.

Strauss, A. 1992. *La trame de la négociation*. Paris: L'Harmattan.

United Nations (UN) 1992. Convention cadre des Nations Unies sur les Changements Climatiques. New York.

UNESCO 2002. *Ecotourisme, écovillage en Afrique de l'ouest : mythe ou réalite? le cas particulier des réserves de la biosphère impliquées dans l'écotourisme : la Pendjari (BENIN) – le Tai (Cote d'Ivoire) – Niokolo Koba (Sénégal) – Région du W (Burkina) – Comoe (Cote d'Ivoire) – Mont Nimba (Guinée)*. Programme l'Homme et la Biosphère (MAB). 1–93.

Vissoh, P.V., Tossou, R.C., Dedehouanou, H., Guibert, H., Codjia, O.C., Vodouhe, S.D. & Agbossou, E.K. 2012. Perceptions et stratégies d'adaptation aux changements climatiques : le cas des communes d'Adjohoun et de Dangbo au Sud-Est Bénin. *Les Cahiers d'Outre-Mer, Revue de Géographie de Bordeaux* 260: 479–492.

Vodounou, J.B.K. & Onibon Doubogan, Y. 2016. Agriculture paysanne et stratégies d'adaptation au changement climatique au Nord-Bénin. *Cybergeo : European Journal of Geography* 794: 1–27.

CHAPTER 16

Climatic factors of soil degradation and adaptation measures of stakeholders in the Oti plain in Benin

I.F. Ouorou Barre
University of Parakou, Parakou, Benin

M.S. Issa, I. Yabi & E. Ogouwale
University of Abomey-Calavi, Cotonou, Benin

ABSTRACT: Ecological deterioration and soil degradation are consequences of climate change. The paper evaluates climatic impacts on soil degradation and peasants' adaptation strategies for sustainable development of agriculture in the Oti plain in Benin. Climatological and pedological data associated with the agricultural statistics of selected crops (maize) have been studied according to current farming practices. Soil samples and their physico-chemical evaluation made it possible to evaluate the state of soil fertility. The soil analyses evidenced that 15% of the studied soils are suitable for certain farming crops. 65% fit moderately and 20% marginally or even unfit to the environmental properties. Most of these soils are shallow and become increasingly uncultivated because of the advanced effect of soil degradation and deterioration of the climate. In response, farmers implement adaptation measures such as soil biological remediation, use of chemical fertilisers, and crop rotation. In the context of the promotion of climate-smart agriculture, accompanying measures deserve to be initiated to support farmers.

16.1 INTRODUCTION AND STUDY AREA

Precipitation is the fundamental climatic element that environmentally conditions agriculture. Rain fed agriculture with traditional bush fires and nomadic livestock keeping is essential for the survival of the population in developing countries (Ogouwale 2004). Due to such practices and demographic pressure, soil used for agriculture is continuously degraded which affects its fertility. In this context of degradation of the productive potential of soils, also climatic variability and climate change has ongoing negative consequences on agricultural production. The population of Benin is growing rapidly and is expected to reach 20 million by 2050 (Ahamide 2015). Therefore, soil has become a critical resource to be protected as a matter of urgency.

Changes in land availability affect more than 2 billion people worldwide, including most of the 852 million people who suffer from hunger, mainly in countries of the global south (Brabant 1996). Soils are essential parts of the environment in the ecosystems. In Benin, nearly 6.4% of soils are experiencing a significant drop in productivity and their ecological functions are partially destroyed. Here, as in all countries of West Africa, although the food crises have not reached the scale of those known by the Sahelian countries which are often confronted with serious food deficits (Issa 1995). Indicators of these food deficits include the imbalance between rapid population growth and declining agricultural productivity due to the ongoing rainfall variability and climate change (Ogouwale 2001). It seems necessary and imperative to reconsider the restoration of agricultural land, and in particular those of the Oti plain in Benin in order to (re)increase agricultural production in the future. It is in order to better appreciate the influence of climatic factors on soil degradation in explaining the difficulties encountered by the farming community of the Oti plain in Benin.

The study area is located in the extreme northwest of Benin between 10°20′56” and 11°30′51” northern latitude and 0°47′52”and 1°55′37” eastern longitude. It covers an area of 10,256 km^2 with the municipalities of Toucountouna, Boukombé, Kérou, Kouandé and Cobly, and the neighbouring countries Burkina Faso (NW) and Togo (W) (Figure 1). The municipalities of Tanguiéta and Matéri are the main crossroads for trade of agricultural products between the Beninise municipalities and also on the inter-state level (see chapter 15 in this volume).

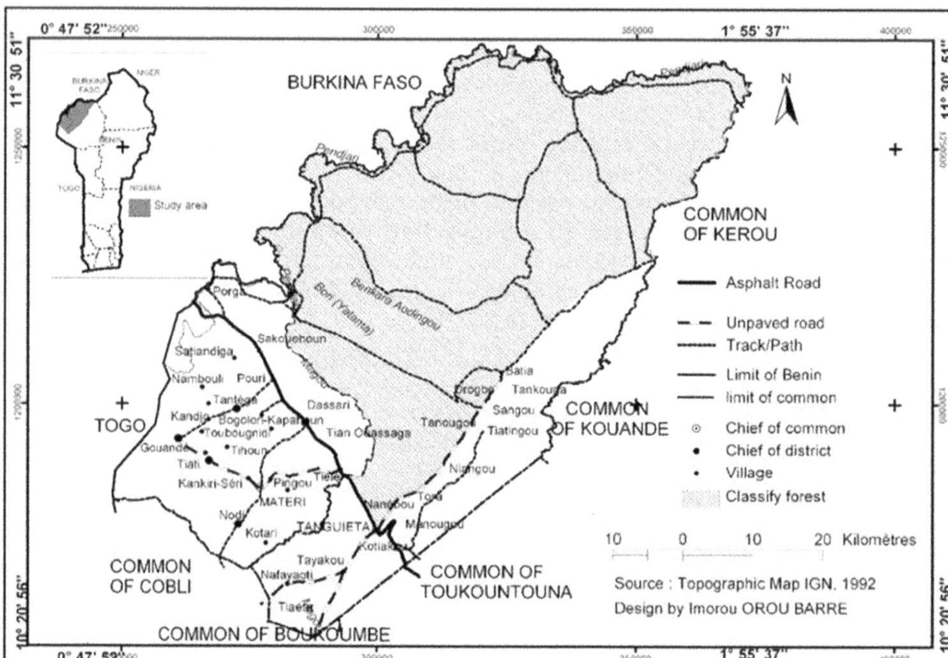

Figure 1. Geographical location of the study area in northwestern Benin.

16.2 DATA AND METHODS

16.2.1 Data

Data used came from documentation and fieldwork. They relate to climatological, pedological, agricultural, socio-economic and socio-ecological sources. Climate data of Tanguiéta and Matéri (precipitation and temperature) helped to evaluate and determine potentials of agricultural activities. The agricultural data consisted essentially of statistics on agricultural yields for food availability. The choice of crops was based on the importance of agricultural production and dietary habits of the people.

The pedological data consisted of soil samples with determination of pH value, Cation Exchange Capacity (CEC), nitrogen (N), phosphorus (P), and potassium (K). Similarly, soil texture, structure and porosity, water retention capacity were determined. The socio-economic and socio-ecological data on adaptation strategies developed by farmers in response to climatic factors of soil degradation were collected by field observations.

16.2.2 Methods

The statistical treatment of climate data was done using EXCEL 2007 and XLSTAT 2008 software that allowed to register decadal, monthly and rainy months, and then transformed into tables and graphs. The average was calculated by the following formula

$$M = \frac{1}{n}\sum^{(x)} i \tag{1}$$

where n = the total number of variables xi. Considered thus n and x can denote either the height or the number of rainy days.

The coefficient of determination (R^2) indicates the utility of the variable "number of rainy days" to explain the variable "seasonal rainfall". The determination of the beginning and end of the rainy season benefited from the method of (Sivakumar 1987). The determination of the pH value was done by using a soil pH-meter (Table 1).

Table 1. Assessment of pH values in soils.

Value of pH	Qualification
< 4,5	Extremely acidic
4,5 – 5,0	Very strongly acidic
5,1 – 5,5	Strongly acidic
5,6 – 6,0	Moderately acidic
6,1 – 6,5	Weakly acidic
6,6 – 7,3	Neutral
7,4 – 7,6	Slightly alkaline
7,7 – 8,4	Moderately alkaline
8,5 – 9,0	Strongly alkaline
> 9,1	Very strongly alkaline

16.2.2.1 *Organic material*

The product of the percentage of soil carbon and conversion factor and the constant 1.724 use in soil science determines the organic matter (OM) content:

$$\%C \cdot 1.724 = OM \tag{2}$$

The assessment of organic matter content was grouped into three classes:

- high content of organic matter + 1.5% OM;
- average content of organic matter between 1 and 1.5% of OM;
- low organic matter content, less than 1% OM.

The assessment of nitrogen (N) content is presented in Table 2.

CEC and saturation values are shown in the assessment grid of Table 3.

The assessment for phosphorus (P) in soils is presented in Table 4.

Fertility classes (I-IV) of soils in the study area have been provided by an overview synopsis presented in Table 5.

Grain size distribution (textural triangle) was determined for clay, silt and sand (Table 6).

Table 2. Assessment of nitrogen content in soils.

Values of nitrogen (N)	Interpretations
+ 3 ‰	very high in nitrogen
2– 3 ‰	high in nitrogen
1 – 2 ‰	moderately high in nitrogen
< 1 ‰	low in nitrogen

Table 3. Assessment of CEC and saturation rate.

Interpretation	Very weak	Weak	Medium	Strong	Very strong
CEC	< 5	5 – 10	10 – 25	25 – 40	> 40
	reserve very weak	weak	medium	strong	very strong
S	< 2	2 – 5	5 – 10	10 – 15	>15
%V = S/T*100	< 15	15 – 40	40 – 60	60 – 90	90 – 100

Table 4. Assessment of phosphorus (P) in soils.

Values of P	Interpretations
> 20 ppm	high phosphorus value
10 – 20 ppm	average phosphorus value
≤ 10 ppm	low phosphorus value

Table 5. Fertility conditions in soils.

Fertility class	Class I	Class II	Class III	Class IV
OM (%)	> 2	1 – 2	1 – 0.05	< 0.05
N (%) at pH 6	> 0.08	0.045 – 0.08	0.045 – 0.03	< 0.03
P (ppm)	> 20	10 – 20	10 – 5	< 5
K (meq/100g of soil)	> 0.4	0.2 – 0.4	0.2 – 0.1	< 0.1
Sum (meq/100g of soil sol)	> 10	5 – 10	5 – 2	< 2
Base saturation (%)	> 60	40 – 60	40 – 15	< 15
CEC (meq/100g of soil)	> 25	10 – 25	10 – 5	< 5

Table 6. Interpretation grid for gain size classes.

Diameter (microns) %	Determined elements
0 – 2 μ	clay
>2 – 20 μ	fine silt
>20 – 5 μ	coarse silt
>50 – 200 μ	fine sand
Y200 – 2000 μ	coarse sand

16.3 RESULTS

16.3.1 Precipitation: Determination of the beginning and the end of the rainy season

The Oti plain is subject to constraints of climate change. The beginning and the end of the cropping season 1971-2010 were determined by the method of Sivakumar (1987), and evidenced a high climatic variability or 'instability' (Figure 2). Based on the Sivakumar method and averaging the months, the beginning of the rainy season is 35% in April; 42.5% in May; 22.5% in June. As for the end of the humid period, it is 9% in September, 88% in October and 3% in November. The average date of the start of the rains is between May 17 and June 21. These underlines a high temporal variability of the beginning and the end of the rainy season. It affects agricultural production with the late and early start of the rains in the Oti plain of Benin.

16.3.2 Temperature

The analysis of air temperatures in the Oti plain shows a rising trend over the period 1971-2010. There is an increase of maximum temperatures of 0.3 °C and for 0.5 °C of the minimum temperatures. These results are consistent with IPCC (2007) which expect that global average temperature could increase by 1.5 to 6 °C by the year 2100. From this observation, the minimum temperatures have a higher upward trend, stronger than the maximum temperatures.

16.3.3 Types of yields according to the growing season

Correlation between agricultural and rainfall indices have been produced (Figure 3).

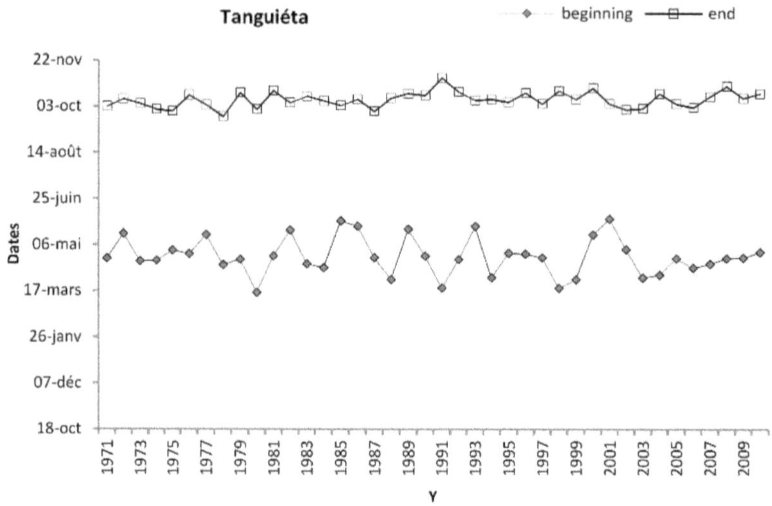

Figure 2. Start and end (dates) of the rainy season at Tanguiéta, Benin, between 1971 and 2010.

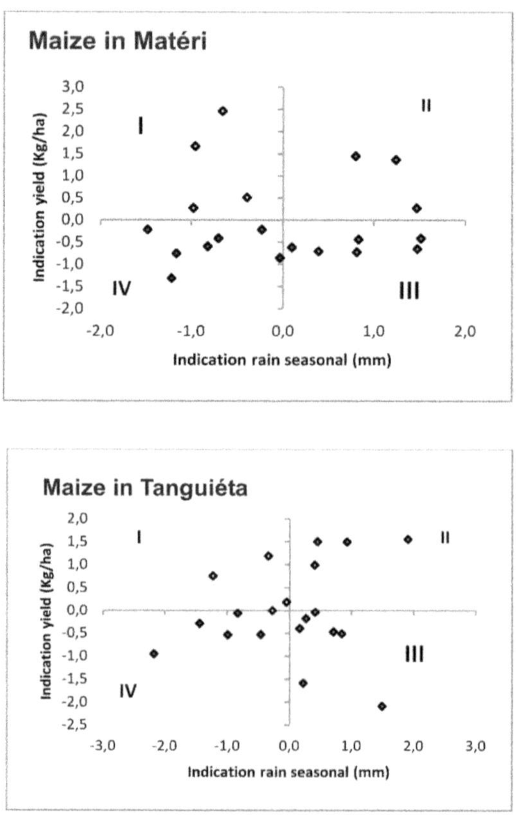

Figure 3. Seasonal yield and rainfall indices of commune of Matéri (top) and Tanguiéta (bottom) (Legend: I = Good yield and excess seasonal rain; II = Poor yield and excess seasonal rain; III = Bad yield and seasonal rain deficit; IV = Good yield and seasonal rain deficit.

Overall, 52.6% of the observed years gave poor performance against 47.4% of years that gave good returns. Rainfall and temperature variations during the rainy seasons have strongly influenced agricultural performance which caused years of poor harvests. In addition, the reduction of soil fertility also caused repercussions in agricultural yields.

Table 7. Proportional seasonal yield of maize at Matéri (Source: fieldwork, November 2012).

Town	VRY – good performance	VRY – bad performance	DS – good performance	DS – bad performance
Matéri	(%)	(%)	(%)	(%)
Maize	31.57	15.78	21.08	31.57
Town	VRY – good performance	VRY – bad performance	DS – good performance	DS – bad performance
Tanguiéta	(%)	(%)	(%)	(%)
Maize	21.05	31.57	21.05	26.31
medium	26.31	23.67	21.06	28.94

VRY= very rainy year; PP = poor performance; DS = dry season; GP = good performance.

The results underline that:

- ferruginous soils are marginally fitted for maize culture;
- hydromorphic soils are moderately fitted for maize culture because of their texture and low organic matter content;
- ferrallitic soils are moderately fitted for maize culture;
- hydromorphous soils are well suitable for maize culture because they have an average content for all physico-chemical characteristics.

Interventions can be carried out to improve the 'fitness' of these soils. For example, it is possible to add organic matter to ferrallitic soils to overcome their limitation and practice contouring on hydromorphic gleyic soils. Fertility of soils showed that 62.5% marginally fit for maize and 37.5% do moderately fit (Figure 4). Appropriate cropping techniques will permit the passage of moderately suitable soils and will improve them concerning the yields of maize.

16.3.4 Classification of soil fertility

The results of the analysis of soil samples are presented in Table 8.

16.3.5 Physico-chemical evaluation of the environment for maize

The results of the assessment of physico-chemical characteristics and the suitability of different soil types for maize cultivation are presented in Table 9.

16.3.6 Anti-erosion measures

Presently, some of the local farmers use the sweet grass *Vetiveria zizanioides* (Syn. *Chrysopogon zizanioides*), originating from tropical Asia, to reduce soil degradation.

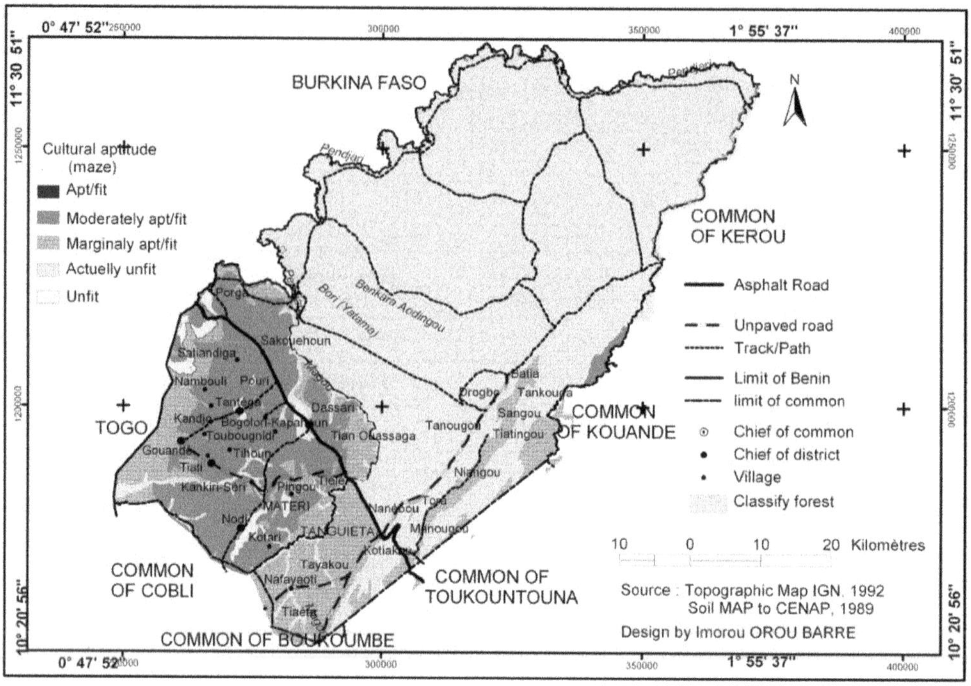

Figure 4. Soil fertility classes ('fitness') for maize in Matéri and Tanguiéta, northwestern Benin.

However, *Vetiveria zizanioides* is applied only by very few farmers, but the plant plays a very important role in mitigating the effect of farmland erosion. Its use is more popular in Boukombé than in the other municipalities of the study area (Figure 5a, b).

Mucuna pruriens (velvet or 'mascate' [french] peas) is a plant belonging to the Fabaceae and also plays an important role in soil conservation. The plant is a good nitrogen fixator, and it protects the soil against erosion and considerably increases crop yield. A field experiment carried out on ferrallitic soils in southern Benin (Azontondé 2000) proved that the *mucuna*-maize system gives an additional seasonal income (+78.4%) compared to traditional crops (Figure 6). An additional surplus of 36% and 48% revenues respectively were obtained when using the NPP (Nitrogen, Phosphorus and Potassium) fertilisation system.

16.4 CONCLUSION

This preliminary geoecological study emphasised on the relationship between climate, soil and land use in northwestern Benin. The correlation between climate variability and its impact on agricultural activity shows that rainfall extremes are rife and have negative repercussions on maize culture. Analyses have shown that soils under cultivation have in most cases lost their agricultural potential. The comparison between the cultivated areas, the slopes and the lowlands indicate the level of soil fertility developed for agriculture. The farmers developed strategies are introduced to raise the fertility of degraded and maintain of their fertility.

Table 8. Classification of soils in relation to the degree of fertility.

Characteristics

Depths: 0 - 20 cm

Pedological units	OM (%)	N (%) to pH 6	Pass Bray 1	K (meq/100g soil)	Sum of cations (meq/100g soil)	Base saturation %	CEC (meq/100g soil)	Fertility level classification	Limiting factors
Soils ferruginous of erosion	4.72	0.1337	46	0.74	17.83	94	19		
Fertility class lithic	I	I	I	I	I	I	II	I-II	S of cations
Ferrallitic soils	1.08	0.02	1.9	0.11	3.98	67.6	6.74		
Fertility class lithic	II	II	IV	III	III	I	III	IV	P, K, S cations, CEC
Revised tropical ferruginous soils hydromorphic	2.46	0.073	12	0.2	7.39	72	10.2		
Fertility class lithic	I	II	II	II	II	I	II	II	
Tropical ferruginous soils (leached)	1.12	0.03	0.6	0.16	3.71	56.6	6.6		
Fertility class lithic	II	II	IV	III	III	II	III	IV	P, K, S cations, CEC
Hydromorphic soils (pseudogley)	2.68	0.07	5	0.25	9.51	43.2	21.96		
Fertility class lithic	I	II	III	II	II	II	II	II-III	
Tropical ferruginous soils leached hydromorphic	1.61	0.042	2	0.16	7.33	72	10.2		
Fertility class lithic	II	II	IV	III	II	I	II	IV	P, K
Gleyic hydromorphic soils	1.94	0.14	2	0.15	10	56.5	19.05		
Fertility class lithic	II	II	IV	III	II	II	II	IV	P, K
Tropical ferruginous soils leached indurated	2.03	0.073	5	0.22	3.33	54	8.8		
Fertility class lithic	I	II	III	II	III	II	III	III	P, S cations, CEC

OM = organic matter; N = nitrogen; P = phosphorus; K = potassium; CEC = Cationic exchange capacity; S Bases = sum of bases; I = no limitation; II = low limitation; III = severe limitation; IV = very severe limitation

Table 9. Evaluation of the physico-chemical characteristics of the environment for maize.

Names Profiles	Soil units	Physico-chemical characteristics of the units											
			W		S			n				Suitability class	
		Topography (%)	Drainage	Flood	Texture	Gravity charge	Prf (cm)	CEC argile	Basesaturation (%)	OM (%)	pH	Current	Potential
OUTAN 1	Undeveloped soils of lithic erosion	3	0	0	1	2	3	1	0	0	0	N1	N1
OUTAN 2	Ferrallitic soils	0	0	0	0	1	0	2	1	2	1	S2	S1
OUTAN 3	Revised tropical ferruginous soils hydromorphic	3	1	0	0	2	0	2	2	1	1	S3	S2
OUTAN 4	Ferruginous soils (leached)	0	1	0	1	2	1	2	1	2	1	S2	S2
OUTAN 5	Hydromorphic soils (pseudogley)	0	2	2	1	1	0	1	2	0	2	S3	S2
OUTAN 6	Tropical ferruginous soils leached hydromorphic	0	1	0	0	1	0	1	0	0	1	S1	S1
OUTAN 8	Gley hydromorphic soils	0	2	3	0	2	0	1	1	1	1	S3	S2
OUTAN 9	Tropical ferruginous soils leached, indurated	0	1	0	1	2	2	2	2	1	1	S3	S2

S1 = apt; S2 = moderately fit; S3 = marginally fit; N 1 = currently unfit; N 2 = unfit; Sat = saturation; Prf = depth;
Source: laboratory results.

Figure 5a. Dense near surface root penetration of *Vetiveria zizanioides* (www.naturalmedicinefacts.info plantvetiveria-zizanioides [accessed 28/04/2020]).

Figure 5b. Anti-erosion measures with strips of *Vetiveria zizanioides* in Boukombé, Benin (Photo: Ouorou Barre, September 2012).

Figure 6. Leaves and fruits of the soil protecting leguminous crop *Mucuna pruriens* (www.botanikmeister.de/mucuna-pruriens-test [accessed 28/04/2020]).

REFERENCES

Ahamide, A., Guengant, J.P. & Kouchoro, G. 2015. La révolution contraceptive, condition necessaire pour réaliser la transition démographique et bénéficier du dividende démographique. Population et développement: quels défis pour le Bénin? *Direction des Politiques et Programmes de Population*, Publication, Novembre, 1–4.

Azontondé, A.H. 2000. *Dynamics of Organic Matter and Nitrogen in the Mucuna-Maïs System on Ferralitic Soil (Bar Land) in Southern Benin.* Doctoral thesis. Montpellier: Université de Montpellier.

Brabant, P. 1996. La dégradation des terres en Afrique. *Afrique Contemporaine* 161: 90–108.

IPCC, 2007. Climate Change, Synthesis Report. Contribution of Working Groups I, II and III to the Fourth Assessment Report of the Intergovernmental Panel on Climate Change [Core Writing Team, Pachauri, R.K and Reisinger, A. (eds.)]. IPCC, Geneva, Switzerland.

Issa, M.S. 1995. *Impact potentiels d'un changement climatique dû au dédoublement du CO_2 atmosphérique sur l'agriculture en République du Bénin.* Mémoire du Diplômes d'Etudes Supérieure Spécialisé (DESS). Université Senghor d'Alexandrie.

Ogouwale, E. 2001. *Vulnérabilité/Adaptation de l'agriculture aux changements climatiques dans le Département des collines.* Mémoire de Maîtrise, Université d'Abomey-Calavi, Bénin.

Ogouwale, E. 2004. *Changements climatiques et sécurité alimentaire dans le Bénin méridional.* Mémoire du Diplômes d'Etudes Approfondies (DEA), Université d'Abomey-Calavi, Bénin.

Sivakumar, M.V.K. 1987. Predicting rainy season in the Southern Sahelian and Sudanian climatic zones of West Africa. *Agricultural and Forest Meteorology* 42: 295–305.

CHAPTER 17

Changes in vegetation cover and land degradation between 1992–2018 evidenced by LANDSAT data in the suburb of Koubri, Ouagadougou, Burkina Faso

J. Ouedraogo & C.E.D. Da

Joseph Ki-Zerbo University, Ouagadougou, Burkina Faso

ABSTRACT: Ouagadougou, the capital of Burkina Faso (~ 2.5 million people), consists of urban, peri-urban and rural areas. In the southern rural district of Koubri land use and vegetation cover changes are linked to population growth, economic activities and land pressure. The study highlights the environmental dynamics between 1992, 2002 and 2018 by multi-temporal mapping and supervised classification using LANDSAT TM (1992), ETM+ (2002) and OLI/TIR (2018) data. Ground truthing, interviews with stakeholders, and questionnaires were used to understand recent land use changes and the causes of land degradation. It was shown that between 1992 and 2018 the surface occupied by bush and grass savannah was reduced from ~65% to ~29% (a decrease of ~37%). By contrast, areas covered by rainfed crops increased from ~32% (1992) to ~43% (2002) and ~58% (2018). The housing area of Koubri grew from 0.08% (1992), to 0.09% (2002) to 28.77% (2018) respectively (onset of urbanisation). Factors contributing to the degradation of ecosystems include climate variability, population pressure, agropastoral practices, overexploitation of natural resources, land speculation, and uncontrolled illegal construction.

17.1 INTRODUCTION

Bare soil without vegetation cover, exposed to water and wind, is sensitive to erosion. Vegetation cover limits stormwater surface runoff and maintains soil structure and fertility; it also reduces near ground wind speeds. The combined effects of increasing anthropogenic pressure on natural resources and climatic conditions lead to disturbance and dysfunctions in the terrestrial ecosystem (Benbrahim et al. 2004). In Africa, several authors have been involved in studying the causes and consequences of the regression of vegetation cover in relation to former variability of climate (Albergel 1987, Runge 2007, Valea & Ballouche 2012, Quattara et al. 2016). Environmental degradation is manifested by several processes that are determined in quantitative or qualitative ways. Destruction of savannah vegetation can cause semidesertic conditions known from ecosystems in arid regions (Taïbi & El Hannni 2004). In sub-Saharan Africa, many countries are sharing transnational watersheds of the Niger, Senegal, Gambia, Volta-Oti, Chari and Logone Rivers, and clearing of tropical forests is an environmental challenge for them. Mobile or nomadic population and the trampling of livestock along watercourses also has negative consequences for the grass cover in West Africa. Ouagadougou, the capital of Burkina Faso, is located south of the Sahel and includes urban, peri-urban and rural areas. As a case study the (still) rural suburb of Koubri was selected: this area south of Ouagadougou is undergoing environmental changes linked to population growth, economic activities and land pressure. The objective is to better understand the dynamics of vegetation cover and land use between 1992, 2002 and 2018 through mapping based on remote sensing data (LANDSAT).

17.2 METHODS

Multitemporal land cover mapping of Koubri (~6.4 km^2) was based on a supervised classification of LANDSAT satellite images: TM for 1992, ETM+ for 2002, and OLI/TIR for 2018. A colour composit (RGB, 4/3/2) was used to map land use units. This band combination allows to distinguish vegetation from land use (band 4, best for vegetation, wetlands and water bodies; band 3, good for plant species; and band 2, for vegetation and housing areas). Similar band combination was successfully applied by Talla (2009) and REPSAHEL (2018). They showed by RGB 4/3/2 processed images the improving of the resilience of Sahelian populations to environmental change.

Two field trips with transects for ground truthing of land use were carried out in the localities of Kouba-Pissi and Koubri-Péélé (Figure 1). Training areas were selected and coordinates taken by a GPS hand receiver. Concerning stakeholders, two focus groups of peasants (seven men and seven women) were addressed to the agricultural sector. Interviews were held at the town hall and at the technical services for agriculture and breeding. They aimed to get information on how access to farmland is organised, and what are the causes of vegetation and landscape degradation. Similarly, a questionnaire was sent to a target population of 124 randomly chosen inhabitants. For the agro-pastoralists' perception of climate variability, it made sense to consult people between 35 and 75 years of age. The technique used to collect data from the population focused on the five villages of Péele, Kouba, Guiguemtenga, Nambé, and Napagtenga (Figure 1). Monthly rainfall data for Koubri were recorded and used to produce climate graphs, as well as the number of rainy days over the time period from 1980 to 2017. Data was obtained from the meteorological station Ouagadougou Somgande (12.35° N; 1.517° W). Therefore, the synoptic station of Ouagadougou Somgande 'represents' an area of one hundred square kilometres.

17.3 STUDY AREA

17.3.1 Location

The topographic map of Ouagadougou ('square degree of Burkina Faso', Figure 1) covers a mainly flat surface of 111 km^2. The geological conditions are characterised by crystalline basement rocks. This area belongs to the most populated regions of Burkina Faso (30% of total population, INSD 2009). The municipalities concerned are Ouagadougou, Ziniare, Koubri, Tanghin-Dassouri, Saaba, Komsilga, Sourgoubila, Sapone, Doulougou, Kombissiri, Gaongo and Bousse. Koubri (12°10′ N; 1°24′ W) consist of 33 villages (~63,781 km^2), located southeast of Ouagadougou, at a distance of 25 km (Figure 1). The spatial extension of the capital Ouagadougou

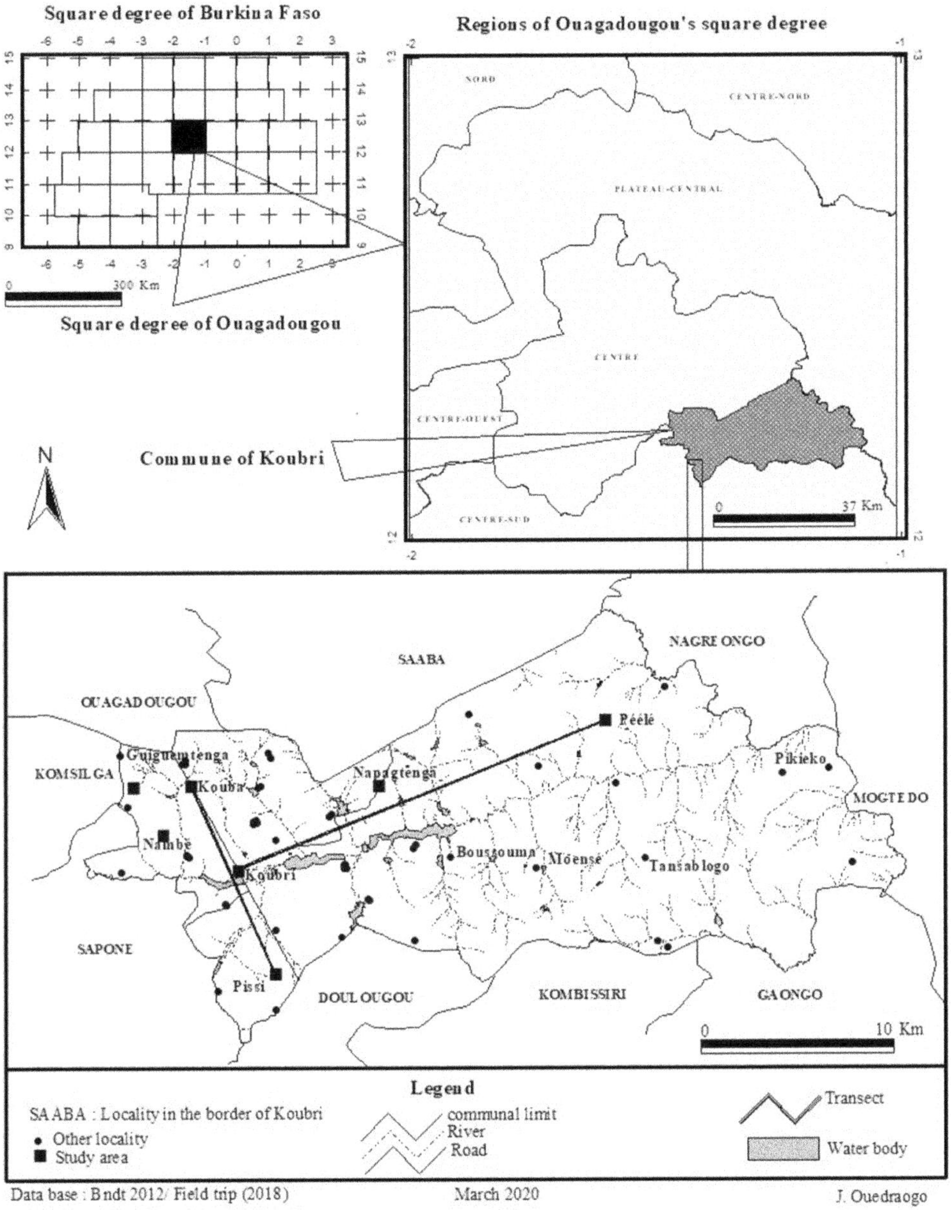

Figure 1. Index map of Burkina Faso and location map of the commune of Koubri close to the capital Ouagadougou.

favours land speculation in the suburbs. Peripheral municipalities are strongly influenced by urban population growth. Koubri experienced a demographic explosion in recent years with land grabbing in its surroundings. Koubri's population increased from 43,928 inhabitants in 2006 to 60,272 inhabitants in 2018 (INSD 2009). Furthermore, Koubri is an agropastoral area where vegetation is still abundant, but with the tendency to be degraded. Natural environment and fields are cleared, sold and transformed by the local population and/or real estate companies. The National topographic data base (BNDT 2012) supported the making of the land use map of Koubri.

17.3.2 Geological and geomorphological features and soils

Crystalline basement rocks shape the suburb of Koubri. Common geological substrates are calc-alkaline granites with biotite, and highly metamorphosed amphibolitic/epidote shales and quartzites. Weathering and erosion of these rocks is caused by water (hydrolysis and runoff), temperature (insolation), wind (deflation), and human activities (agriculture and livestock).

These type of weathering and erosion agents favour the establishment of the Sudano-Sahelian landscape. Annual precipitation is less than 1200 mm. Geomorphological features are planation surfaces with inselbergs, tabular mountains protected by lateritic crusts, extended slopes ('glacis') connecting to extended basins and depressions ('mare').

Inselbergs are located on the watersheds and the partly iron indurated flat slopes (gradient of only 2%) and represent the land use areas for rainfed agriculture. The presence of gullies and linear fluvial cuts ('ravines') give evidence to the importance of water erosion. Erosion processes on crystalline rocks are closely linked to different soils (e.g. shallow to deeply weathered and leached tropical ferruginous soils) used on the 'glacis' slopes by traditional agriculture for the production of cereals. Further on, and of minor importance, low lying hydromorphic soils, eutrophic brown soils and alluvial to colluvial soils occur in geomorphic depressions (BUNASOLS 1982).

17.3.3 Vegetation cover

Savannah woodland is a typical vegetation type with trees of seven metres or more in height with a high and continuous herbaceous mat of at least 80 cm (Guillamet & Koechlin 1971). During fieldwork in Koubri it was revealed that shrubby savannah dominates the study area. Savannah species have been identified in the transects (Figure 1). Savannah woodland includes *Adansonia digitata, Bombax costatum, Khaya senegalensis, Lannea acida, Lannea microcarpa, Parkia biglobosa, Prosopis africana, Sclerocarya birrea,* and *Vitellaria paradoxa*. The shrub savannah includes Combretaceae (*Combretum micranthum, Combretum molle, Combretum nigricans*) and *Acacia seyal, Balanites aegyptiaca, Guiera senegalensis, Piliostigma reticulatum, Piliostigma thonningii* with a continuous grass mat of *Andropogon gayanus, Cymbopogon schoenanthus* and *Loudetia togoensis.* Along the river courses plant species are *Acacia albida, Acacia macrostachya, Acacia nilotica, Acacia seyal, Diospyros mespiliformis, Mitragyna inermis, Parkia biglobosa, Pterocarpus lucens* and *Saba senegalensis*.

17.3.4 Hydrology

The hydrographic network is dense and dentritically shaped, and part of the Nakambé catchment (Massili sub-watershed). The main water course is the Nyarlé River with lower runoff than the Nakambé (formerly 'White Volta'), with short periods of high water and floodings (July-August) and a long dry period in between (often dried out). The commune of Koubri has access to several dams and water bodies (~ 40) where residents carry out agro-pastoral and fishing activities.

17.3.5 Population dynamics and socio-economic activities

The population of the commune of Koubri grew from 43,928 inhabitants in 2006 to 60,272 inhabitants in 2018 (INSD 2009). The increase in population is leading to a quest for productive land in Koubri and similarly for housing plots for Ouagadougou's urban population. In peri-urban and urban areas, the high demand for housing and farm area is at the root of land speculation. Economic activities are dominated by rainfed cultivation on 'glacis', in depressions and on hillsides, followed by market gardening in the dry season around water bodies. In addition, trade, livestock, fishing, handicrafts and a few industrial units, mineral water and dairy production are of importance (Dembele 2010).

17.4 RESULTS AND DISCUSSION

Agricultural and residential land use are factors in the decline of vegetation cover. Specifically, these are settlement and housing areas, sealed surfaces and bare soils, water bodies and rainfed fields. The quasi natural vegetation cover in Koubri is savannah woodland, gallery forest, shrubby savannah and grassy lands.

17.4.1 Land use and vegetation cover units

Land use and vegetation cover dynamics between 1992, 2002 and 2018 is illustrated in Table 1.

17.4.1.1 Housing area

The housing area has increased strongly from 52.58 ha (1992) to 58.13 ha (2002) and 483.63 ha in 2018 (Table 1). Trees and shrubs are cut down and replaced by buildings and urban infrastructure. On agricultural land, only non-timber species remain. These shrubs are also pruned during periods with lack of firewood.

Table 1. Evolution of land use and vegetation cover in hectares (ha) between 1992, 2002 and 2018 (Source: LANDSAT images processing statistics).

Land use and vegetation cover	Year	1992	2002	2018
Housing area		52.58	58.13	483.63
Bare soil		249.58	426.41	54.61
Water body		593.86	1701.69	1311.47
Rainfed cultivation		20,180.22	27,122.87	34,042.58
Gallery forest		861.56	1039.28	772.46
Savannah woodland		22.23	178.69	8781.44
Shrub savannah/grassy area		41,821.74	33,177.09	18,356.69
Total (ha)		63,781.77	63,781.77	63,781.77

17.4.1.2 Bare soil

Vegetation cover protects the soil against erosion. Bandre & Da (2004) noted that grass cover reduces the kinetic energy of raindrops, lowering rainwater runoff and maintaining soil porosity. Exposure of the soil through abusive land clearing, wood cutting (75% of respondents), trampling of livestock (45%), and bush fires (25%) accentuates water and wind erosion. Such anthropogenic activities lead to superficial denudation, leaching of soils, and gullying ('badland' development). All the people interviewed stated that they no longer have any productive land left. They are forced to carry out 'development projects' aimed at increasing rainfed cultivation yields on the degraded soils. Encrusted and bare soil surfaces are highly dynamic (0.36 % in 1992; 0.67 % in 2002 to 0.09 % in 2018) because stakeholders are opting for water and soil conservation as well as agroforestry techniques in order to improve the agronomic capacity of degraded areas. The techniques applied in the communal area are mainly stony cords, Zaï (a traditional form of soil mulching) and tree fruits culture with *Citrus limon*, *Citrus tangelo*, and *Mangifera indica*. In addition to low productive of bare soils, there are rocky granite outcrops used for quarry stone production (armourstone) in erosion control.

17.4.1.3 Water body

The surface area of the water bodies in the study area is on the rise due to the construction of water reservoirs and dams starting in the 2000 (from 0.96% in 1992 to 2.67% in 2002) and then declining to 0.61% in 2018. These reservoirs are used for market gardening and animal watering. The decrease in the volume of these water reservoirs is mainly due to water and wind erosion. Agropastoral activities (rainfed and market gardening cultivations, trampling of livestock) practiced upstream, are the main causes of soil erosion which favours the continuous silting of artificial dams.

17.4.1.4 Rainfed cultivation

Rainfed cultivated areas increased significantly in the Koubri municipal land from 32.01%, 43.13% to 53.36% between 1992, 2002 and 2018 as they occupy the majority of the working population. Rainfed cultivation is dominated by cereals (rice, sorghum, maize, millet), oilseeds and leguminous crops. It is a subsistence activity, very unproductive with the quantities of cereals harvested being less than one ton per hectare and growing season (FAO 2019).

17.4.1.5 Gallery forest

Riparian vegetation along the water courses also declined, especially between 2002 and 2018, by 1039.28 to 772.46 ha respectively (Table 1). Similarly, the tree savannah in the study area increased from 0.28 % in 2002 to 13.76 % in 2018 following various tree planting campaigns with woody species such as *Acacia albida, Acacia macrostachya, Acacia nilotica, Acacia seyal, Blighia sapida, Diospyros mespiliformis, Mangifera indica, Mitragyna inermis, Parkia biglobosa, Pterocarpus lucens* and *Saba senegalensis.*

17.4.1.6 Savannah woodland

The area occupied by savannah woodland increased between 1992 (0.3%) and 2018 (13.76%) in the environs of Koubri. This change in vegetation cover is due to the conservation of species considered useful in the farm area by producers. Taxonomically, the tree stratum is dominated by the species: *Anogeissus leiocarpus; Bombax costatum, Lannea acida; Khaya senegalensis, Lannea microcarpa; Parkia biglobosa; Tamarindus indica; Sclerocarya birrea; Tamarindus indica,* and *Vitellaria paradoxa.*

17.4.1.7 Shrub savannah and grassy areas

It has been revealed that between 1992, 2002 and 2018, the shrubby-grassy savannah occupies 41,821.7 ha, 33,177.09 ha and 18,356.69 ha respectively (Table 1), representing a regression rate of -36.8% for the benefit of fields and housing area. The shrub stratum is dominated by *Acacia seyal, Balanites aegyptiaca, Combretum glutinosum, Combretum micranthum, Combretum nigricans, Dichroctacia glomenta, Guiera senegalensis, Piliostigma thonningii, Ximenia americana,* and *Zizuphus mauritiana.* As for the herbaceous layer, it is dominated by *Andropogon gayanus, Cymbopogon schoenanthus* and *Loudetia togoensis.*

17.4.2 Farmers' perception of environmental degradation

According to the field surveys, the factors that degrade plant formations in peri-urban areas are in order of importance, excessive logging, climate variability, land speculation, poor agropastoral practices, uncontrolled construction, rapid urbanisation, bush fires and other uses of wood (Figure 2).

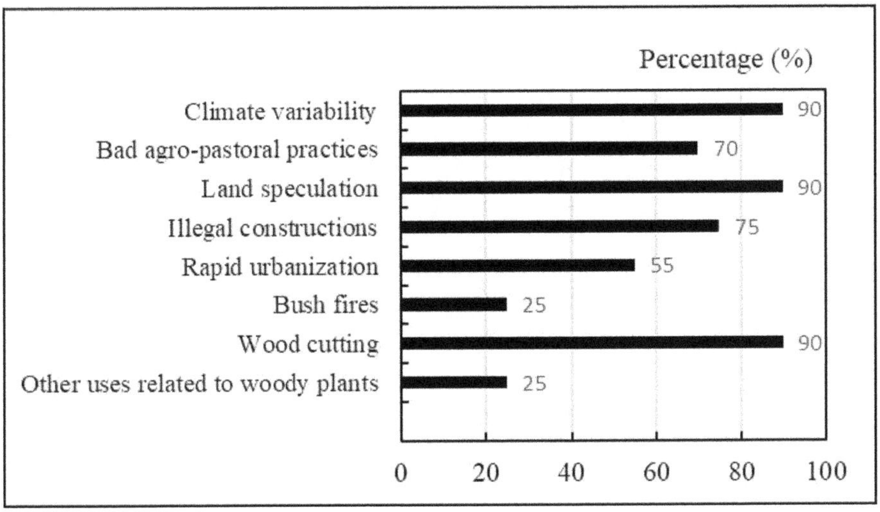

Figure 2. Main factors of degradation of vegetation cover in Koubri by stakeholder's feedback.

17.4.3 Factors responsible for the decline of vegetation cover

17.4.3.1 Slope inclination

Slope inclination ranges from 2% to 24% and altitudes between 220 to 420 m asl in the Koubri area. Slopes' lengths over schists are greater than in crystalline rocks such as granodiorite, rhyolite, pegmatite, and basalt. The shallow depth of the geological substratum conditions the establishment of thin and shallow soils exposed to various erosion processes by water and wind. Degradation of biodiversity is also a result of declining vegetation cover and soil erosion. Steepness and length of slopes also determines runoff processes in a generally levelled landscape (Sanou 1992; Da 1993). The quantity of particles carried by runoff water is estimated via the slope's characteristics. This was well evidenced by the work of Zouagui et al. (2018) and Follain et al. (2018) in the Moulay Bouchta watershed in Morocco and on the Herault wine-growing lands in Switzerland. In sub-Saharan Africa, this point of view is shared by Roose (1985) from arid to semi-humid tropical environments (Sudano-Sahelian savannahs in West Africa). Kabore (1999) did comparable studies in the province of Kourittenga (Burkina Faso) on the problem of the degradation of the vegetation cover of three 'terroirs' of Tensobentenga. Avakoudjo et al. (2014) discussed factors of land use and vegetation cover dynamics in 'W' park in northern Benin.

17.4.3.2 Climate variability

Climate variability is partly summarized by the interannual irregularity of rainfall in the study area between 1980 and 2017. In order to better analyses this interannual irregularity of rainfall at the Ouagadougou Somgande meteorological station, a graph was produced (Figure 3). Analysis of rainfall data indicates a temporal

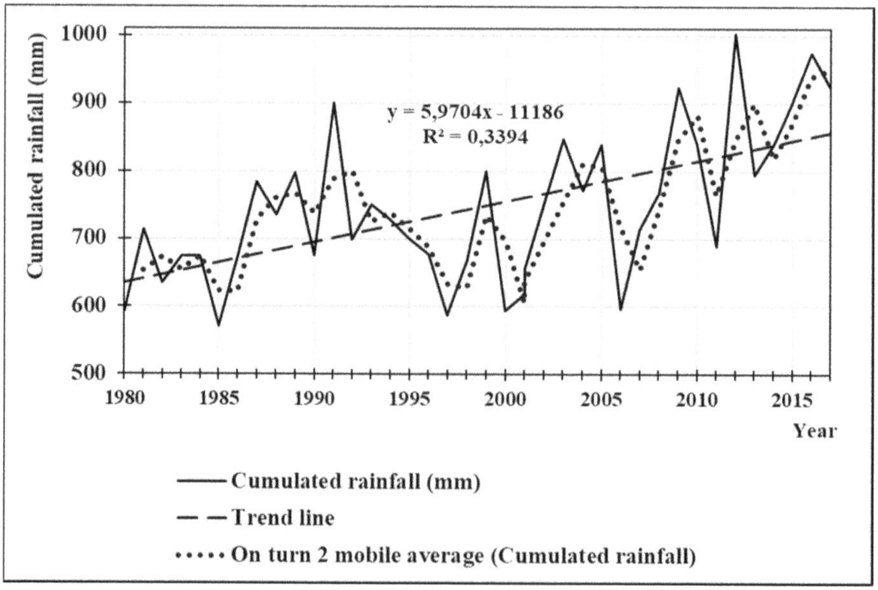

Figure 3. Interannual irregularity of total rainfall at Ouagadougou Somgande meteorological station.

distribution of rainfall with an increasing trend and irregularities between years. Between 1980 and 1985, rainfall experienced a 21.8 mm deficit, from 1986 to 1991 an increase of 227.7 mm, a decrease of 104.3 mm for the period 1992-2000 and a current upward trend (306.2 mm) from 2001 to 2017. Zoungrana (2010) and Dipama (2004) confirm this assertion respectively in the Ziga basin, in the Central Plateau Region and in the Kompienga catchment area.

The average annual rainfall at Ouagadougou Somgande station is 745.7 mm over the 1980-2017 period. The highest rainfall was recorded in 2012, 2016 and 2017 with 1003 mm; 974.1 mm and 924.4 mm. Years with low precipitation of less than 600 mm were registered in 2006, 2000, 1997, 1985 and in 1980. The 1980-1985 period had a negative impact on local vegetation cover. From 1986 to 1991, rainfall increased and marked an episode of wetter years. Between 1992 to 2000, once again marked rainfall deficits were recognisable. A transitional period (2010-2012) followed by again increase of rainfall. From 2001 to the present day, occasional floods, similar to those in 2009, 2012 and 2016, occurred. Hygrophilous and wetland species appeared as they are most vulnerable to interjacent, shorter periods of drought (e.g. *Mimosa pigra* and *Mitragyna inermis*). According to the respondents' impression (15%), rainfall decreased during this period.

The total number of rainy days was characterised by an amount of rain in the range of 5 mm or more, and it varied between 53 and 88 days (in average 67 days), thus imposing short-cycle cereal varieties of the farmers, resulting in a decrease in biomass. In the urban to rural environs of Ouagadougou the number of rainy days was in significantly increasing (regression coefficient R^2= 0.0607) between from 1980 to 2017 (Figure 4).

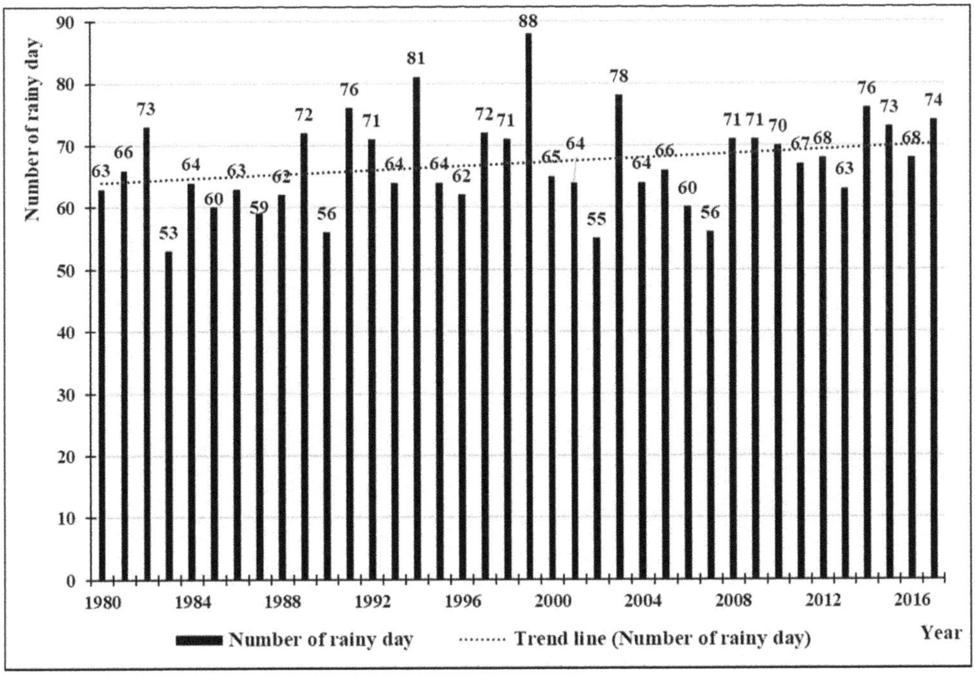

Figure 4. Chronological development of the number of rainy days at Ouagadougou Somgande station 1980-2017.

17.4.3.3 Wood cutting

Excessive logging (75% of respondents) has led to a significant reduction in the rate of vegetation cover. The exposure of bush fires (25%) further reduces the amount of vegetation cover.

According to reactions during field work, the most popular trees for wood bundles are *Acacia macrostachya, Khaya senegalensis*, and *Vitellaria paradoxa*. The search for firewood for cooking leads to an excessive clearing (75%) of the landscape. In combination with the rapid growth of population, it exacerbates the fragility and sensitivity of savannah ecosystems (Figure 5). This is supported by Bandre & Da (2004) and Kankpénandja (2016) working on the dynamics of vegetation cover in north-central Burkina Faso and in the Oti plain in northern Togo. According to Rouamba (2008), population growth has led to an increase in the capital's demand for wood energy since the 1970s (see chapter 18 in this volume).

17.4.3.4 Trampling of livestock

According to DSGU (2014) the communal area of Koubri is also an agropastoral sector dominated by ruminants such as cattle, goats, and sheep. The extensive nomadic to semi-nomadic system of animal breeding is operated by the Peulh herdsmen. Trampling of livestock along the banks of watercourses, around water

Figure 5. Illegal tree cutting of *Khaya senegalensis* in Guiguimtenga, southeast of Ouagadougou (Photo: J. Ouedraogo).

reservoirs, and in places ploughing too, contribute to the degradation of soils (texture) and the disappearance of the herbaceous stratum at the end of the dry season. Animal hooves are impacting the landscape's surface, leading to degradation of the grass cover. This is the starting point of water erosion via the splash effect. These observations are in line with Bougere & Hugot (1979) and Baritse (1986) in the Burkinabe Sahel, and in the Savannah Region of northern Togo.

17.4.3.5 Land pressure

The lack of space for housing, the search for single-family homes and rising land prices in the urban central of the capital Ouagadougou are forcing city dwellers to look for available and low-cost space in the suburbs. The latter takes place towards the south of the capital Ouagadougou at the expense of agricultural and natural areas. The "consumption" of peri-urban spaces by the urban population through setting up buildings and agropastoral activities testifies the scarcity of spaces in the urban perimeter. Land speculation (90%) has as its main consequence the regression of vegetation cover.

According to INSD (2009) the population of the capital Ouagadougou has almost doubled over the last 20 years: from 709,736 inhabitants to 1,475,830 inhabitants. Nowadays, the number of urban dwellers in Ouagadougou is estimated at about 2,500,000 inhabitants with a surface area of 2805 km^2 (Zida 2009). The population of the municipality of Koubri increased from 43,928 inhabitants between 2006 and 60,272 in 2018 (+37.2%). The average density of the municipality was about 69 inhabitants/km^2 while at the national level it was 50 inhabitants/km^2 (INSD 2009). To the land pressure (90%) it must be added animal surcharge (90%). In Guiguimtenga a herd of cattle grazes less far from the concessions, favouring the destruction of the herbaceous mat at the end of the dry season in April (Figure 6).

The most devastating bush fires are those during the mid-dry season from January to February. This is the period during which the dry monodirectional Harmattan wind predominates, with a northeast to southwest blowing speed of 4 and 6 meters per second. Combined with the dry herbaceous carpet, bush fires spread quickly while decimating small shrubs resulting in soil exposure. These fires are difficult to control and can cross slopes and burn for several days. Only pyrophytic species such as *Annona senegalensis, Bombax costatum, Mitragyna inermis, Piliostigma thonningii*, and *Vitellaria paradoxa* are better adapted and therefore spared from the bushfires (Bandre & Da 2004). The pyrophytic groups belongs to Combretaceae and Sapotaceae, and they become more and more important (Ballouche 2019).

Runge (2007) also attests that nowadays African rain forest is destroyed by many disturbances like bush fires, clearing for agriculture and for wood collection. Drylands in Africa are often used for extensive livestock grazing by domestic animals (Baumer 1983). In the north of Ivory Coast, heavy cattle loads are condemning the forests to rapid degradation and even long-term disappearance due to bush fires started by pastoralists. These, while regenerating the pastures, destroy the young woody trees. and prevent the regeneration of forests. In the absence of fire, livestock favours the impoverishment of the flora and diminishes its economic wealth because it destroys the regeneration of the species that are most 'hungry' (Louppe 1996).

Valea & Ballouche (2012) consider these savannah fires as a source of regeneration of the grass stratum favourable to pastoral activities.

Bush fires are therefore a normal and indispensable ecological factor for the maintenance of the Sudanese savannahs. As soon as the savannahs are grazed, it feeds bush fires that are already less aggressive against woody trees. But trampling

Figure 6. Illegal construction and degradation of grass cover by informal cattle grazing at the end of the dry season (April) in Guiguimtenga, southeast of Ouagadougou (Photo: J. Ouedraogo).

by cattle probably favours the fixation of seeds on the ground and the tilling of grasses (Boutrais 1983).

Bush fires in the savannah lands (e.g. Mount Nimba in Guinea) occur as a result of pastoral, agricultural or poaching activities. Bush fires are less frequent in submontane savannah and while they do not adversely affect grassy cover, they will modify woodland edges and undergrowth in forests and ravines. Early season fires are the only way of reconciling plaster management and conservation in the savannah ecosystem without negative impacts for the vegetation cover and fauna (Poilecot & Loua 2009).

The Sudano-Sahelian and Sahelian areas of West Africa are among the regions where populations and ecosystems are most sensitive and vulnerable to climatic hazards. The rapid increase in land requirements due to rapid population growth and spontaneous urbanisation increases pressure on natural resources and causes their degradation (Hountondji 2008). In the Guirkouh forest in southern Chad, Guiguindibaye (2014) attests that fire has been more destructive in shrub savannah where herbaceous biomass is more abundant. Herbaceous plants showed variable responses to the passage of fire. In some species such as *Andropogon gayanus*, fire stimulates regrowth. At the level of the vegetation cover, fire causes damage to individuals but seems to favour their foliage and regeneration. In southeast Togo the results revealed significant changes, both natural and anthropogenic, who achieved water resources and their buffer areas, and vegetation cover. During 25 years, vegetation cover around wetlands decreased by ~26% and aquatic plants showed a surface increase of ~13% (Ahlonko et al. 2015).

17.4.3.6 Other uses of woody plants

Other uses of woody plant species concern traditional medicine and the search for wood for construction of houses, woody pallisades (fences) and especially for craftsmen (sculpture). Some of the most sought-after species are *Balanites aegyptiaca*, *Diospyros mespiliformis*, *Khaya senegalensis*, *Mitragyna inermis*, and *Vitellaria paradoxa* (Valea & Ballouche 2012). The use of grasses, bark and woody flowers, in the long term, contributes to the regression of the woody and herbaceous stratum. But for other considerations such as tradition, species such as *Sclerocarya birrea*, *Tamarindus indica* are conserved. This observation is in line with those of Traore et al. (2011) in the southwestern part of Burkina Faso. However, some aspects of traditional beliefs (prohibitions) are favourable to the conservation of plant resources that protect species from burning. There is evidence that the combination of field inventories and ethnobotanical surveys are essential as a reliable approach to identifying threatened species.

17.5 CONCLUSION

The study of the factors responsible for the decline in vegetation cover in Koubri, a peri-urban suburb of Ouagadougou, based on LANDSAT images, revealed profound transformations in the physical and biophysical environment over the last 30 years. The shrubby/grassy savannah is now replaced by buildings, rainfed crops and quarries (extraction of clay and granite). This degradation of the vegetation cover also has its origins in population growth and pressure on degraded ferruginous soils. The factors contributing to the decline of woody and herbaceous species are essentially intensive logging, land speculation and climate variability. From this regression of plant cover many geomorphological, hydromorphological, and hydro-pedological consequences occur.

REFERENCES

Ahlonko, K.B., Selifa, S. & Kouami, K. 2015. Identification of environmental changes and land use in the Logoon ecosystems of Togo Southeastern. *Open access library journal* 2: 9.

Albergel, J. 1987. *Genesis and predetermination of floods in Burkina Faso per m^2 per km^2: study of hydrological parameters and their evolution*. Doctoral thesis: 342. Paris: University of Paris XI.

Avakoudjo, J., Mama, A., Toko, I., Kindomihou, V. & Sinsin, B. 2014. Dynamics of land use in the W National Park and its periphery in north-west Benin, *In J. Biol. Chem. Sci* 8(6): 2608–2625.

Ballouche, A. 2019. Ordinary, degraded or patrimonial legacies in west African savannah landscapes. *Landscape in action in the tropics, Agrocampus Angers* 21:1–21.

Bandre, E. & Da, D.E.C. 2004. Impacts of vegetation cover dynamics on silting of Lake Dem in Burkina Faso in Priorities and Development Practices in Sub-Saharan Africa. *Proceedings of the Enreca/Ird-Social Sciences International Symposium from 19 to* 20 December 2002: 137–148. Ouagadougou: University of Perugia.

Baritse, L. 1986. *Slopes and slope systems, the example of Northern Togo*. Doctoral thesis: 197. Paris: Sorbonne University.

Baumer, M. 1983. Grazing land in arid zones: reflections for a management of the environment. *Mediterranean forest*, t.v., 2: 173–184.

Benbrahim, F.K., Ismaili, M., Benbrahim, F. S. & Tribak, A. 2004. Problems of environmental degradation through desertification and deforestation: impact of the phenomenon in Morocco. *Secheresse* 15(4): 307–314.

BNDT 2012. National topographic data base, Geographical institute of Burkina (IGB), degree square of Ouagadougou, Digital Support.

Bougere, J. & Hugot, G. 1979. Trampling of livestock, a factor in the degradation of pastures, such as the dune of Tin Edjar (Haute Volta). *Revue photo-inter* 1(5-6b): 33–45. Paris: édition technique.

Boutrais, J. 1983. *Sudanese breeding. From savannah trails to ranches (Cameroon-Nigeria)*. Bondy: ORSTOM.

BUNASOLS 1982. Study, agro-economic in the Koubri area. Ouagadougou: 126.

Da, D.E.C.1993. Crusts and micro-steps, the natural environment of Burkina Faso. *Collection Cret/ "Landlocked Countries" 7*: 46–60 Paris.

Dembele, J. 2010. Mapping socio-rural areas: a planning support tool for management in agriculture (Burkina Faso). *Agricultural Water Management Solutions Project/Bill and Melinda Gates Foundation.*

Dipama, J.-M. 2004. Peasant perception of environmental degradation around the Kompienga dam lake. In *Priorities and development practices in sub-Saharan* Africa, *proceedings of Enreca-Ird /social sciences international symposium*, University of Perugia: 149–164.

DSGU, Directorate of Statistics and Urban Management 2014. State of the major estates of the rural commune of Koubri 2012-2014 (62). Ouagadougou.

FAO 2019. *Analysis of food losses: causes and solutions. Case study on sorghum, maize in Burkina Faso*. Rome: FAO.

Follain, S., Fayolle, E., Noll, D. & Quiquerez, A. 2018. Erosion of wine soils and adaptation to change: vulnerability, diagnosis and adaptation strategies. From the adaptation of practices to the reorganization of vineyard. *La revue des oenologues* 166: 12–15.

Guiguindibaye, M. 2014. *Biological and ecological impacts of control on vegetation and soil in the Guikouh forest, southern Chad.* Doctoral thesis: 169. Ouagadougou: University of Ouagadougou.

Guillaumet, J-L & Koechlin, J. 1971. Contribution to the definition of vegetation types in tropical regions (example of Madagascar). *Candolla* 26(2): 263–277.

Hountondji, Y-C.H. 2008. Environmental dynamics in the Sahelian and Sudanian zones of West Africa: Analysis of changes and assessment of land cover degradation. Doctoral thesis: 152. Liège: University of Liège.

INSD, National Institute of Statistics and Demography 2009. Population projections from 2007 to 2020. Ouagadougou.

Kabore, M. 1999. *The dynamics of vegetation cover and its socio-economic impact on the populations of three terroirs in the department of Tensobentenga, Kourittenga province*. Master's thesis: 83. Ouagadougou: University of Ouagadougou.

Kankpénandja, L. 2016. *Morphogenesis and land management in the Oti sub-watersheds*. Doctoral thesis: 315. Lomé: University of Lomé.

Louppe, D. 1996. Research in agroforestry in Ivory Coast Northern (Role of woody trees in the agroforestry space), CIRAD-Forest/Korhogo: 15.

Poilecot, P. & Loua, S-N. 2009. Bush fires in the savanna lands of mount Nimba, Guinea. *Woods and forests of the tropics* 301(3): 51–66.

Quattara, A., Kouame, A., Tiebre, M-S., Cisse, A. & N'guessan, K. 2016. Flora diversity and plant use in the Sudanese zone of northwestern Côte d'Ivoire. *Journal of Animal and Plant Sciences* 31: 4828–4840.

REPSAHEL 2018. Land use mapping: a practical guide. *Saharan and Sahel observatory*: 85.

Roose, E. 1985. Land Degradation and Development in West Africa. *Bull. Rech. Agrom Gembloux* 20 (3/4): 505–537.

Rouamba, H. 2008. Environmental degradation in peri-urban villages: the case of Saaba in the outskirts of Ouagadougou. Master's thesis in geography, University of Ouaga-dougou: *124*. Ouagadougou.

Runge, J. 2007. Of deserts and forests. Late Quaternary landscape and climate history of Central Africa. *Geo-Eco-Top* 31: 1–18.

Sanou, D.C. 1992. Runoff and erosion in small catchment areas: the case of Imiga-Tinbin, Ministry of Water (National Office of Wells and Drilling, INSHUS), University of Ouagadougou: 64. Ouagadougou.

Taïbi, A-N. & El Hannni, M. 2004. Identification and monitoring of environmental degradation processes in the Oued Dadès basin (Morocco) by satellite and aerial remote sensing. *Télédétection* 4(1): 543–553.

Talla, N. 2009. Digital image processing, course material. Ensai-N'Gaoundere/Cameroon: 85.

Traore, L., Ouedraogo, I., Ouedraogo, A., & Thiombiano, A. 2011. Perception, uses and vulnerability of woody plant resources in southwest Burkina Faso. *Int J. Biol Chem Sci* 5(1): 258–278.

Valea, F. & Ballouche, A. 2012. Bushfires in West Africa: Environmental concepts or environmental management tool? The example of Burkina Faso. *Territoire d'Afrique* 3: 36–47.

Zida Y. 2009. Central Region monograph: 154. Ouagadougou.

Zouagui, S., Sabir, M., Naimi, M., Chikhaoui, M., & Benmansour, M. 2018. Modelling water erosion risk by universal loss equation in the western rif: case of the Moulay Boutcha catchment area (Morocco). *European Scientific Journal* 14(8): 524–544.

Zoungrana, T.P. 2010. Adaptation strategies of rural producers to climate variability in the Ziga basin, Central Burkina Faso. *Annales de l'Université de Ouagadougou* A(11): 585–606.

CHAPTER 18

Food insecurity and household adaptability in rapidly changing informal districts of Ouagadougou, Burkina Faso

S. Rouamba

Norbert Zongo University, Koudougou and Joseph Ki-Zerbo University, Ouagadougou, Burkina Faso

ABSTRACT: Like many West African cities the population of the informal districts of Ouagadougou, Burkina Faso's capital (1,500,000 inhabitants estimated in 2006, INSD 2009), are vulnerable to food insecurity. Ongoing population growth and climate change lead to the extension of rural lands and the emergence of informal neighbourhoods. These marginal agricultural areas are home for the capital's poorest people (nearly 25% of the urban population of Ouagadougou). Access to food is becoming increasingly uncertain for them. In a context where the impact of climate change on food security in urban areas is poorly addressed with little disparate and abundant documentation, the objective of this study is to analyse the determinants of food insecurity in the suburb of Polesgo. Both quantitative and qualitative methods have been used for data collection. The application of FAO techniques allowed to analyse the determinants of food insecurity and to understand household responses. This evidenced that the problem has become a daily reality for the people of Polesgo and that 3/5 of households are regularly experiencing food insecurity. The determinants are both conjunctural and structural. The purchasing power is very low: 77.5% of households in Polesgo have a monthly income of less than FCFA 25,000 (~ € 38). Agricultural production has become very low and market prices for some cereals have fluctuated widely over the past decade. Faced with this food insecurity aggravated by climate change, adaptation strategies are developed by populations, including livestock farming (bovine, ovine, porcine, caprine, poultry), and income-generating activities (carpentry, welding, mechanics, small trade), and the rationalisation of family spending.

18.1 INTRODUCTION

Since the early 2000s, the tools, protocols and mechanisms governing food and nutrition crisis management have evolved significantly (WFP, World Food Programme, FAO, Food and Agriculture Organization). They are subsequent to episodes of disastrous crises (Janin 2015), like the droughts of the 1970s and 1980s caused famines in sub-Saharan Africa. Many West African countries are experiencing food crises that are both cyclical and structural (CILSS 2008). These different crises in the context of climate change have significantly increased the number of undernourished people in the population of sub-Saharan Africa. One in three people is exposed to an average food deficit of 300 kilocalories per day for a standard of 2,400 kilocalories, both in urban and rural areas (FAO 2006). The number of undernourished people in Africa increased from 181.7 million between 1990-1992 to 232.5 million during the period 2014-2016

(FAO 2015). The average daily food availability per person still remains below the minimum threshold (300 at less than 2400 kilocalories per day per person), despite the increasing use of food aid by African States delivered by the World Food Programme (PAM 2005).

In their attempts to respond to the scourge of hunger, policy-makers, researchers and stakeholders in the development sector have focused for years on rural environment, which for them is full of people who are vulnerable in terms of food. Most of the documents related to the subject deal with the issue of food security in rural areas, while urban dwellers are also exposed to food insecurity linked to structural, economic and social factors (FAO 2015).

Since 2005, demonstrations against the high living costs in many capitals, including Ouagadougou, have shown a new social and urban dimension of hunger. It now appears that food insecurity is no longer a topic of rural areas alone. It has become a major concern in urban areas (CILSS 2008). The rate of urbanisation is very high and urban population in West Africa has already reached 472 million inhabitants (2017), and will double in the next 25 years. It will reach one billion inhabitants in 2040. In 2025 African cities will have 187 million additional inhabitants (World Bank 2017). World Bank also documented in 2000 that 71% of poor people lived in rural areas, compared to 29% in urban areas. The poor survive on less than one dollar per day, which does not allow them to reach normal dietary energy availability (Ghersi 2015). In Ouagadougou the phenomenon of 'rurbanisation' (urbanisation of rural areas around the city) is triggered by ongoing climatic change that are affecting the country. Poor distribution of rainfall over time and space (droughts) enhances the rural exodus. This massive migration of people from the back country to Ouagadougou contributes to the spatial extension of the city. In these precarious neighbourhoods, about 25% of the population of the capital is concentrated where the conditions of access to food seem more and more problematic (INSD 2009). As a result, in some unplanned and informal neighbourhoods, food insecurity strongly influences the daily lives of these population.

18.2 STUDY AREA

Polesgo district is located on the northern periphery of Ouagadougou, ten kilometres from the capital's centre, beyond the 'green belt' that surrounds the urban area (Figure 1). The green belt is a vegetated strip which was set up in 1976 by the government with the support of the German forestry mission. Planned to cover 2100 ha, this reforested strip was intended to protect the capital from drying winds (Harmattan) and to trap the dust coming from the Sahara, and to mitigate the stripping of the soil by runoff water towards the city's dams (Figure 1).

In the current division of the Commune of Ouagadougou, Polesgo is attached to the Nongr-Mâasom District (District 4) and is part of Sector 20. Polesgo district is bordered to the southwest by the waste treatment and recycling centre of Ouagadougou and by the Tuudbweogo cemetery, located on the access road to the city centre. It is bounded to the south-east by the green belt and the industrial zone of Kossodo, and from the north and to the west by agricultural land. Polesgo covers an area of 134.96 ha (~1.35 km^2) and housing about 3600 households (INSD 2009).

18.3 METHODS

This work is based on qualitative and quantitative research techniques: extensive literature review, data collection, data treatment and analysis. Collection tools include

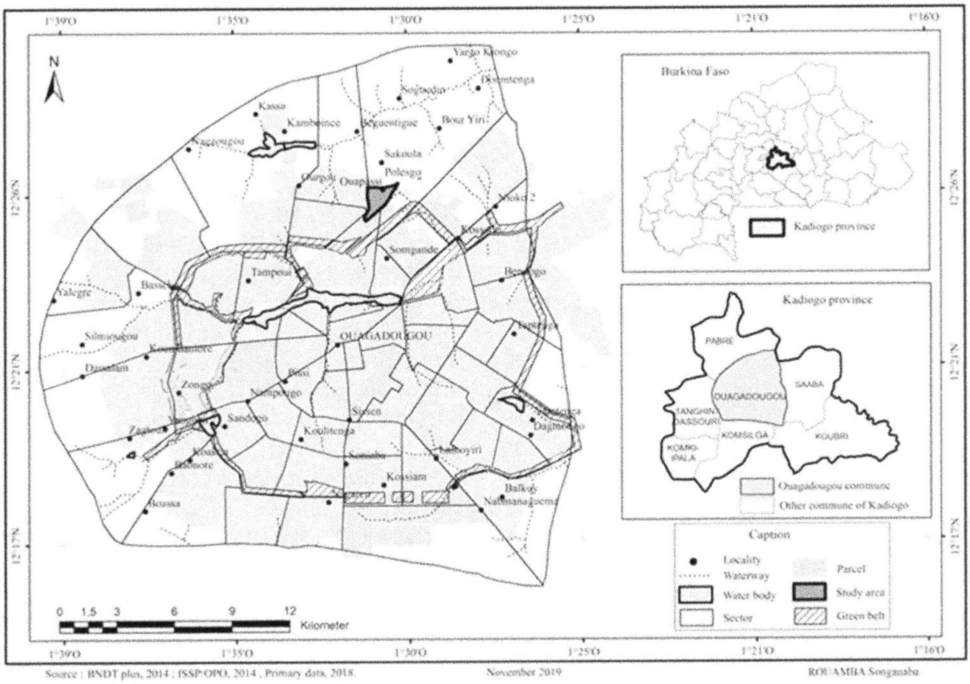

Figure 1. Location map of the informal suburb of Polesgo north of Ouagadougou.

questionnaires, interview guides, direct observations and the Food Insecurity Score (FSI). It was derived from the answers to nine questions in a standardised questionnaire on the household's experience of food insecurity, using the method recommended by FANTA (Food and Nutrition Technical Assistance). This method was developed by Coates et al. (2006); Webb Coates et al. (2006), and Swindale & Bilinsky (2006). The respondent is the person primarily in charge of food in the household. For each response, a score is assigned based on the frequency with which the corresponding event was experienced by the household in the previous months: 0 (never), 1 (rarely), 2 (sometimes) or 3 (often). The FSI corresponds to the sum of the scores assigned to the nine questions and therefore varies from 0 to 27, with "0" corresponding to a situation where no food insecurity was experienced by the household and "27" reflecting maximum, dramatic insecurity. The food insecurity category for each household is assigned according to a grid proposed by FANTA (Coates et al. 2006). The category in which a household is classified corresponds to the most precarious category identified by at least one answer to one of the questions. A household is classified as "food secure" if all responses are in green boxes; it is classified as "low food insecurity" if at least one response is in a yellow box, without any being in an orange or red box, and so on.

A stratification scale of the different adaptation strategies "survival" is also constructed using the FANTA method (Coates et al. 2006). These different survival strategies depend on the levels of insecurity in which households find themselves. There are four different levels of severity of survival strategies. The use of these strategies makes it possible to identify the level of household risks through so-called level 1, 2, 3 and 4 strategies. The more the household uses level 4 strategies, the more exposed it is to food insecurity where reversibility is difficult. Conversely, the more the household uses level 1 and 2 strategies, the higher the probability of reversibility.

- Level 1 survival strategies: these are the most common strategies used by at-risk or food-insecure households. They result in the consumption of less preferred foods and a decrease in food intake. Level 1 strategies present a low risk of livelihood depletion.
- Level 2 survival strategies: these are strategies used by food-insecure households. They include the following strategies: purchase of food on credit, unusual exodus. These strategies present a medium risk of erosion and disruption of livelihoods.
- Level 3 survival strategies: These strategies cover all households facing food insecurity. The aim is to limit adult consumption in favour of the young, to make active members of the household eat at the expense of the inactive. These strategies preserve the living conditions of children and assets, while conversely, they degrade those of adults and the inactive. This shows that food insecurity is experienced differently at the household level.
- Level 4 survival strategies: These are the most dangerous distress strategies. These include selling production goods, avoiding health services, spending the night or day without food. These strategies undermine livelihoods, nutritional status and increase long-term vulnerability.

The secondary data was obtained and the primary data was collected from 120 randomly selected households. QGIS was applied for the processing of spatial information and to produce thematic maps; EXCEL was used for the statistical processing of data (climatic, quantitative) to produce tables and graphs.

18.4 RESULTS AND DISCUSSION

18.4.1 Socio-economic characteristics of food-insecure households

Food insecurity can be observed in this neighbourhood through the characteristics of households and their level of insecurity. Some households show features that predispose them to difficulties in accessing sufficient food. The status of the head of household could indicate food insecurity within the household. In Polesgo 7.5% of households are managed by divorced or widowed women. Most of them are abandoned or driven out with their children by their family in-laws:

> *"I lost my husband and my family in-laws drove me out. That's why I came to sit here with my children" (statement by K.M., widow head of household in Polesgo, August 2017).*

For this stakeholder, women who are alone in the management of their families cannot afford to bear the household expenses. Another important food insecurity issue characterises households in Polesgo: the level of education of heads of households. More than the half (54.2%) of them have no level of education, and 30% have obtained a very low education level (primary school). This situation constitutes a major handicap for the family economy, because heads of households (84.2%) will experience difficulties in obtaining adequate jobs capable to providing their families' food needs. Moreover, the various main activities carried out testify to the precariousness of this local economy: 87.5% of heads of households work in the informal sector and 12.5% in agriculture. In addition, the population of Polesgo is quite heterogeneous as (74.2%) of households are migrants. A high proportion of migrants can have a negative effect on social cohabitation and make it difficult for development partners interventions and provide social assistance in the neighbourhood.

The precarious socio-economic living conditions of households justify their inability to satisfy the food needs of their family members. The nature of housing is an undeniable factor in assessing the economic level of households in a locality. In Polesgo 87.5% of households live in dwellings built with precarious materials (mud bricks). The same is true for almost all sanitation facilities (toilets), unfortunately 81.3% of which are disposed end up in the shallows or on the street. The energy used in this district also remains precarious: the cooking method is traditional: 71% of households use firewood and 40% use battery-powered flashlights for lighting. The predominance of land bricks in buildings, the method of waste disposal and the precariousness of the energy used testify to the low purchasing power of the populations of Polesgo. This has negative impacts on food security.

18.4.2 Determinants of food insecurity in Polesgo

The results based on nine items in a standardized questionnaire according to the method recommended by FANTA are as follows: It was proved that food insecurity has become a daily reality for the inhabitants of Polesgo and 3/5 of households are food insecure. Widow households (5.8%) remain the most vulnerable to food insecurity. Widows or single mothers with dependent children are among the vulnerable social groups (FAO 1999). The determinants of food insecurity are both cyclical and structural, including the low purchasing power of populations linked to poverty, the virtual non-existence of structural and community aid and agonizing agricultural activity in a context of spatial and climatic dynamics. Precisely, these structural and cyclical problems deteriorate the level of food security of African cities' households. This is confirmed by Courade (1996), Martin-Prevel et al. (2000), Maire & Delpeuch (2004) and Janin (2009).

Socio-economically the determinants of food insecurity are present in this part of Ouagadougou. Within the population there is a significant proportion of inactive people or 2/3 of household members. However, the results of surveys show that the inactive constitute a real burden for the food autonomy of households, especially when these households have many mouths to feed (an average of six members per household). Moreover, there is a predominance of a large fringe of households with very low purchasing power: in Polesgo 77.5% of households have a monthly income of less than FCFA 25,000 (~ € 38). This low household income is justified by the education level of the heads of households (83.2%). With such low levels of education, it is quite legitimate for the heads of households to face difficulties in accessing jobs. The analysis of the monetary expenditures of these households evidenced that food expenditures occupy an important place with 42.6%.

In economic terms, agriculture, which was the main food-providing activity, only employs only 13% of the population due to climate change and 'rurbanization'. The informal neighbourhoods of Ouagadougou are constantly handicapped by ongoing climate change. This change is manifested by an increase in floods, droughts, heat waves and a disruption of the rainy season (the beginning of the season is increasingly late and the end of the season is increasingly early, Rouamba 2017). Agriculture is hindered by rainfall irregularities which reduce yields. Populations testify that before the 2000s, the abundance and regularity of rainfall enabled them to produce good harvests. Their cereal products were abundant, and they have no longer bought food after the end of the agricultural season. Nowadays, things have changed: producers are starving (Rouamba 2017). These results are similar to those of Duchemin et al. (2009) cited in Avard (2015) and FAO (2014), who find that urban agriculture (within or fringe), also called “city farming”, helps improve economic conditions as well as the health of the poor and the vulnerable in a developing nation. In this case it contributes to food availability, provides employment and income and can contribute to the food security and nutrition of urban dwellers (FAO 2014). Unfortunately, as confirmed by Von Braun (2008) adverse climate conditions significantly disrupt production and reduce farmers’ assets.

In addition to the decrease in agricultural yield due to climate change, there is the transformation of agricultural space into informal neighbourhoods (residential areas). Between 1960 and 1984 the informal districts of Ouagadougou, which were dominated by agricultural activities, occupied 71.4% of the urbanized area. In 2008, these districts represented only 24.2% of the urban surface area, and were dominated by informal housing (Ouattara 2010, Boyer & Delaunay 2009, Rouamba 2017).

Like in other informal districts of Ouagadougou, Polesgo is permanently fed by a population flow originating from the rural exodus, the unemployed and the most deprived of the city of Ouagadougou, at a rate of 11% per year against 3% for the formal city. In Polesgo, the area occupied by new buildings has tripled in 3 years (from 9 ha to 27 ha, Figure 2).

With such a rapid increase, an uncontrolled occupation of space and 98% of buildings built with simple materials (clay bricks), this district is more vulnerable to the effects of climate change. The rapid growth of cities in sub-Saharan Africa and especially the precariousness in which human settlements evolve around these cities

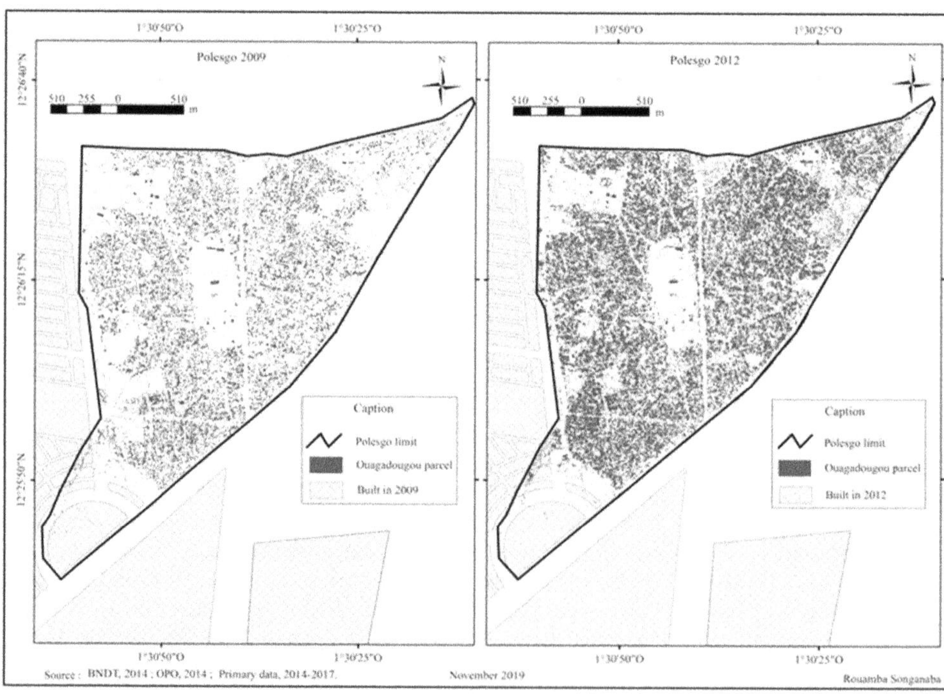

Figure 2. Growth of buildings at Polesgo district (Ouagadougou) from 2009 (left) until 2012 (right). This map shows the dynamics of land occupation by people seeking housing. As housing is very expensive in Ouagadougou, the poorest people rush to peri-urban farmland to build their homes.

and their corollaries are also addressed in the work of Jaglin (1995), Martin-Prevel et al. (2000) and Maire & Delpeuch (2004).

In addition to the decline in yields and loss of agricultural land aggravated by climate change, there is the problem of access to water resources in Polesgo. The increase and the intensification of climatic phenomena such as heat waves, droughts and increased evapotranspiration reduce the availability of water resources. Surface water and groundwater used by people in agriculture, livestock, construction and households are rapidly drying up (Dipama 2014, Rouamba & Ouedraogo 2016, Rouamba 2017). Over the last sixty years (1950-2013), there has been an increase in maximum and minimum temperatures in Ouagadougou. However, high temperatures, especially in the dry season, increase the hygrometric capacity of the air, resulting in the premature drying of water points by evaporation (Dipama 2014, Yaka et al. 2015). While drinking water provided by the National Water and Sanitation Office (ONEA) and some private actors are used for these various activities, they come mainly from surface water. The availability of this drinking water depends on rainfall, which has a downward trend during the same period (1950-2013). Thus, in the dry season, especially in periods of high heat (March-April-May), water becomes a competitive resource in Polesgo (Rouamba & Ouedraogo 2016).

With regard to these results it is noted that 87% of Polesgo households make greater use of the market for obtaining food. Prices of some foodstuff on the local markets have fluctuated significantly over the last decade. For example, the prices of most staple foods, increased slightly from 2016 to 2017 (Table 1).

It sometimes reflects the inability of administrative and political authorities to effectively regulate the prices of basic foodstuff. This translates into higher prices for food products that traders buy from farmers which increases food insecurity in informal households. This is pointed out by Kalinda & Songolo (2017), that many African countries have encouraged an increase in agricultural productivity without making the same effort to improve food prices on the markets, which has resulted in high transport costs and significant post-harvest losses while maintaining the vicious circle of poverty caused by food security that has remained inaccessible. According to Chongo (2017), it is the task of local authorities in developing countries to set up laws to improve the sustainability of urban food security. The solutions would lie in trade policy reforms aimed, in developed countries, at reviewing strategies related to agriculture and biofuels, and stopping new policies of market distortion that are mutually harmful (Von Braun 2008).

The price fluctuation for cereals is followed by increasing prices for water. As water becomes less and less available in Ouagadougou, it will become more and more expensive. The resource is becoming scarce and expensive in Polesgo. Consumers do travel long distances to get water (1 to 5 km), and sometimes they spend whole days around water points. Waiting time can reach 20 hours during periods of water shortage, especially in April or May (Figure 3). Difficult access to water in the undeveloped

Table 1. Food prices on the Polesgo market (in FCFA/€).

Specification	Price per kg in 2016	Price per kg in 2017
Corn	112.5 (~ € 0,17)	137.5 (~ € 0,21)
Small millet	200 (~ € 0,31)	237.5 (~ € 0,36)
White sorghum	106.25 (~ € 0,16)	150 (~ € 0,23)
Cowpea	312.5 (~ € 0,48)	262.5 (~ € 0,40)

(Source: Primary data for September 2017).

districts which confirms the comments of people in these districts, that water problems persist throughout the dry season until the beginning of the rains. At that time, women hardly sleep in their homes anymore, but rather at fountains and boreholes.

Water harvesting becomes an activity in its own right to the detriment of income-generating activities in Polesgo. Moreover, the price of water is becoming exorbitant for poor households: among retailers, it varies from 250 to 300 FCFA (~ € 0.38 to 0.46) per barrel of 200 liters in normal periods (July to February), against 500 to 2000 FCFA (~ € 0.76 to 3.05) in times of water shortage (March to June). Between the normal and shortage periods, the price of water among resellers increased fivefold in Polesgo (Rouamba 2017). This arbitrary and uncontrolled setting of water prices in precarious neighbourhoods shows that reflections on water prices, the ability of households to pay and the access of the poorest to a service that has become profitable have not yet resulted in economically and socially viable proposals (Baron & Tidjani Alou 2011). This situation justifies the difficulties that populations have in getting rid of food insecurity. This is what Lesauvage (1997), Maxwell & Frankenberger (1992), Maxwell et al. (2000) and Briand (2001) show: the precariousness and professional situation of heads of households characterise families vulnerable to food insecurity in urban areas.

This food insecurity is further aggravated in Polesgo where the level of assistance and social mutual aid remains very low because of the predominance of migrants (74.2%).

The photo (Figure 3) illustrates how difficult access to water is in the informal settlements of Ouagadougou. It confirms what the populations of these neighbourhoods

Figure 3. Water queue at a well of an informal settlement in the environs of Ouagadougou (Photo: S. Rouamba, May 2014).

say: water problems persist throughout the dry season until the rains begin. At this time, women hardly sleep in the homes anymore, but rather at the standpipes and boreholes.

18.4.3 Household food insecurity in Polesgo

The analysis of the level of food insecurity in Polesgo also revealed household behaviour with regard to insecurity indicators. The issue of concern about access to food is a sentiment that drives a large part of the population in the Polesgo district. With regard to concern about access to food, 57.3% of households are experiencing or have already experienced this stress. For the consumption of non-preferred food, 58% of households sometimes or often opt for cheaper and more accessible food products that are not necessarily preferable. Most often, the proportion of populations that consume a single food repeatedly (a food monotony) is significant: 77% have experienced this situation and 22% experience food monotony on a recurrent basis. The same is true for the consumption of avoided foods: in the specific case of the Polesgo district, 21.7% of households experienced it on a recurrent basis. Among the indicators of food insecurity, a decrease in the quantity and number of meals is evidenced. At this level, 44.2% of households reduce the amount of food when access to food becomes a problem, and 83.3% reduce the number of meals from three to one meal a day. At worst, some families go to bed hungry (50%); some households do not have food (50%) and others are forced to fast all day, 23% of households experience this situation on a recurrent basis.

A link between the function of the head of household and food insecurity shows that the households most affected are those whose head works mainly in the informal sector. It is also noted that the most vulnerable households (67%) are those with the largest numbers of children. Analyses show that households experiencing food shortages represent 69.2%. Of these households 67% have more than 5 members each. In these large households (44%), members go to bed without eating. These results on food security in the Polesgo are confirmed by Bricas (2017) who says that hungry populations are those who don't have access to means of food production and sufficient purchasing power to feed themselves properly. This is also confirmed by FAO (1999) and Janin (2006) who state that in West African cities, populations in informal settlements are vulnerable in terms of food.

18.4.4 Strategies to address food insecurity in Polesgo

Facing food insecurity, adaptation strategies are developed by the population, including livestock farming, income-generating activities, restrictive measures in family expenditure. Livestock farming is a form of standing savings adopted by some households to address the issue of food accessibility. This breeding mainly concerns poultry as well as sheep and goats. Like poultry farming, sheep and goat farming is an insurance mechanism for the purchase of cereals if the crisis is not acute. For households this activity is therefore an alternative to the financial constraints. The populations use the cereal banks at social prices set up by the government (120 FCFA/~ € 0.18 per kg against 180 FCFA/~ € 0.27 per kg among traders). Unfortunately, people do not always have access to it because of stock shortages. This leads some households to resort to financial borrowing (14%) and cereal borrowing (4%) to remedy their food deficit.

Dropping health services for food needs is a resilience strategy that 18% of households, especially the poorest, are developing. This strategy can put the lives of children and the elderly in the household at risk because all expenditure is directed towards food needs at the expense of health care. Food fasting and unusual migration are coping strategies used by the population of Polesgo against food insecurity

(Figure 4). Fasting, i.e. spending the night or day without eating, is a strategy that 43% of households use to cope with food insufficiency. Exodus only becomes an indicator of food insecurity when it is unusual: in Polesgo 10% of households practice this temporary migration to find food.

Food consumption strategies are also observed: they include the obligation to eat certain less preferred foods, reducing food intake, limiting adult consumption in favour of young children, making active household members eat at the expense of the inactive and buying food on credit (Figure 5). The consumption of certain foods such as bran of cereal, wild plant species, which 70% of the households surveyed claim to have used becomes an indicator of food insecurity in the event of an obligation.

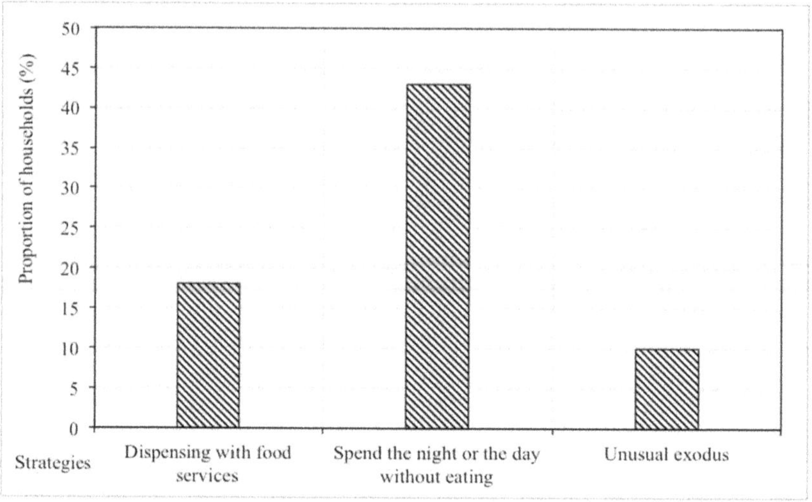

Figure 4. Household practice of non-food strategies (Source: Primary data, August 2017).

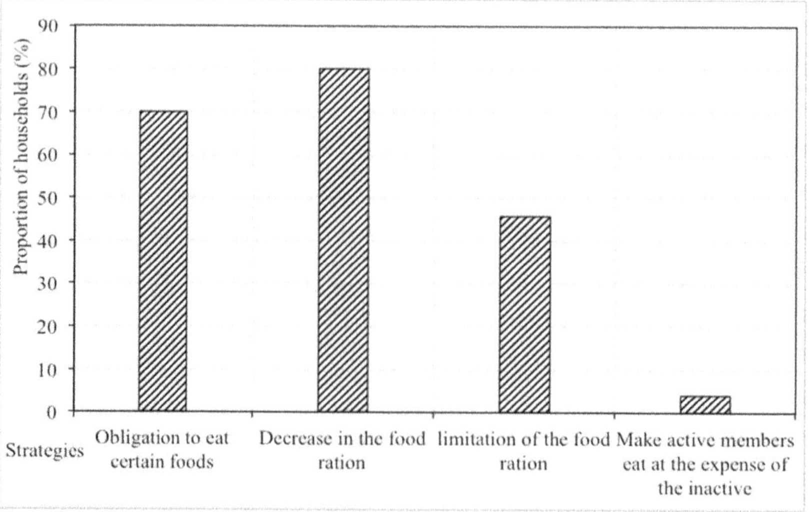

Figure 5. Household practice of food strategies (Source: Primary data, August 2017).

Also 80% of households in times of food insecurity adopt as a strategy, the reduction of food and the variety of food. Some households (45.8%) limit adult food consumption in favour of the youngest in times of food shortage. In some (minority) cases, the strategy of feeding active household members at the expense of the inactive is used. This practice mainly concerns breakfast and is generally adopted by agricultural households (4.2%). These strategies are generally reactive, ad hoc and as old as the structural adjustment programmes of the 1990s. These strategies are influenced by the almost constant poverty of urban households in informal settlements. Courade (1996) shows that the rise in food prices and the decline in purchasing power in the 1990s weakened food security in some African cities. The combination of these two contrasting situations forces poor households to adjust the quantities of food consumed or to consume certain unwanted or avoided foods. The persistence of food insecurity in urban areas demonstrates the limits of the various stakeholders in securing sustainable household food security. That's why Maire & Delpeuch (2004) argue that cities or economies are still largely dominated by poverty phenomena with huge disparities between households, it is necessary to go beyond the sole criterion of food supply to cities, and find specific solutions to problems of financial accessibility, within the framework of poverty reduction policies at the level of local communities or even households.

18.5 CONCLUSION

Food accessibility in urban areas illustrated by the showcase of the Polesgo district of Ouagadougou, gave evidence to a number of indicators of food insecurity but also to the production of an assessment of the levels of food strategies. The latter made it possible to prioritize the different levels of vulnerability in the area. 29% are in extreme vulnerability compared to 9% for the non-vulnerable, 33% for the medium vulnerability, and 29% are at risk of vulnerability. The analysis of FAO's food insecurity items shows that 3/5 of households are food insecure. The study also showed that the causes of food insecurity are both cyclical and structural. Almost all heads of households claim that this phenomenon has been aggravated in recent years by the weakening of traditional village mechanisms for risk prevention and management, but also by the transformation of their spatial environment, which places populations in a high level of vulnerability and leads to a high level of household decapitalisation.

Facing food insecurity, populations resort to several strategies in several forms: economic, food or no-food strategies. Following the analysis of these strategies, it appears that they remain deficient with regard to the different degrees of risk of exposure of populations. It therefore seems urgent to get out of this spiral of food insecurity and vulnerability that exposes the communities of Polesgo to both environmental and health risks. The possibilities of developing the area are possible and must be reflected in a better targeting of interventions and strategies compatible with this area. Finally, the analysis of food insecurity indicators and adaptation strategies allows either a cyclical monitoring of this phenomenon and its vulnerability, or a structural approach allowing a better approach to the causes of vulnerability.

REFERENCES

Avard, E. 2015. *Northern Greenhouses: An alternative local food provisioning strategy for nunavik.* Quebec: Laval University.

Baron, C. & Tidjani Alou, M. 2011. L'accès à l'eau en Afrique subsaharienne: au-delà des modèles, une pluralité d'innovations locales. *Mondes en développement* 155(3): 7–22.

Boyer, F. & Delaunay, D. 2009: *Ouaga 2009 - Peuplement de Ouagadougou et Développement urbain.* Ouagadougou: IRD - Ambassade de France.

Briand, V. 2001. L'insuffisance du concept de pauvreté pour comprendre l'insécurité alimentaire urbaine. Analyse dans le cas de Bouaké en Côte d'Ivoire. *Colloque Universités, coopération et solidarité internationales, Pauvreté et développement durable.* Paris: Université Montesquieu-Bordeaux IV-Unesco.

Bricas, N. 2017. Villes et alimentation, quels enjeux? In Debru, J., Albert, S., Bricas, N. & Conaré, D. (eds.) 2017, *Politiques alimentaires urbaines. Actes de la rencontre internationale sur les expériences en Afrique, Amérique latine et Asie en matière de marchés, restauration collective et connexion urbain/rural,* 16-18 November 2015: 17–23. Montpellier.

Chongo, M.C. 2017. Des marchés pour réduire l'insécurité alimentaire. In Debru, J., Albert, S., Bricas, N. & Conaré, D. (eds.) 2017, *Politiques alimentaires urbaines. Actes de la rencontre internationale sur les expériences en Afrique, Amérique latine et Asie en matière de marchés, restauration collective et connexion urbain/rural,* 16-18 November 2015: 45-52. Montpellier.

CILSS (Comité Inter-Etats de Lutte contre la Sècheresse dans le Sahel) 2008. *Mesure de la vulnérabilité alimentaire en milieu urbain sahélien: Résultats de l'étude de Ouagadougou.* Ouagadougou: CILSS.

Coates, J. Anne, S. & Paula, B. 2006. *Household Food Access Scale (HFIAS) for Measurement of Food Access: Indicator Guide.* Washington: FANTA Project.

Courade, G. 1996. Entre liberalisation et ajustement structurel: la securite alimentaiere dans un etau. *Cahiers Agricultures* 5: 221–7.

Dipama, J-M. 2014. Approche spatiale du phénomène du réchauffement climatique à l'échelle du Burkina Faso et perceptions des populations. *Climat et développement* 16: 36–49.

Duchemin, E. Wegmuller, F. & Legault, A.-M. 2009. Urban Agriculture: Multidimensional Tools for Social Development in Poor Neighbourhoods. *Field Actions Science Report,* 2(1-8): 43–52.

FAO (Food and Agriculture Organization of the United Nations) 1999. *La sécurité alimentaire urbaine et la commercialisation des aliments: un défi lancé aux villes et aux collectivités locales.* Rome: FAO.

FAO (Food and Agriculture Organization of the United Nations) 2006. Sécurité alimentaire. *Notes d'orientation n°2 juin 2006, Division de l'économie agricole et du développement (ESA) de la FAO.* The Netherlands: FAO.

FAO (Food and Agriculture Organization of the United Nations) 2014. *Food for the Cities: Production Systems – UPA.* URL: http://www.fao.org/fcit/upa/en/(accessed 29/ 11/2014).

FAO (Food and Agriculture Organization of the United Nations) 2015. *Vue d'ensemble régionale de l'insécurité alimentaire en Afrique: Des perspectives plus favorables que jamais.* Accra: FAO.

Ghersi, G. 2015. *Nourrir 9 milliards d'hommes.* Paris: Agropolis-Museum.

INSD (Institut National de la Statistique et de la Démographie) 2009. Monographie de la commune urbaine de Ouagadougou. Ouagadougou: INSD/Recensement général de la population et de l'habitation de 2006 (RGPH-2006).

Jaglin, S. 1995. *Gestion urbaine partagée à Ouagadougou: pouvoirs et périphéries (1983-1991).* Paris: Karthala/IRD.

Janin, P. 2006. La vulnérabilité alimentaire des Sahéliens: concepts, échelles et enseignements d'une recherche de terrain. *L'Espace géographique* 4(35): 355–366.

Janin, P. 2009. The complexity of the risk of food insecurity in the Sahel. *Annales de Géographie* 4(668): 430–448.

Janin, P. 2015. Nouvelles analyses des crises alimentaires en Afrique de l'Ouest. *BASE* 19(3): 290–300.

Kalinda, S. & Songolo, C. 2017. Politiques alimentaires urbaines durables via les infrastructures de marché et la logistique. In Debru, J., Albert, S., Bricas, N. & Conaré, D. (eds.), *Politiques alimentaires urbaines. Actes de la rencontre internationale sur les expériences en Afrique, Amérique latine et Asie en matière de marchés, restauration collective et connexion urbain/rural*, 16-18 November 2015: 30–38. Montpellier.

Lesauvage, S. 1997. *Mise au point d'outils pour le rappel des 24 heures en milieu urbain sénégalais*. Montpellier: ORSTOM, Université de Montpellier.

Maire, B. & Delpeuch, F. 2004. La transition nutritionnelle, l'alimentation et les villes dans les pays en développement. *Cahiers Agricultures* 13: 23–30.

Martin-Prevel, Y., Maire, B. & Delpeuch, F. 2000. Nutrition, urbanisation et pauvrete en afrique subsaharienne. *Médecine Tropicale* 60(2): 179–191.

Maxwell, D. & Frankenberger, T. 1992. *Household Food Security: Concepts, Indicators, Measurements. A Technical Review*. New York/Rome: IFAD/Unicef.

Maxwell, D., Levin, C., Armar-Klemesu, M., Ruel, M., Morris, S. & Ahiadeke, C. 2000. *Urban Livelihoods and Food and Nutrition Security in Greater Accra, Ghana*. Washington: IFPRI.

Ouattara, A. 2010. Les enjeux de la dynamique des extensions périurbaines à Ouagadougou (Burkina Faso). *Vertigo, revue électronique en science de l'environnement* 10(2): 9.

PAM (Programme Alimentaire Mondial) 2005. *Analyse et cartographie de la vulnérabilité structurelle à l'insécurité alimentaire en milieu rural au Burkina Faso*. Ouagadougou: UNDP.

Rouamba, S. & Ouédraogo, F.C. 2016. Les extrêmes climatiques, un problème d'accès à l'eau dans les quartiers informels de Ouagadougou. In E.W. Vissin, H.S. Totin Vodounon & C.S. Houssou (eds.), *Risques et catastrophes climatiques: vulnérabilités et adaptation en Afrique de l'ouest 1: Climat et environnement. Actes du colloque en hommage aux professeurs Michel Boko et Fulgence Afouda*, 27-30 septembre 2016: 362–374. Cotonou: Université d'Abomey-Calavi/Laboratoire Pierre Pagney: Climat, Eau, Ecosystèmes et Développement (LACEEDE).

Rouamba, S. 2017. *Variabilité climatique et accès à l'eau dans les quartiers informels de Ouagadougou*. Ouagadougou: Université Joseph Ki ZERBO.

Swindale, A. & Bilinsky, P. 2006. Development of a universally applicable household food insecurity measurement tool: process, current status, and outstanding issues. *Journal of nutrition* 136: 1449–1452.

Von Braun, J. 2008. *Hausse des prix alimentaires et actions stratégiques proposées: que faire, par qui et comment*. Washington: International food policy research institute.

Webb Coates, J., Frongillo, E.A., Rogers, B.L., Swindale, A. & Bilinsky, P. 2006. Measuring household food insecurity: why it's so important and yet so difficult to do. *Journal of nutrition* 136: 1404–1408.

World Bank 2017. *Rapport sur l'urbanisation en Afrique: pour soutenir la croissance il faut améliorer la vie des habitants et des entreprises dans les villes*. URL: http://www.banquemondiale.orgfrnews press-release20170209world-bank (accessed 29/ 10/2018).

Yaka, P., Rouamba, S. & Ouédraogo, A.E. 2015. Variabilité climatique en zone périurbaine de Ouagadougou. In A.B. Soura, S. Dos Santos & F.C. Ouédraogo 2015 (eds.), *Climat et ressources en eau en zones informelles de Ouagadougou*: 39–55. Ouagadougou: Presses universitaires de Ouagadougou.

For Product Safety Concerns and Information please contact our EU representative GPSR@taylorandfrancis.com
Taylor & Francis Verlag GmbH, Kaufingerstraße 24, 80331 München, Germany

www.ingramcontent.com/pod-product-compliance
Ingram Content Group UK Ltd.
Pitfield, Milton Keynes, MK11 3LW, UK
UKHW050754180926
475701UK00007B/1052
* 9 7 8 0 3 6 7 6 1 8 2 2 3 *